Springer Proceedings in Mathematics & Statistics

Volume 31

For further volumes:
http://www.springer.com/series/10533

Springer Proceedings in Mathematics & Statistics

This book series features volumes composed of select contributions from workshops and conferences in all areas of current research in mathematics and statistics, including OR and optimization. In addition to an overall evaluation of the interest, scientific quality, and timeliness of each proposal at the hands of the publisher, individual contributions are all refereed to the high quality standards of leading journals in the field. Thus, this series provides the research community with well-edited, authoritative reports on developments in the most exciting areas of mathematical and statistical research today.

Athanasios Migdalas • Angelo Sifaleras
Christos K. Georgiadis • Jason Papathanasiou
Emmanuil Stiakakis
Editors

Optimization Theory, Decision Making, and Operations Research Applications

Proceedings of the 1st International Symposium and 10th Balkan Conference on Operational Research

Springer

Editors
Athanasios Migdalas
Department of Mathematical and Physical Sciences
Faculty of Engineering
Aristotle University of Thessaloniki
Thessaloniki, Greece

Christos K. Georgiadis
Department of Applied Informatics
University of Macedonia
Thessaloniki, Greece

Emmanuil Stiakakis
Department of Applied Informatics
University of Macedonia
Thessaloniki, Greece

Angelo Sifaleras
Department of Technology and Management
Economical and Social Sciences
University of Macedonia
Naoussa, Greece

Jason Papathanasiou
Department of Marketing and Operations Management
University of Macedonia
Edessa, Greece

ISSN 2194-1009 ISSN 2194-1017 (electronic)
ISBN 978-1-4614-5133-4 ISBN 978-1-4614-5134-1 (eBook)
DOI 10.1007/978-1-4614-5134-1
Springer New York Heidelberg Dordrecht London

Library of Congress Control Number: 2012951435

Printed on acid-free paper

Springer is part of Springer Science+Business Media (www.springer.com)

We dedicate this book to the memory of
Prof. Konstantinos Paparrizos

Preface

It is a pleasure to introduce this book entitled Optimization Theory, Decision Making, and Operational Research Applications, of the series Springer Proceedings in Mathematics, which contains selected, refereed Proceedings from the 1st International Symposium and 10th Balkan Conference on Operational Research (BALCOR 2011).

This conference is an established biennial event attended by a large number of Operational Research (OR) scientists, instructors, and students, not only from the Balkan countries but also usually from all over Europe. In order to promote the conference and to attract people not only from Europe, but also from all over the world, it was decided to extend it with the 1st International Symposium on Operational Research in Thessaloniki. According to the number of different countries of the participants, we believe that this goal was achieved.

The general aim of the conference is to facilitate the exchange of scientific and technical information related to OR and to promote international cooperation especially among the Balkan countries. The conference was co-organized by the Branch of Macedonia-Thrace of the Hellenic Operational Research Society (HELORS) and the University of Macedonia, Economic and Social Sciences. It took place in the Holiday Inn Thessaloniki Hotel, Greece, 22–24 September 2011. The conference chair was Prof. Athanasios Migdalas, supported by the seven board members of the Branch of Macedonia—Thrace of HELORS, Prof. G. Arabatzis, Prof. C.K. Georgiadis, Dr. L. Karamitopoulos, Prof. V. Kostoglou, Dr. J. Papathanasiou, Dr. A. Sifaleras, and Dr. E. Stiakakis. Around 110 papers were presented in 32 streams, including combinatorial optimization, stochastic optimization, multibojective optimization, computational operational research, parallel processing in OR, multicriteria decision making, data envelopment analysis, inventory management, project management, etc. BALCOR 2011 was attended by participants from around 20 different countries, both from the Balkans and from France, Israel, Russia, Brazil, China, Algeria, Iran, etc.

Three plenary talks were given by eminent researchers in the field of optimization. Professor Vangelis Th. Paschos opened the conference by addressing the possible trade-off between polynomial approximation and exact computation.

He demonstrated how to use the ideas from both fields in order to design approximation algorithms for several combinatorial problems. Professor Nenad Mladenovic presented an overview of variable neighborhood search (VNS) meta-heuristic, and surveyed VNS-based approaches for solving the Travelling Salesman Problem (TSP). Finally, Prof. Leo Liberti presented theory and applications of combinatorial methods in Euclidean distance geometry to the protein visualization problem.

Apart from the funds provided by the co-organizers, University of Macedonia, Economic and Social Sciences, the conference attracted the sponsorship of well-known optimization software companies, such as LINDO Systems, Inc., Banxia Software Ltd., and Marathon Data Systems. These companies provided single-user professional software licenses of LINGO, What's Best, LINDO API, Frontier Analyst, or Decision Explorer as the best paper awards. They were given to the authors of four papers, selected by the Program Committee as these present interesting optimization models and methods.

This volume is dedicated to the memory of Prof. Konstantinos Paparrizos who passed away on Tuesday, December 6, just a few months after the BALCOR 2011 conference.

Professor Konstantinos Paparrizos was born in 1949 in Kozani, Greece. He earned his B.Sc. degree with honors in Mathematics from the Aristotle University of Thessaloniki, Greece (1972) and an M.Sc. and PhD both of them with honors in Operations Research (Minor in Computer Science) from the Case Western Reserve University, Cleveland OHIO, USA, in 1981 and 1983, respectively. He was a full-time professor at the Department of Applied Informatics, University of Macedonia, Economic and Social Sciences, Greece, Thessaloniki and also a professor in the Hellenic Open University, since 2008.

The late Prof. K. Paparrizos was a founding member of the Branch of Macedonia-Thrace of the Hellenic Operational Research Society. He was also a member of INFORMS and of the International Honorary Society OMEGA RHO, Case Western Reserve University Student Chapter. He served as a member of the Editorial Board of the *Operational Research: An International Journal* and the *Yugoslav Journal of Operations Research*. He also served as a member of the scientific and organizing committee of several Balkan Conferences on Operational Research in the past, as also in the recent BALCOR 2011 in Thessaloniki.

Professor K. Paparrizos was an established expert in the area of optimization algorithms. His research interests included mathematical programming, linear programming and network flows, design and analysis of algorithms, and data structures. He had significant research contributions in the design and analysis of exterior point simplex-type algorithms for linear and network optimization problems and also published several papers in leading international peer-reviewed journals (e.g., Mathematical Programming).

Thessaloniki, Greece

Athanasios Migdalas
Angelo Sifaleras
Christos K. Georgiadis
Jason Papathanasiou
Emmanuil Stiakakis

Contents

Contributors

Dimitra Alexiou Department of Spatial Planning and Development Engineering, School of Engineering, Aristotle University of Thessaloniki, Veria, Greece

Samira Almoustafa Brunel University, Uxbridge, UB8 3PH, UK

Henry Amankwah Department of Mathematics & Statistics, University of Cape Coast, Ghana

Alain Bretto Université de Caen, GREYC CNRS UMR-6072, Campus II, Bd Marechal Juin BP 5186, 14032 Caen cedex, France

Djamal Chaabane USTHB University, B.P. 32, Bab-Ezzouar, El-Alia 16111, Algiers, Algeria

Daniel Ciuiu Technical University of Civil Engineering, Bucharest, Bd. Lacul Tei No. 122-124, Sector 2, Bucharest, Romania

Romanian Institute for Economic Forecasting, Calea 13 Septembrie No. 13, Sector 5, Bucharest, Romania

Mohamed El Ghami Høgskolen i Nesna, Nesna University College, N-8700 Nesna, Norway

Aristotelis Giannakos Picardie University, Amiens, France

Nikolaos D. Goumagias University of Macedonia, Egnatia Av. 156 Thessaloniki, Greece

Evangelos Grigoroudis Technical University of Crete, Department of Production Engineering and Management, University Campus, Kounoupidiana, GR73100 Chania, Greece

Alexander Grusho Institute of Informatics Problems, RAS, Vaviliva St., 44, build 2, 119333 Moscow, Russia

Said Hanafi Lamih-Universite de Valenciennes, France

M'hand Hifi Picardie University, Amiens, France

Kaj Holmberg Department of Mathematics, Linköping University, Sweden

Dimitrios Hristu-Varsakelis University of Macedonia, Egnatia Av. 156 Thessaloniki, Greece

Jelena Jocković Department of Physics and Mathematics, Faculty of Pharmacy, University of Belgrade, Vojvode Stepe 450, 11000 Belgrade, Serbia

A. Karakitsiou Technological Educational Institute of Serres, Terma Magnesias, 62100 Serres, Greece

Stefanos Katsavounis Department of Production and Management Engineering, School of Engineering, Demokritos University of Thrace, Xanthi, Greece

Rezika Kheffache Faculty of Sciences, Mouloud Mammeri University, Algeria

Vassilis Kostoglou Department of Informatics, Alexander Technological Educational Institute of Thessaloniki, 57400, Thessaloniki, Greece

Panayiotis Kotzanikolaou Department of Informatics, University of Piraeus, Karaoli & Dimitriou 80, 185 34 Piraeus, Greece

Torbjörn Larsson Department of Mathematics, Linköping University, Sweden

Carlile Lavor Department of Applied Math. (IMECC-UNICAMP), State University of Campinas, 13081-970, Campinas - SP, Brazil

Leo Liberti LIX, Ecole Polytechnique, 91128 Palaiseau, France

Kristian Lundberg Department of Mathematics, Linköping University, Sweden

Yannis Marinakis Decision Support Systems Laboratory, Department of Production Engineering and Management, Technical University of Crete, 73100 Chania, Greece

Magdalene Marinaki Industrial Systems Control Laboratory, Department of Production Engineering and Management, Technical University of Crete, 73100 Chania, Greece

Ruxandra Marinescu-Ghemeci University of Bucharest, Str. Academiei, 14, Bucharest, Romania

Konstantinos Margaritis Department of Applied Informatics, University of Macedonia, 156 Egnatia Str., 54006 Thessaloniki, Greece

Fatma Mebrek ENSTP-KOUBA, Algiers, Algeria

George Minos Department of Aquaculture and Fisheries Technology, Alexander Technological Educational Institute of Thessaloniki, 63200, N. Moudania, Greece

Nenad Mladenović Brunel University, Uxbridge, UB8 3PH, UK

Theodoros Ntouskas Department of Informatics, University of Piraeus, Karaoli & Dimitriou 80, 185 34 Piraeus, Greece

Ioannis Paparrizos Computer Science Department, Columbia University, New York, NY, USA

Vangelis Th. Paschos LAMSADE, CNRS UMR 7243, Université Paris-Dauphine & Institut Universitaire de France, France

Nikolaos Ploskas Department of Applied Informatics, University of Macedonia, 156 Egnatia Str., 54006 Thessaloniki, Greece

Nineta Polemi Department of Informatics, University of Piraeus, Karaoli & Dimitriou 80, 185 34 Piraeus, Greece

Rachid Ouafi University of Technology, Bab Ezzouar, Algeria

Nils-Hassan Quttineh Department of Mathematics, Linköping University, Sweden

Nikolaos Samaras Department of Applied Informatics, University of Macedonia, 156 Egnatia Str., 54006 Thessaloniki, Greece

Andreea Mădălina Stancu Institute of Mathematical Statistics and Applied Mathematics, of the Romanian Academy, Calea 13 Septembrie Nr. 13, Sector 5, Bucharest, Romania

Ioan M. Stancu-Minasian Institute of Mathematical Statistics and Applied Mathematics, of the Romanian Academy, Calea 13 Septembrie Nr. 13, Sector 5, Bucharest, Romania

Srdjan S. Stanković Faculty of Electrical Engineering, University of Belgrade, Serbia

Milorad J. Stanojević Faculty of Transport and Traffic Engineering, University of Belgrade, Serbia

Dragoslav D. Šiljak School of Engineering, Santa Clara University, Santa Clara, USA

Cerasela Tanasescu ESSEC Business School, 1 avenue Bernard Hirsch Cergy, France

Björn Textorius Department of Mathematics, Linköping University, Sweden

Elena Timonina Moscow State University, Leninskie Gory, GSP-2, 119992 Moscow, Russia

Emmanouil Tolis Department of Informatics, Alexander Technological Educational Institute of Thessaloniki, 57400, Thessaloniki, Greece

Marianna Tsitsiloni Technical University of Crete, Department of Production Engineering and Management, University Campus, Kounoupidiana, GR73100 Chania, Greece

Constantin Zopounidis Technical University of Crete, Department of Production Engineering and Management, University Campus, Kounoupidiana, GR73100 Chania, Greece

BALCOR 2011

Conference Chair

A. Migdalas, Aristotle University of Thessalonike (Greece)

Organizing Committee

G. Arabatzis, Democritus University of Thrace (Greece)
C.K. Georgiadis, University of Macedonia (Greece)
L. Karamitopoulos, ATEI of Thessaloniki (Greece)
V. Kostoglou, ATEI of Thessaloniki (Greece, O.R. Branch Macedonia-Thrace chairman)
J. Papathanasiou, University of Macedonia (Greece)
A. Sifaleras, University of Macedonia (Greece)
E. Stiakakis, University of Macedonia (Greece)

International Scientific Committee

M. Cangalovic, University of Belgrade (Serbia)
P. Capros, National Technical University of Athens (Greece)
M. Demange, ESSEC Business School Romania Foundation (France)
A. Georgiou, University of Macedonia (Greece)
D. Giannakopoulos, Technological Educational Institution of Piraeus (Greece)
N. Farmakis, Aristotle University of Thessaloniki (Greece)
J. Figueira, Technical University of Lisbon (Portugal)
M. Iosifescu, Vice-President of the Romanian Academy (Romania)
M. Jacimovic, University of Montenegro (Montenegro)

V. Kovacevic-Vujcic, University of Belgrade (Serbia)
V. Manos, Aristotle University of Thessaloniki (Greece)
M. Martic, University of Belgrade (Serbia)
N. Matsatsinis, Technical University of Crete (Greece, O.R. Society chairman)
I. Mierlus Mazilu, Technical University of Civil Engineering (Romania)
N. Mladenovic, University of Belgrade (Serbia) and Brunel University (UK)
D. Nace, Universite de Technologie de Compiegne (France)
S. Papachristos, University of Ioannina (Greece)
K. Paparrizos, University of Macedonia (Greece)
B. Papathanasiou, Aristotle University of Thessaloniki (Greece)
P. Pardalos, University of Florida (USA)
V. Paschos, Universite Paris Dauphine (France)
V. Preda, University of Bucharest (Romania)
J. Psarras, National Technical University of Athens (Greece)
D. Radojevic, Mihajlo Pupin Institute (Serbia, O.R. Society chairman)
N. Samaras, University of Macedonia (Greece)
S. Sburlan, Naval Academy Mircea cel Batran Constanta (Romania)
Y. Siskos, University of Pireaus (Greece)
R. Slowinski, Poznan University of Technology (Poland)
S. Stanic, University of Banja Luka (Republic of Srpska, Bosnia and Herzegovina)
D. Teodorovic, University of Belgrade (Serbia)
R. Trandafir, Technical University of Civil Engineering (Romania)
C. Tsouros, Aristotle University of Thessaloniki (Greece)
A. Tsoukias, Universite Paris Dauphine (France)
M. Vujocevic, University of Belgrade (Serbia)
C. Zopounidis, Technical University of Crete (Greece)

Referees

G. Arabatzis	S. Kostopoulou	K. Petridis
G. Aretoulis	E. Livanis	E. Pimenidis
U. Bilsel	S. Liu	L. Pitsoulis
B. Delibasic	Ch. Malesios	N. Ploskas
M. Doumpos	B. Mamalis	J. Psarras
A. Georgiou	Y. Marinakis	R. Ravindran
S. Gujar	N. Matsatsinis	N. Samaras
J. Hernandez	A. Mendoza	D. Thilikos
A. Karakitsiou	G. Minos	C. Tsouros
V. Kostoglou	I. Mourtos	G. Zioutas
C. Zopounidis	D. Zoros	

Moderately Exponential Approximation: Bridging the Gap Between Exact Computation and Polynomial Approximation

Vangelis Th. Paschos

Abstract This paper proposes a way to bring together two seemingly "foreign" domains that are the polynomial approximation and the exact computation for **NP**-hard problems. We show how one can match ideas from both areas in order to design approximation algorithms achieving ratios unachievable in polynomial time (unless a very unlikely complexity conjecture is confirmed) with worst-case complexity much lower (though super-polynomial) than that of an exact computation.

1 Introduction

The two most known paradigms to come up with **NP**-hard problems are either the exact computation (i.e., the computation of optimal solutions for them) or the heuristic resolution, i.e., the development of fast algorithms that hopefully compute near-optimal solutions. Notable part of the heuristic paradigm is the so-called *polynomial approximation* where one tries to devise polynomial algorithms computing feasible solutions that are close to optimal under an a priori criterion called *approximation ratio*.

Both exact computation and polynomial optimization are very active areas in theoretical computer science and combinatorial optimization. Dealing with the former, very active research has been conducted around the development of optimal algorithms with nontrivial worst-case complexity. As an example, let us consider MAX INDEPENDENT SET. Any optimal algorithm can solve it with complexity $O^*(2^n)$ (where $O^*(\cdot)$ is as $O(\cdot)$ ignoring polynomial factors), where n is the order of G (i.e., the cardinality of V) by exhaustively examining all the subsets in 2^V and by taking the largest among them that forms an independent set; hence,

---

V.Th. Paschos (✉)
LAMSADE, CNRS UMR 7243 - Université Paris-Dauphine and Institut Universitaire de France, 103, boulevard Saint-Michel, 75005 Paris, France
e-mail: paschos@lamsade.dauphine.fr

A. Migdalas et al. (eds.), *Optimization Theory, Decision Making, and Operations Research Applications*, Springer Proceedings in Mathematics & Statistics 31,
DOI 10.1007/978-1-4614-5134-1_1,

an interesting question is if we can compute a maximum independent set with complexity $O^*(\gamma^n)$, for $\gamma < 2$. More about such issues for several combinatorial problems can be found in [42] and in the more recent book by Fomin and Kratsch [29]. This area has known a renew of the researcher's interest not only due to numerous pessimistic results in polynomial approximation but also due to the fantastic increase of the computational power of modern computers.

On the other hand, dealing with polynomial approximation, very intensive research since the beginnings of 1970s has led to numerous results exhibiting possibilities but also limits to the approximability of **NP**-hard problems. Such limits are expressed as statements that a given problem cannot be approximated within a certain approximation level (for instance, within a constant approximation ratio) unless a very unlikely complexity condition (e.g., **P** = **NP**) holds. A very rich landscape of the polynomial approximation area can be found in [2, 34, 39, 41]. Since the beginning of 1990s, and using the celebrated PCP theorem [1], numerous natural hard optimization problems are proved to admit more or less pessimistic inapproximability results. For instance, MAX INDEPENDENT SET is inapproximable within the approximation ratio better than $n^{\varepsilon-1}$, unless **P** = **NP**, [43].

These two areas have remained foreign for long time and until very recently. Researchers in any of them produced results fitting the corresponding paradigm and without links with the other one. Staring point of this work is the idea that both of them can be linked by mutual exchanges of tools and concepts, in order that new results and ideas are established handling solution mechanisms of **NP**-hard problems. For instance, it is interesting to efficiently approximate such problems by devising algorithms that achieve approximation ratios that cannot be achieved in polynomial time, with a worst-case complexity that is significantly lower (though super-polynomial) than the complexity of a exact computation. This issue is called *moderately exponential approximation* in what follows.

2 What Is Moderately Exponential Approximation?

An optimization problem Π is in **NPO** if the decision version of Π is in **NP**. Formally, an **NPO** problem Π is defined as a four-tuple $(\mathscr{I}, \mathrm{sol}, m, \mathrm{opt})$ such that: $\mathscr{I}$ is the set of instances of Π and it can be recognized in polynomial time; given $x \in \mathscr{I}$, $\mathrm{sol}(x)$ denotes the set of feasible solutions of x; for every $y \in \mathrm{sol}(x)$, $|y|$ is polynomial in $|x|$; given any x and any y whose length is polynomial in $|x|$, one can decide in polynomial time if $y \in \mathrm{sol}(x)$; given $x \in \mathscr{I}$ and $y \in \mathrm{sol}(x)$, $m(x,y)$ denotes the value of y for x; m is polynomially computable and is commonly called feasible value, or objective value; finally, $\mathrm{opt} \in \{\max, \min\}$ denotes the optimization goal for Π. The set of **NP** optimization problems forms the class **NPO**. Given an instance x of an **NPO** problem $\Pi = (\mathscr{I}, \mathrm{sol}, m, \mathrm{opt})$, and a feasible solution y for x, we denote by $\mathrm{opt}(x)$ the value of an optimal solution of x.

For an approximation algorithm $\mathtt{A}$ computing a feasible solution y for x with value $m_{\mathtt{A}}(x,y)$, its approximation ratio on y is defined as $\rho_{\Pi}^{\mathtt{A}}(x,y) = m_{\mathtt{A}}(x,y)/\mathrm{opt}(x)$.

The approximation ratio $\rho_{\Pi}^{\mathtt{A}}$ of $\mathtt{A}$ is then defined as the worst[1] over any instance $x \in \mathscr{I}$, of $\rho_{\Pi}^{\mathtt{A}}(x,y)$. In what follows, whenever it is understood, references to problem Π and/or $\mathtt{A}$ will be dropped.

Polynomial approximation is a very active area since the beginning of the 1970s. The celebrated paper by Johnson [36] is considered as the startpoint of this research program that has dominated a large part of the research conducted in complexity theory. On the other hand, exact solution of combinatorial problems is a natural requirement that remains in the heart of the research in combinatorial optimization and in operational research more generally. These two approaches are complementary in the sense that, informally, the former gives priority to fast computation of feasible solutions against optimality, while, for the latter, priority is given to solutions' optimality against speed of such computation.

If the area of exact computation is in the heart of combinatorial optimization since the beginnings of this domain, the main concerns of a large majority of its researchers were rather about the design of clever solution algorithms (mainly based upon tree-search procedures, dynamic programming, etc.) than the precise estimation of their running-time. Before the middle of 1990s, when a broad research program around such concerns has been built, fairly little research has been dedicated to this issue.

On the other hand, numerous open questions, posed since the beginnings of polynomial approximation as, for example, the approximation of MAX INDEPENDENT SET[2] within constant ratio, have received strongly negative answers with the proof of the famous PCP theorem (carrying over a novel and fine characterization of **NP** [1]). Similar answers, known as *inapproximability* or *negative results* in polynomial approximation theory, have been provided for numerous other paradigmatic optimization problems, as MIN SET COVER,[3] MIN VERTEX COVER,[4] MIN COLORING,[5] etc. Informally, an inapproximability result is a statement that a problem is not approximable within ratios better than some approximability level unless something very unlikely happens in complexity theory (for example, $\mathbf{P} = \mathbf{NP}$, or the Exponential Time Hypothesis (ETH), saying that no problem in **NP** can be solved in sub-exponential time, is disproved, or ...). For instance, we know today that under several more or less strong complexity hypotheses:

- MAX INDEPENDENT SET or MAX CLIQUE is inapproximable within ratios $\Omega\left(n^{-1}\right)$ [43]

[1]The min if Π is a maximization problem, the max, otherwise.

[2]Given a graph $G(V,E)$, MAX INDEPENDENT SET consists of finding a set $S \subseteq V$ of maximum size such that for any $(u,v) \in S \times S$, $(u,v) \notin E$.

[3]Given a ground set C of cardinality n and a system $\mathscr{S} = \{S_1,\dots,S_m\} \subset 2^C$, MIN SET COVER consists of determining a minimum size subsystem $\mathscr{S}'$ such that $\cup_{S \in \mathscr{S}'} S = C$.

[4]Given a graph $G(V,E)$, MIN VERTEX COVER consists of finding a set $C \subseteq V$ of minimum size such that, for every $(u,v) \in E$, either u, or v belongs to C.

[5]Given a graph $G(V,E)$, MIN COLORING consists of determining a minimum-size partition of V into independent sets.

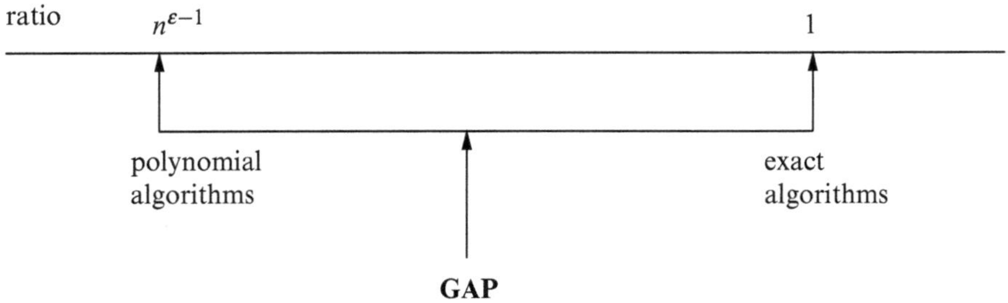

Fig. 1 The approximability gap for MAX INDEPENDENT SET

- MIN VERTEX COVER is inapproximable within ratio $2-\varepsilon$ for any fixed constant $\varepsilon < 1$ [37]
- MIN SET COVER is inapproximable within ratios $o(\log n)$ [28]
- MIN INDEPENDENT DOMINATING SET is inapproximable within ratios $o(n)$ [32]
- MIN TSP is inapproximable within better than exponential ratios, [39];
- MIN COLORING is inapproximable within ratios $o(n)$ [43]

These results exhibit large gaps between what it is possible to do in polynomial time and what becomes possible in exponential time. Let us take, once again, the case of MAX INDEPENDENT SET. As mentioned above, it is proved in [43] that this problem is inapproximable within ratio better than $O(n^{\varepsilon-1})$, unless $\mathbf{P} = \mathbf{NP}$ (note that any approximation algorithm trivially achieves approximation ratio $O(n)$ in polynomial time). We so are faced with a huge gap impossible to be bridged in polynomial time (Fig. 1).

Hence, a natural question is how much time takes the computation of an r-approximate solution, for $r \in [n^{\varepsilon-1}, 1[$? Of course, we have a lower bound to this time (any polynomial to the size of the instance) and also an upper bound (the running time of exact computation). But:

- Can we devise, for some ratio r, an r-approximate algorithm with an improved running time located somewhere between these bounds?
- Is this possible for *any* ratio r, i.e., can we specify a global relationship between running time and approximation ratio?

In this paper, we try to bring answers to these questions by matching ideas and results from exact computation and from polynomial approximation (mainly around approximation preserving reductions). This issue has been also marginally handled in [5] for minimum coloring and more systematically in [10]. It is handled in [11, 20] for MIN SET COVER, in [19, 31] for MIN BANDWIDTH, in [12] for MIN INDEPENDENT DOMINATING SET, in [8] for dominating clique problems, in [6] mainly for STEINER TREE and for TSP, in [27] for MAX SAT, in [21] for capacitated dominating set, ... Moderately exponential approximation has been also handled in [17, 18, 24], though in a different setting and with different objectives oriented towards development of fixed-parameter algorithms (see [23] for more details about fixed parameter tractability). In the same setting, we quote the paper by Brankovic and Fernau [15] that improves a result of [9, 14] on parameterized approximation of MIN VERTEX COVER. A different but very interesting kind of trade-off between exact computation and polynomial time approximation is settled in [40]. Note finally that trade-offs between approximation ratio and running time have already been studied for polynomially solvable problems (but with practically long running times) such as the MAX MATCHING[6] problem.

In what follows in this paper, we sketch some basic techniques for moderately exponential approximation and illustrate them on some paradigmatic combinatorial optimization problems.

Before closing this section we give some notations that will be used later. Let $T(\cdot)$ be a super-polynomial and $p(\cdot)$ be a polynomial, both on integers. In what follows, using notations in [42], for an integer n, we express running-time bounds of the form $p(n) \cdot T(n)$ as $O^*(T(n))$ by ignoring, for simplicity, polynomial factors. We denote by $T(n)$ the worst-case time required to solve the considered combinatorial optimization problem with n variables. We recall (see, for instance, [26]) that, if it is possible to bound above $T(n)$ by a recurrence expression of the type $T(n) \leq \sum T(n - r_i) + O(p(n))$, we have $\sum T(n - r_i) + O(p(n)) = O^*(\alpha(r_1, r_2, \ldots)^n)$, where $\alpha(r_1, r_2, \ldots)$ is the largest zero of the function $f(x) = 1 - \sum x^{-r_i}$.

Given a graph $G(V,E)$, we denote by n the size of V, by $\alpha(G)$ the size of a maximum independent set of G and by $\tau(G)$ the size of a minimum vertex cover of G. Also, we denote by $\Delta(G)$ the maximum degree of G. Given a subset V' of V, $G[V']$ denotes the subgraph of G induced by V'. Sometimes, for a graph G, we denote by $V(G)$ its vertex-set.

3 Generating a "Small" Number of Candidate Solutions or Exhaustively Searching a Small Part of Instance

The key idea of this technique consists of generating a "small" number of candidate solutions for a given problem and of finally choosing the best of them for the final solution of the problem.

For instance, assume that the problem to solve is the MAX INDEPENDENT SET, consider a graph $G(V,E)$ of order n and run the following algorithm:

- Generate all the $\sqrt{n}$-subsets (subsets of cardinality $\sqrt{n}$) of V
- If one of them is independent, then output it
- Otherwise output a vertex at random

---

[6]Given a graph $G(V,E)$, a matching is a set $E' \subseteq E$ that they have no common endpoints; the MAX MATCHING problem consists of determining a matching of maximum size.

It is easy to see that the approximation ratio of this algorithm is $n^{-1/2}$. Indeed, if algorithm's output is done at the second item, i.e., an independent set of size $\sqrt{n}$ is discovered, then, since $\alpha(G) \leq n$, the approximation ratio achieved is at least $\sqrt{n}/n = n^{-1/2}$. On the other hand, if no independent set is found at the second step, then $\alpha(G) \leq \sqrt{n}$ and the approximation ratio guaranteed in third step is at least $1/\sqrt{n} = n^{-1/2}$, impossible for polynomial algorithms according to Håstad [33].

The complexity of the algorithm above is roughly bounded above by $O^*(\binom{n}{\sqrt{n}}) = O^*(2^{\sqrt{n}\log n})$, that is subexponential and much lower than the best known exact complexity for MAX INDEPENDENT SET that is $O^*(1.2125^n)$ (see [7, 13]).

The technique of generation of a small number of candidate solutions has also been used in [21] for approximately solving the CAPACITATED DOMINATING SET[7] problem, that is a generalization of the well-known MIN DOMINATING SET problem. More precisely, the following theorem is proved there.

Theorem 1 ([21]). *There exists an approximation algorithm for the* CAPACITATED DOMINATING SET *problem that for any fixed constant* $c \in (0, 1/3)$ *runs in time* $O((c^c(1-c)^{1-c})^{-1})^n$. *For* $c \leq 1/4$, *its approximation ratio is at most* $(1/4c) + c$, *while for* $c \geq 1/4$ *its approximation ratio is at most* $2 - 3c$.

Another technique that has given interesting results in moderately exponential approximation is a kind of exhaustive search in a small part of the instance. Roughly speaking, this technique consists of constructing a part (or the whole) of a solution by exhaustive search in a "small" part of the input-instance and of completing this solution (if necessary) by some polynomial algorithm.

Let us give an example of this technique to MIN INDEPENDENT DOMINATING SET.[8] Consider the following algorithm:

1. Compute every independent dominating set of at most size n/r
2. If such a set exist then return it
3. Otherwise, return some maximal independent set (for example, using the maximum degree greedy MAX INDEPENDENT SET-algorithm)

Note that *for any* $r \geq 3$, *it is possible to enumerate all independent dominating sets (i.e., maximal independent sets) of size at most* n/r *with running time* $O^*(r^{n/r})$, [16].

Theorem 2 ([12]). *For any* $r \geq 3$, *it is possible to compute an* r*-approximation of* MIN INDEPENDENT DOMINATING SET *with running time* $O^*(2^{n\log_2 r/r})$.

[7] Given a graph $G(V,E)$ with each of its vertex v equipped with a number $c(v)$ that represents the number of the other vertices that v can dominate, a set $S \subset V$ is a capacitated dominating set if there exists a function $f_S : V \setminus S \rightarrow S$ such that $f_S(v)$ is a neighbor of v for each $v \in V \setminus S$ and $|f_S^{-1}(v)| \leq c(v)$; the goal of CAPACITATED DOMINATING SET is to determine a capacitated dominating set of the smallest possible size.

[8] Given a graph $G(V,E)$, MIN INDEPENDENT DOMINATING SET consists of finding the smallest independent set of G that is maximal for the inclusion.

4 Divide-and-Approximate

The key idea of this technique consists of splitting, in some appropriate way, the initial instance in a series of "small" sub-instances whose sizes are functions of the ratio that is to be achieved, of optimally solving the problem on each of them and, finally, of "fastly" composing a solution in the initial instance by the solutions of the sub-instances.

We illustrate this technique, once more, on MAX INDEPENDENT SET. Consider a graph G of order n and fix a rational $\rho \leq 1$ (the ratio that one wishes to achieve). Since $\rho \in \mathbb{Q}$, it can be written as $\rho = p/q$, $p,q \in \mathbb{N}$, $p \leq q$. Consider now the following algorithm, called with parameters G and ρ and denoted by `MEXPIS`:

1. Arbitrarily partition G into q induced subgraphs $G_1,\ldots,G_q$ of order (except eventually for G_q) n/q
2. Build the q subgraphs $G'_1,\ldots G'_q$ that are unions of p consecutive subgraphs $G_{i+1},\ldots,G_{i+p}$, $i = 1,\ldots,q$ (where of course $G_{q+1} = G_1$)
3. Optimally solve MAX INDEPENDENT SET in every G'_i, $i = 1,\ldots,q$
4. Output the best of the solutions computed in step 3

Theorem 3 ([9, 14]). *Assume that there exists an exact algorithm* `A` *for* MAX INDEPENDENT SET *with worst-case complexity $O^*(\gamma^n)$ for some $\gamma \in \mathbb{R}$, where n is the order of the input-graph, for* MAX INDEPENDENT SET. *Then for any $\rho \in \mathbb{Q}$, $\rho \leq 1$, there exists a ρ-approximation algorithm for* MAX INDEPENDENT SET *that runs in time $O^*(\gamma^{\rho n})$.*

As the basic argument of the proof of Theorem 3 in [9, 14] is based upon the heredity of the solution, Theorem 3 holds for several hereditary properties. A graph property π is said to be *hereditary* if every subgraph of G satisfies π whenever G satisfies π. Furthermore, π is nontrivial if it is satisfied for infinitely many graphs and it is false for infinitely many graphs. A graph-problem is said hereditary if its feasible solutions are the subsets of vertices such that the corresponding induced subgraph verifies some hereditary property. Under this definition, "clique," "planar graph," "bipartite graph," etc. are hereditary properties, and the problems of determining a maximum order induced subgraph that is a clique, or a planar graph, or a bipartite graph, or it is k-colorable, are hereditary problems. Theorem 3 applies to all of such problems.

Note also that any improvement to the basis γ of the exponential for the running time of the exact algorithm for MAX INDEPENDENT SET is immediately transferred to the running time claimed by Theorem 3.

We now show that the result of Theorem 3 can be also used for approximating MIN VERTEX COVER in a moderately exponential way. There exists a very close and well-known relation between a vertex cover and an independent set in a graph $G(V,E)$, [3]: if S is an independent set of G, then the set $V \setminus S$ is a vertex cover of G. The same complementarity relation holds obviously for a maximum independent set S^* and the set $C^* = V \setminus S^*$ that is a minimum vertex cover of G.

For the moderately exponential approximation of MIN VERTEX COVER, we use the seminal result by Nemhauser and Trotter [38] characterizing the polytope of MAX INDEPENDENT SET (or, equivalently, of MIN VERTEX COVER). Before, for readability, let us recall the integer linear program of MAX INDEPENDENT SET (denoted also by is) as well as the mathematical program of its linear programming relaxation (LP-relaxation), denoted by MAX INDEPENDENT SET-R. Given a graph G, denoting by A its incidence matrix:

$$\text{MAX INDEPENDENT SET} = \begin{cases} \max \ \underline{1} \cdot \underline{x} \\ \quad A\underline{x} \leq \underline{1} \\ \quad \underline{x} \in \{0,1\}^n \end{cases}$$

$$\text{MAX INDEPENDENT SET-R} = \begin{cases} \max \ \underline{1} \cdot \underline{x} \\ \quad A\underline{x} \leq \underline{1} \\ \quad \underline{x} \in (\mathbb{Q}^n)^+ \end{cases}$$

Obviously, solution of MAX INDEPENDENT SET-R can be done in polynomial time by any continuous linear-programming algorithm.

Theorem 4 ([38]). *The basic optimal solution of the LP-relaxation of* MAX INDEPENDENT SET *is semi-integral, i.e., it assigns to the variables values from* $\{0,1,1/2\}$. *Let* V_0, V_1 *and* $V_{1/2}$ *be the subsets of* V *associated with 0, 1 et 1/2, respectively. There exists a maximum independent set* S^* *such that* $V_1 \subseteq S^*$ *and* $V_0 \subseteq C^* = V \setminus S^*$.

From Theorem 4 and its proof in [38], the following corollary holds.

Corollary 1. $\alpha(G[V_{1/2}]) \leq |V_{1/2}|/2$ *and* $\tau(G[V_{1/2}]) \geq |V_{1/2}|/2$. *Also, denoting by* S' *and* C' *an independent set and a vertex cover of* $G[V_{1/2}]$, $S = V_1 \cup S'$ *is an independent set of* G *and* $C = V \setminus S = V_0 \cup C'$ *is a vertex cover of* G.

The following lemma links approximabilities of MAX INDEPENDENT SET and MIN VERTEX COVER.

Lemma 1 ([9, 14]). *If* MAX INDEPENDENT SET *is approximable within approximation ratio* ρ, *then* MIN VERTEX COVER *is approximable within ratio* $2-\rho$.

Let us note that when tackling approximation of MAX INDEPENDENT SET and of MIN VERTEX COVER, we can restrict ourselves to subgraph $G[V_{1/2}]$, instead of the whole G (since the sets V_0 and V_1 can be computed in polynomial time).

Using Corollary 1 and Lemma 1, MIN VERTEX COVER can be approximately solved by the following algorithm called MEXPVC:

1. Solve the LP-relaxation of MAX INDEPENDENT SET to obtain sets V_1, V_0, and $V_{1/2}$ (this step runs in polynomial time);
2. Set $G = G[V_{1/2}]$ and run $\texttt{MEXPIS}(\texttt{G}, \rho)$;
3. Output $V \setminus (V_1 \cup \texttt{MEXPIS}(\texttt{G}, \rho))$.

Combination of Theorem 3 and Lemma 1 immediately derives the following result.

Theorem 5 ([9, 14]). *For any $\rho \leq 1$, Algorithm* MEXPVC *computes a $(2-\rho)$-approximation of* MIN VERTEX COVER *with running time $O^*(\gamma^{\rho n})$.*

In other words, any approximation ratio $r \in [1,2[$ for MIN VERTEX COVER can be attained by Algorithm MEXPVC, with complexity $O^*(\gamma^{(2-r)n})$. This result is improved in [9, 14].

Using divide-and-approximate, Cygan et al. [20] proposes moderately exponential approximation algorithms for MIN SET COVER (in its weighted version, where weights are assigned on the set of $\mathscr{S}$ and the objective becomes to find a minimum-weight set cover). More precisely, the following results are proved there.

Theorem 6 ([20]). *Assume that there exists an exact algorithm for weighted* MIN SET COVER *running in time $O^*(c^m)$ for some constant $c > 1$. Then, for any $r > 1$, there exists an r-approximation algorithm running in $O^*(c^{m/r})$.*

5 Approximately Pruning the Search Tree

The most common tool used to devise exponential algorithm with nontrivial worst case complexity consists of pruning the search tree [42]. The key idea of the technique of approximate pruning of the search tree consists of performing a branch-and-cut by allowing a "bounded error" in order to accelerate the algorithm, i.e., of making the instance-size decreasing quicker than in exact computation by keeping the produced error "small".

Consider a simple search tree-based algorithm for solving MAX INDEPENDENT SET, which consists of recursively applying the following rule (see, for instance, [42]):

1. If $\Delta(G) \leq 2$, then output a maximum independent set
2. Else, branch on a vertex v with degree at least 3 as follows:
   (a) Either take v and solve MAX INDEPENDENT SET in the subgraph surviving after the removal of v and its neighbors
   (b) Or do not take v, and solve MAX INDEPENDENT SET in the subgraph surviving after the removal of v

Step 1 can be done in polynomial time. On the other hand, when branching, we have to solve a subproblem of size either $n - \Delta(v) - 1 \leq n - 4$, or $n - 1$. This leads to a running time $T(n) \leq T(n-1) + T(n-4) + p(n)$, for some polynomial p, which comes up to $T(n) \leq O^*(1.381^n)$.

We now explain how one can get a 1/2-approximation algorithm based upon a modification of the above algorithm, with running time much better than $O^*(1.381^n)$. The idea is that, when a branching occurs, in case 2a an optimum solution built via the above algorithm takes v. In this case, if we only seek a 1/2-approximate solution, then roughly speaking, an approximation algorithm can make an error on another vertex (not taking it in the solution while an optimal solution takes it). Indeed, vertex v taken in both solutions compensates "at half" this error. So, when applying the branching, in case 2a we can remove any other vertex of the graph. We then get a subproblem of size $n-5$ instead of $n-4$. More generally, consider an edge (v_i, v_j) in the surviving graph (or even a clique K). Since an optimal solution can take at most one vertex of a clique, then when branching in case 2a, we can remove vertices v_i and v_j (resp., the whole clique K).

A second improvement deals with step 1. Indeed, we do not need to deal with cases where the optimum can be polynomially reached, but with cases where a 1/2-approximate solution can be found in polynomial time. For instance, MAX INDEPENDENT SET can be approximately solved in polynomial time within approximation ratio $5/(\Delta(G)+3)$ [4]. Hence, if $\Delta(G) \leq 7$, then MAX INDEPENDENT SET is 1/2-approximable in G. This leads to the following algorithm:

1. If $\Delta(G) \leq 7$, then run the algorithm by Berman and Fujito [4]
2. Else, branch on a vertex v with degree at least 8 as follows:
   (a) Either take v, and solve MAX INDEPENDENT SET in the subgraph surviving after the removal of v, of its neighbors and of two other adjacent vertices v_i, v_j
   (b) Or do not take v, and solve the problem in the subgraph surviving after the removal of v

It is easy to recursively verify that the algorithm above guarantees an approximation ratio 1/2. Concerning its running time, during step 2a we remove 11 vertices (note that if there is no edge (v_i, v_j) to be removed, the surviving graph is an independent set per se); hence, $T(n) \leq T(n-1) + T(n-11) + p(n)$. This leads to $T(n) = O^*(1.185^n)$.

Note that the above algorithm can be generalized to find a $(1/k)$-approximation algorithm (for any $k \in \mathbb{N}$) in time $T(n) \leq T(n-1) + T(n-7k+3) + p(n)$. Obviously, improved running times would follow from considering, for example, either modifications of algorithms more sophisticated than the one presented in this section or a more efficient counting technique such as the one presented in [30].

We now apply this technique to MIN SET COVER. We set $\Delta = |S^*|$, where $S^* = \text{argmax}\{|S| : S \in \mathscr{S}\}$ and $d = m+n$. Let us note that in [25], using semi-local optimization techniques, a $(1/2) + \ln\Delta$-approximation algorithm is given. Consider the following Algorithm MEXPSC, parameterized by the ratio q one wishes to guarantee:

- Fix $q \in \mathbb{N}^*$ and compute the largest integer p such that $(1/2) + \ln\Delta \leq q$
- While C remains uncovered do:
  1. If there exists an item of C that belongs to a single subset $S \in \mathscr{S}$, then add S to the solution
  2. If there exist two sets S, R in $\mathscr{S}$ such that S is included into R, then remove S without branching
  3. If all the residual subsets have cardinality at most p, then run the algorithm by Duh and Fürer [25] in order to compute a q-approximation of the optimal solution in the surviving instance

4. Determine q sets $S_1, \ldots, S_q$ from $\mathscr{S}$ such that $\cup_{i \le q} S_i$ has maximum cardinality and perform the the following branching: either add every S_i to the solution (and remove $\cup_{i \le q} S_i$ from C), or remove all of them

Theorem 7 ([11]). *For any integer $q \ge 1$, Algorithm* MEXPSC *computes with running time $O^*(\alpha^d)$ a q-approximation of* MIN SET COVER, *where α is the solution of:*

$$x^{q(2+p)} - x^{q(1+p)} - 1 = 0 \tag{1}$$

and p is the largest integer such that $\ln p + 1/2 \le q$.

Corollary 2 ([11]). *For any integer $q \ge 1$, Algorithm* MEXPSC *computes a q-approximation of* MIN SET COVER *in $O^*(2^{m/q})$.*

The result of Theorem 7 is improved in several ways in [11]. Also, several differential approximation [22] moderately exponential algorithms are given.

The same basic technique of approximately pruning the search tree, although using more involved technical arguments is used in [19] for the BANDWIDTH problem. The following result is proved there.

Theorem 8 ([19]). *For any positive integer r, there exists a $(4r-1)$-approximation algorithm for* BANDWIDTH *running in $O^*(2^{n/r})$-time.*

6 Randomization

The results seen in Sect. 4, mainly those of Theorems 3 and 5, were of the form: *if a problem Π is solvable to optimality in time $O^*(\gamma^n)$, for some $\gamma > 1$, then it is approximable within ratio r in time $O^*(\gamma^{rn})$*. Can we do better? In other words, is it possible to get ratios r in time better than $O^*(\gamma^{rn})$ for problems handled by theorems similar Theorems 3 and 5? Indeed, this is very frequently possible by randomization. The key idea of this technique is the following:

- Randomly split the graph into subgraphs in such a way that the problem at hand is to be solved in graphs G'_i of order $r'n$ with $r' < r$
- Compute the probability $\Pr[r]$ that $|S^* \cap \mathrm{sol}(G'_i)| \ge r|S^*|$ (in other words, get an r-approximation with probability $\Pr[r]$)
- Repeat splitting $N(r)$ times so that $\Pr[r] \to 1$ (in other words, get an r-approximation with probability $\sim$1 in time $N(r)\gamma^{r'n}$)

For instance, dealing with MAX INDEPENDENT SET, it can be shown that by splitting into sub-instances of (smaller) size βn, with $\beta < r$, one can achieve approximation ratio r by iterating the splitting a very large (exponential) number of times.

Theorem 9 ([9, 14]). *For any $\rho < 1$ and for any β, $\rho/2 \le \beta \le \rho$, it is possible to find an independent set that is, with probability $1 - \exp(-cn)$ (for some constant c), a ρ-approximation for* MAX INDEPENDENT SET, *with running time $O^*(K_n \gamma^{\beta n})$, where:*

$$K_n = \frac{n\binom{n}{\beta n}}{\binom{n/2}{\rho n/2}\binom{n/2}{\beta n-\rho n/2}}$$

It is shown in [9, 14] that an optimal choice of β decreases the overall running time of the derived algorithm by an exponential term.

In [9, 14], the result of Theorem 9 is further improved and extended to the case of MIN VERTEX COVER. Also, randomization also works for MIN SET COVER (seen in Sect. 5).

7 Final Remarks

What kind of results can be expected in the area of (sub)exponential approximation? All the algorithms given in this paper have exponential running times when a *constant* approximation ratio (unachievable in polynomial time) is seeked. On the other hand, for several problems that are hard to approximate in polynomial time (like MAX INDEPENDENT SET, MIN COLORING, ...), subexponential time can be easily reached for ratios growing (to infinity) with the input size (this is the case of MAX INDEPENDENT SET seen in Sect. 3). An interesting question is to determine, for these problems, if it is possible to devise a constant approximation algorithm working in subexponential time. An easy argument shows that this is not always the case.

For instance, the existence of subexponential approximation algorithms (within ratio better than 4/3) is quite improbable for MIN COLORING since it would imply that 3-COLORING can be solved in subexponential time, contradicting so the ETH, [35]. We conjecture that this is true for any constant ratio for MIN COLORING, and that the same holds for MAX INDEPENDENT SET.

Anyway, the possibility of devising subexponential approximation algorithms for **NP**-hard problems, achieving ratios forbidden in polynomial time or of showing impossibility of such algorithms is an interesting open question that deserves further investigation.

Acknowledgements Research supported by the French Agency for Research under the DEFIS program TODO, ANR-09-EMER-010.

References

1. Arora, S., Lund, C., Motwani, R., Sudan, M., Szegedy, M.: Proof verification and intractability of approximation problems. J. Assoc. Comput. Mach. **45**(3), 501–555 (1998)
2. Ausiello, G., Crescenzi, P., Gambosi, G., Kann, V., Marchetti-Spaccamela, A., Protasi, M.: Complexity and Approximation. Combinatorial Optimization Problems and Their Approximability Properties. Springer, Berlin (1999)

3. Berge, C.: Graphs and Hypergraphs. North Holland, Amsterdam (1973)
4. Berman, P., Fujito, T.: On the approximation properties of independent set problem in degree 3 graphs. In: Proceedings of the International Workshop on Algorithms and Data Structures, WADS'95. LNCS, vol. 955, pp. 449–460. Springer, Berlin (1995)
5. Bjorklund, A., Husfeldt, T.: Inclusion-exclusion algorithms for counting set partitions. In: Proceedings of the FOCS'06, pp. 575–582 (2006)
6. Boria, N., Bourgeois, N., Escoffier, B., Paschos, V.Th.: Exponential approximation schemata for some network design problems. Cahier du LAMSADE 303, LAMSADE, Université Paris-Dauphine. Available at http://www.lamsade.dauphine.fr/sites/default/IMG/pdf/cahier-303. (2011)
7. Bourgeois, N., Escoffier, B., Paschos, V.Th., van Rooij, J.M.M.: Fast algorithms for MAX INDEPENDENT SET. Algorithmica vol. 62, 1–2, pp. 382–415 (2012)
8. Bourgeois, N., Della Croce, F., Escoffier, B., Paschos, V.Th.: Exact algorithms for dominating clique problems. In: Dong, Y., Du, D.Z., Ibarra, O. (eds.) Proceedings of the International Symposium on Algorithms and Computation, (ISAAC'09). LNCS, vol. 5878, pp. 4–13. Springer, Berlin (2009)
9. Bourgeois, N., Escoffier, B., Paschos, V.Th.: Efficient approximation of combinatorial problems by moderately exponential algorithms. In: Dehne, F., Gavrilova, M., Sack, J.R., Tóth, C.D. (eds.) Proceedings of the Algorithms and Data Structures Symposium, (WADS'09). LNCS, vol. 5664, pp. 507–518. Springer, Berlin (2009)
10. Bourgeois, N., Escoffier, B., Paschos, V.Th.: Efficient approximation of MIN COLORING by moderately exponential algorithms. Inform. Process. Lett. **109**(16), 950–954 (2009)
11. Bourgeois, N., Escoffier, B., Paschos, V.Th.: Efficient approximation of MIN SET COVER by moderately exponential algorithms. Theoret. Comput. Sci. **410**(21–23), 2184–2195 (2009)
12. Bourgeois, N., Escoffier, B., Paschos, V.Th. Patt-Shamir B., Ekim T.: Fast algorithms for MIN INDEPENDENT DOMINATING SET. In: Proceedings of the Colloquium on Structural Information & Communication Complexity, (SIROCCO'10). LNCS, vol. 6058, 247–261 Spinger (2010)
13. Bourgeois, N., Escoffier, B., Paschos, V.Th., van Rooij, J.M.M.: A bottom-up method and fast algorithms for MAX INDEPENDENT SET. In: Kaplan, H. (ed.) Proceedings of the Scandinavian Symposium and Workshops on Algorithm Theory, (SWAT'10). LNCS, vol. 6139, pp. 62–73. Spinger, Berlin (2010)
14. Bourgeois, N., Escoffier, B., Paschos, V.Th.: Approximation of MAX INDEPENDENT SET, MIN VERTEX COVER and related problems by moderately exponential algorithms. Discrete Appl. Math. **159**(17), 1954–1970 (2011)
15. Brankovic, L., Fernau, H.: Combining two worlds: parameterized approximation for vertex cover. In: Cheong, O., Chwa, K.Y., Park, K. (eds.) Proceedings of the International Symposium on Algorithms and Computation, (ISAAC'10). LNCS, vol. 6506, pp. 390–402. Springer, Berlin (2010)
16. Byskov, J.M.: Enumerating maximal independent sets with applications to graph colouring. Oper. Res. Lett. **32**(6), 547–556 (2004)
17. Cai, L., Huang, X.: Fixed-parameter approximation: conceptual framework and approximability results. In: Bodlaender, H.L., Langston, M.A. (eds.) Proceedings of the International Workshop on Parameterized and Exact Computation, (IWPEC'06). LNCS, vol. 4169, pp. 96–108. Springer, Berlin (2006)
18. Chen, Y., Grohe, M., Grüber, M.: On parameterized approximability. In: Bodlaender, H.L., Langston, M.A. (eds.) Proceedings of the International Workshop on Parameterized and Exact Computation, (IWPEC'06). LNCS, vol. 4169, pp. 109–120. Springer, Berlin (2006)
19. Cygan, M., Pilipczuk, M.: Exact and approximate bandwidth. Theoret. Comput. Sci. **411**(40–42), 3701–3713 (2010)
20. Cygan, M., Kowalik, L., Wykurz, M.: Exponential-time approximation of weighted set cover. Inform. Process. Lett. **109**(16), 957–961 (2009)

21. Cygan, M., Pilipczuk, M., Wojtaszczyk, J.O.: Capacitated domination faster than $o(2^n)$. In: Kaplan, H. (ed.) Proceedings of the Scandinavian Symposium and Workshops on Algorithm Theory, (SWAT'10). LNCS, vol. 6139, pp. 74–80. Springer, Berlin (2010)
22. Demange, M., Paschos, V.Th.: On an approximation measure founded on the links between optimization and polynomial approximation theory. Theoret. Comput. Sci. **158**, 117–141 (1996)
23. Downey, R.G., Fellows, M.R.: Parameterized Complexity. Monographs in Computer Science. Springer, New York (1999)
24. Downey, R.G., Fellows, M.R., McCartin, C.: Parameterized approximation problems. In: Bodlaender, H.L., Langston, M.A. (eds.) Proceedings of the International Workshop on Parameterized and Exact Computation, (IWPEC'06). LNCS, vol. 4169, pp. 121–129. Springer, Berlin (2006)
25. Duh, R., Fürer, M.: Approximation of k-set cover by semi-local optimization. In: Proceedings of the STOC'97, pp. 256–265 (1997)
26. Eppstein, D.: Improved algorithms for 3-coloring, 3-edge-coloring, and constraint satisfaction. In: Proceedings of the Symposium on Discrete Algorithms, (SODA'01), pp. 329–337 (2001)
27. Escoffier, B., Paschos, V.Th., Tourniaire, E., Manindra A., S.B. Cooper and Angsheng Li.: Approximating MAX SAT by moderately exponential algorithms. Cahier du LAMSADE 303, LAMSADE, Université Paris-Dauphine. Available at http://www.lamsade.dauphine.fr/sites/default/IMG/pdf/cahier-304 Springer, LNCS, vol. 7287, 202–213 (2012)
28. Feige, U.: A threshold of $\ln n$ for approximating set cover. J. Assoc. Comput. Mach. **45**, 634–652 (1998)
29. Fomin, F.V., Kratsch, D.: Exact Exponential Algorithms. EATCS. Springer, Berlin (2010)
30. Fomin, F.V., Grandoni, F., Kratsch, D.: Measure and conquer: domination – a case study. In: Caires, L., Italiano, G.F., Monteiro, L., Palamidessi, C., Yung, M. (eds.) Proceedings of the ICALP'05. LNCS, vol. 3580, pp. 191–203. Springer, Berlin (2005)
31. Fürer, M., Gaspers, S., Kasiviswanathan, S.P.: An exponential time 2-approximation algorithm for bandwidth. In: Proceedings of the International Workshop on Parameterized and Exact Computation, (IWPEC'09). LNCS, vol. 5917, pp. 173–184. Springer, Berlin (2009)
32. Halldórsson, M.M.: Approximating the minimum maximal independence number. Inform. Process. Lett. **46**, 169–172 (1993)
33. Håstad, J.: Clique is hard to approximate within $n^{1-\varepsilon}$. Acta Mathematica **182**, 105–142 (1999)
34. Hochbaum, D.S. (ed.): Approximation Algorithms for NP-Hard Problems. PWS, Boston (1997)
35. Impagliazzo, R., Paturi, R., Zane, F.: Which problems have strongly exponential complexity? J. Comput. Syst. Sci. **63**(4), 512–530 (2001)
36. Johnson, D.S.: Approximation algorithms for combinatorial problems. J. Comput. Syst. Sci. **9**, 256–278 (1974)
37. Khot, S., Regev, O.: Vertex cover might be hard to approximate to within $2-\varepsilon$. In: Proceedings of the Annual Conference on Computational Complexity, (CCC'03), pp. 379–386 (2003)
38. Nemhauser, G.L., Trotter, L.E.: Vertex packings: structural properties and algorithms. Math. Program. **8**, 232–248 (1975)
39. Paschos, V.Th.: Complexité et approximation polynomiale. Hermès, Paris (2004)
40. Vassilevska, V., Williams, R., Woo, S.L.M.: Confronting hardness using a hybrid approach. In: Proceedings of the Symposium on Discrete Algorithms, (SODA'06), pp. 1–10 (2006)
41. Vazirani, V.: Approximation Algorithms. Springer, Berlin (2001)
42. Woeginger, G.J.: Exact algorithms for NP-hard problems: a survey. In: Juenger, M., Reinelt, G., Rinaldi, G. (eds.) Combinatorial Optimization - Eureka! You shrink!. LNCS, vol. 2570, pp. 185–207. Springer, Berlin (2003)
43. Zuckerman, D.: Linear degree extractors and the inapproximability of max clique and chromatic number. In: Proceedings of the STOC'06, pp. 681–690 (2006)

Multistart Branch and Bound for Large Asymmetric Distance-Constrained Vehicle Routing Problem

Samira Almoustafa, Said Hanafi, and Nenad Mladenović

Abstract In this chapter we revise and modify an old branch-and-bound method for solving the asymmetric distance-constrained vehicle routing problem suggested by Laporte et al. in 1987. It is based on reformulating distance-constrained vehicle routing problem into a travelling salesman problem and use of assignment problem as a lower bounding procedure. In addition, our algorithm uses the best-first strategy and new tolerance-based branching rules. Since our method was fast but memory consuming, it could stop before optimality is proven. Therefore we introduce the randomness, in case of ties, in choosing the node of the search tree. If an optimal solution is not found, we restart our procedure. In that way we get multistart branch-and-bound method. As far as we know instances we solved exactly (up to 1,000 customers) are much larger than instances considered for other VRP models from the recent literature. So, despite its simplicity, this proposed algorithm is capable of solving the largest instances ever solved in the literature. Moreover, this approach is general and may be used in solving other types of vehicle routing problems.

1 Introduction

The vehicle routing problem VRP is defined as follows: finding vehicle tours to connect the depot to n customers with m vehicles, such that every customer is visited exactly once; every vehicle starts and ends its tour at the depot. It is an NP-hard problem [10, 27]. There are many kinds of VRPs. For an overview of

S. Almoustafa (✉) • N. Mladenović
Brunel University, Uxbridge, Middlesex UB8 3PH, UK
e-mail: samira.al-moustafa@brunel.ac.uk; nenad.mladenovic@brunel.ac.uk

S. Hanafi
Lamih-Universite de Valenciennes, France
e-mail: said.hanafi@univ-valenciennes.fr

A. Migdalas et al. (eds.), *Optimization Theory, Decision Making, and Operations Research Applications*, Springer Proceedings in Mathematics & Statistics 31,
DOI 10.1007/978-1-4614-5134-1_2,

VRPs variants and applications, we refer to [4, 5, 16, 17, 26, 29]. We consider in this chapter distance-constrained VRP (DVRP) where the total travelled distance by each vehicle in the solution is less than or equal to the maximum possible travelled distance. If the distance from node i to node j is different from node j to node i, we call this problem asymmetric (ADVRP) otherwise the symmetric DVRP is defined. Surprisingly ADVRP is not studied as other types of VRPs. To the best of our knowledge, there are a few papers discuss this problem see [20].

A variety of integer programming formulations have been proposed for VRPs, including the so-called two- and three-index formulations, the set partitioning formulation and various formulations based on extra variables representing the flow of one or more commodities (see e.g. survey and formulation comparisons in [23]). On the other hand, recent solution techniques are mostly based on branch-and-cut or on branch-and-cut-and-price (see [4, 28] and recent survey by [5]).

Three types of algorithms are used to solve any VRP. The first type consists of exact algorithms. It has been studied by [5, 6, 18, 19, 22], etc. This type of methods is time-consuming. The second type consists of classical *heuristics* such as greedy, local search, relaxation-based, etc. It has been also studied by many researchers, e.g., [7, 21, 29]. They produce an approximate solution faster, when compared to the first type, but without guarantees of optimality. The third type consists of heuristics that are based on some *metaheuristic* rules. Such metaheuristics or framework for building heuristics are Simulated annealing, Tabu search, Genetic algorithms [29], Variable neighborhood search [11], etc.

In this chapter we suggest a new simple algorithm for solving ADVRP that is based on Branch-and-Bound (B&B) method. As in [20], the ADVRP is first transformed to the Travelling salesman problem (TSP). The lower bounds are obtained by relaxation of subtour elimination and maximum distance constraints. Thus the Assignment problem (AP) is solved in each node of the B&B tree. We use the best-first-search strategy and adapted tolerance-based rules for branching. That is, the next node in the tree is one with the smallest relaxed objective function value. In the case of tie, we use two tie-breaking rules: (1) the last one in the list; (2) the random one among them. We found that our B&B-based method is very fast but memory consuming. That is why we suggested multistart B&B method (MSBB-ADVRP). It simply uses random tie-breaking rule in selection of the next subproblem. Computational results show that we are able to provide exact solutions for instances with up to 1,000 nodes. The size of problems could be even larger if more powerful computer (with larger memory) is used. As far as we know those instances are much larger than instances considered for other similar VRP models and exact solution approaches from the recent literature. For example, in the recent paper by Baldacci and Mingozzi [3], several VRP problem types are studied and sophisticated exact solution methods tested. The largest instances solved had 199 customers. Therefore, our simple algorithms are capable of solving the largest instances ever solved in the literature.

The structure of this chapter is as follows. In Sect. 2 we present mathematical programming formulations of ADVRP. In Sect. 3 we discuss details of deterministic branch-and-bound method for ADVRP. In Sect. 4 we present our multistart approach for solving ADVRP. Section 5 contains details regarding data structure that we used in our implementation. Computational results are provided in Sect. 6. In Sect. 7 we give conclusion and future research directions.

2 Mathematical Programming Formulations for ADVRP

In this section we give two mathematical programming formulations of ADVRP. The first one, so-called flow-based formulation, is used for comparison purposes in computational results sect. The second is based on transformation of ADVRP to asymmetric travelling salesman problem (TSP). We use its relaxation in our B&B exact method that will be described in Sect. 3.

Let $N' = \{1,2,\ldots,n-1\}$ denote the set of customers and $V' = N' \cup \{0\}$ denote the set of nodes where 0 is index of the depot. A set of arcs is denoted by A', $A' = \{(i,j) \in V' \times V' : i \neq j\}$. The travelled distance from customer i to customer j and the number of vehicles are denoted by d'_{ij} and m, respectively. The maximum distance allowed is denoted by $D_{\max}$. The shortest distance between customer i and customer j is denoted by c_{ij}. The decision binary variable x_{ij} is defined as follows:

$$x_{ij} = \begin{cases} 1 & \text{if the arc } (i,j) \text{ belongs to any tour;} \\ 0 & \text{otherwise.} \end{cases} \tag{1}$$

2.1 Flow-Based Formulation

For the sake of comparison, we use another formulation of ADVRP with polynomial number of variables and constraints, without copying depots. This is achieved by introducing the new set of variables z_{ij}. They present the shortest length travelled from the depot to customer j, where i is the predecessor of j. The formulation of ADVRP, which will be later used with CPLEX solver (CPLEX-ADVRP), is given below [13]:

$$f(S) = \min \sum_{(i,j)\in A'} c_{ij}x_{ij} \tag{2}$$

subject to

$$\sum_{i\in N'} x_{ij} = 1 \qquad \forall\ j \in N' \tag{3}$$

$$\sum_{j\in N'} x_{ij} = 1 \qquad \forall\, i \in N' \tag{4}$$

$$\sum_{i\in N'} x_{i0} = m \tag{5}$$

$$\sum_{j\in N'} x_{0j} = m \tag{6}$$

$$\sum_{(i,j)\in A'} z_{ij} - \sum_{(i,j)\in A'} z_{ji} - \sum_{j\in V'} c_{ij}x_{ij} = 0 \qquad \forall\, i \in N' \tag{7}$$

$$z_{ij} \le (D_{\max} - c_{j0})x_{ij} \qquad \forall\, j \neq 0, \forall (i,j) \in A' \tag{8}$$

$$z_{ij} \ge (c_{ij} + c_{0i})x_{ij} \qquad \forall\, i \neq 0, \forall (i,j) \in A' \tag{9}$$

$$z_{0i} = c_{0i}x_{0i} \qquad \forall\, i \in N' \tag{10}$$

$$x_{ij} \in \{0,1\} \qquad \forall\, (i,j) \in A'. \tag{11}$$

Obviously, there are polynomial number of variables and constraints. This model is known as flow-based model since constraint (7) is typical flow constraint. It says that the distance from node i to any other node j on the tour should be equal to the difference between distance from depot to i and distance from depot to j. Constraint (8) presents that the total distance from depot to customer j and the shortest distance from customer j to depot directly are less than or equal to the maximum distance allowed. In addition, according to constraint (9) the total distance from depot to customer j should be greater than or equal to the distance from the depot to customer i plus the distance from customer i to customer j. Constraint (10) gives the initial value for z_{0i} which is equal to the distance from depot to customer i. Last constraint (11) introduces the decision variables x_{ij} as binary variables.

2.2 TSP Formulation

The TSP formulation may be obtained by adding $m-1$ copies of the depot to V' [22]. Now there are $n+m-1$ nodes in the new augmented directed graph $G(V,A)$, where

$$V = V^{'} \cup \{n, n+1, \ldots, n+m-2\}.$$

The distance matrix D is obtained from D' by the following transformation rules, where $i,j \in V$:

$$d_{ij} = \begin{cases} d'_{ij} & \text{if } (0 \le i < n, 0 \le j < n, i \neq j) \\ d'_{0j} & \text{if } (i \ge n, 0 < j < n) \\ d'_{i0} & \text{if } (0 < i < n, j \ge n) \\ \infty & \text{otherwise} \end{cases} \tag{12}$$

Then the formulation of TSP [8] is given below (13)–(17) as follows:

$$f(S) = \min \sum_{(i,j)\in A} d_{ij}x_{ij} \tag{13}$$

where x_{ij} satisfies these conditions

$$\sum_i x_{ij} = 1 \qquad \forall\, j \in V \tag{14}$$

$$\sum_j x_{ij} = 1 \qquad \forall\, i \in V \tag{15}$$

$$\sum_{i,j \in U} x_{ij} \leq \mid U \mid -1 \qquad \forall\, U \subset V, |U| \geq 2, \tag{16}$$

$$x_{ij} \in \{0,1\} \qquad \forall\, i,j \in V. \tag{17}$$

$$+\ distance\ \ constraints \tag{18}$$

The constraints (14) and (15) ensure that *in* and *out* degree of each node are equal to 1. The constraint (16) eliminates subtours, where U is any subset of V. The constraint (17) is integrality constraint. To formulate ADVRP in addition to (13)–(17), we need to add distance constraint (18), which check that the total distance for each tour should be less than maximum distance allowed ($D_{\max}$).

The weak point of this formulation is exponential number of constraints in (16), since the number of subsets U is exponential. However, in our B&B method that will be explained in the next sect., this set of constraints will be relaxed.

DVRP may be also seen as a special case of VRP with time windows constraints [23]. In the vehicle routing problem with time windows (VRPTW), it takes a time $t_{ij} \geq 0$ to traverse arc (i,j). When $i \in N'$, the quantity t_{ij} includes any time required to service i. For each $i \in N'$, service must begin within the time window $[e_i, l_i]$, where $0 \leq e_i \leq l_i \leq \infty$. We will also allow that each vehicle be required to leave the depot at time e_0 or afterwards, and arrives back at the depot by time l_0 or earlier. A vehicle is permitted to wait at a vertex, either before or after serving a customer. The DVRP can be viewed as a special case of the VRPTW by setting $t_{ij} = d_{ij}$, $e_i = d_{0i}$ and $l_i = D - d_{i0}$.

3 Branch and Bound for ADVRP

The branch and bound (B&B) is an exact method for solving integer programming problem. It consists of enumerating all possible solutions within the so-called search tree and pruning subtrees when better solutions than the current one (upper bound) could not be found. B&B rules are briefly given below:

- The initial feasible solution of good quality is usually obtained by heuristic and its objective function value is initial upper bound (UB). If the heuristic solution is not known, then the upper bound is set to infinity ($UB = \infty$)
- The original problem is placed at the root of the branch-and-bound or search tree. All other nodes represent subproblems. In solving subproblems some variables are fixed and some constraints are ignored (relaxed), so that the lower bound is obtained

- The search strategy defines the way in which we choose the next node for branching. There are three basic branching strategies: breadth first search, depth first search and hybrid search which is also called *best first search strategy*. In this chapter we implement this strategy

3.1 Lower Bounds

To apply B&B we need to have lower bounding procedure that should be applied in each node of the search tree. Of course, there are many ways to relax model (13)–(18). The more constraints are included, the better (higher) lower bound is obtained. We use AP as lower bounding procedure of the TSP formulation given in (13)–(15), (17), i.e., we relax all tour elimination and maximum distance constraints (16) and (18). Although the quality of the lower bound is not high, the benefit is in using very fast exact *Hungarian* method [15] for solving AP. Here we use its implementation described in [12]. The complexity of AP at root node is in $O(n^3)$ [31]. Another advantage is that the relaxed AP solution is already integer.

Proposition 1. *Any feasible* AP *solutions for problem (13)–(15), (17) consists of a set of cycles, i.e., a sequence of arcs starting and ending at the same vertex with the number of arcs in each cycle* $\omega \geq 2$.

Proof. It is clear from (14) and (15) that the degree of each vertex in S is equal to 2. It has one incoming and one out coming arc. If the matrix D was symmetric and n even, then those cycles would contain just 2 vertices and therefore $\omega = 2$. However, in all other cases, ω is obviously larger than 2: if vertex i is assigned to j, then j is not necessary assigned to i. □

There are three types of cycles obtained by AP relaxation:

- *A served cycle*—contains exactly one depot
- *An unserved cycle*—contains no depot
- *A path*—contains more than one depot

In the last case, each path may be divided into served cycles. Therefore the number of served cycles is equal to the number of depots in the path. Subsequently, the term *tour* is used to denote either a served cycle or unserved cycle or a path; the term *depot* is used to denote either the original depot or a copy of the depot. A tour is called *infeasible* if its total distance is larger than $D_{\max}$ or if it contains no depot.

3.2 Tolerance-Based Branching Rule

Since the sets of constraints (16) and (18) are relaxed, the AP solution may have many infeasible tours. If the tour is infeasible, it must be destroyed, i.e., one arc should be excluded (deleted). We exclude an arc from the current infeasible solution S by giving to it large value (∞) and then resolve AP relaxation again. Really, in the new solution, such an arc will not appear, since we minimize AP objective function. There are several ways to remove an arc from S. We can try all possible removals (one at the time) and collect all objective function values obtained from solving new corresponding AP 's.

However, in this chapter we use the concept of tolerance. Tolerance is one of the sensitivity analysis techniques (for more details on sensitivity analysis, see [14,24]). The definition of tolerance is used as a branching rule within B&B method in [30] for solving ATSP. We first extend this idea for solving ADVRP.

The difference between the value of the objective function before and after the exclusion of an arc in the current solution is called upper tolerance (UT) of the arc [9]. The arc to be removed corresponds to the smallest objective function obtained. Therefore, in our ADVRP tolerance-based B&B (TOL-ADVRP), the arc which has the smallest upper tolerance is chosen to be excluded. Some preliminary results of this approach has been given in [1].

Another possibility of destroying infeasible tour is to exclude the arc with the largest cost. B&B method that uses such a branching rule we call COST-ADVRP. However, based on extensive computational analysis, the results obtained with COST-ADVRP were of slightly worse quality than those obtained by our TOL-ADVRP. That was the reason why in computational analysis sect. we give TOL-ADVRP results. In TOL-ADVRP when there are more than one subproblem in the list of active subproblems (that have the smallest objective value) we choose the last among them to branch next. In other words, our tie-breaking rule is deterministic.

3.3 Algorithm

The main steps of TOL-ADVRP algorithm are given in Algorithm 2.1, where we use the following notation:

S—the relaxed solution obtained by AP or $S = \{T_1, T_2, \ldots, T_{M(S)}\}$ presented as a set of tours where $M(S)$ is the number of tours in S
$d(T_k)$—the total travelled distance in a tour k where $d(T_k) = \sum_{(i,j)\in T_k} d_{ij}$
$t(T_k)$—the number of arcs in a tour k
$f(S) = \sum_{k=1}^{M(s)} d(T_k)$—the value of an optimal solution to AP
S^*—an optimal solution to ADVRP
L—the list of active subproblems (or unfathomed nodes in a search tree), it is updated during the execution of the code
LB—the lower bound. It is the smallest value of the objective function to AP among those in L
UB—the upper bound value to $ADVRP$
$APcnt$—counts the number of nodes in B&B tree (how many times AP subroutine is called)
$Maxnodes$—the maximum number of nodes allowed in B&B tree. In order to prevent termination with no memory message, we use it as stopping condition. Here we set $Maxnodes = 100,000$

Algorithm 2.1: (TOL-ADVRP) algorithm with tolerance-based branching

ProcedureTOL-ADVRP $(n, m, D_{max}, D, Maxnodes, \beta, S^*, ind)$;
$UB \leftarrow m \times D_{max}$, $APcnt \leftarrow 1$, set iteration counter $i \leftarrow 1$;
Solve AP to get its solution $S = \{T_l | l = 1, \ldots, M\}$;
$L = \{1\}$- the list contains the root node;
Calculate $d(T_l)$ for every tour of $T_l \in S$;
if (S feasible) **then** ($S^* = S$ is an optimal solution; *ind*=1; stop);
while $(APcnt < Maxnodes)$ **do**
 Branching. Choose subproblem $b_i \in L$ with the smallest value of the objective function; in the case of tie, choose one from the list with respect to β value;
 Best first. Find the ratio $d(T_k)/t(T_k)$ for every infeasible tour $k = 1, .., M'$ where M' is the number of infeasible tours and choose the tour k^* with the largest ratio;
 Tolerance (expanding search tree). Calculate upper tolerances for all arcs in this k^* tour by solving $t(T_{k^*})$ times AP problem to get solutions S_r where $r = 1, \ldots, t(T_{k^*})$. Expand search tree with those $t(T_{k^*})$ subproblems and update $APcnt$ as: $APcnt = APcnt + t(T_{k^*})$;
 Feasibility check. Check feasibility of all new (expanded) nodes;
 Update. Update UB (if necessary), and L based on the current UB value as: $L \leftarrow \{r | f(S_r) < UB\} \setminus \cup \{b_j | j = 1, .., i\}$ where S_r are infeasible solutions;
 Optimality conditions. If ($L = \emptyset$ and $UB \neq m \times D_{max}$) **then** (S^* is the optimal solution where $f(S^*) = UB$; *ind*=1; stop). Otherwise, **If** ($L = \emptyset$ and $UB = m \times D_{max}$) **then** (there is no feasible solution; *ind*=4; stop);
 $i = i + 1$;
end
Termination. If ($UB \neq m \times D$) **then** (S is the new incumbent; *ind* =2; return), otherwise (no memory; *ind*=3; return) ;

β—the type of tie-breaking rule that has two values:

$$\beta = \begin{cases} 0 & \text{deterministic (last in } L\text{)} \\ 1 & \text{at random} \end{cases}$$

ind—the variable that covers all possible outputs of the algorithm. The basic algorithm may stop with the following outputs:

$$ind = \begin{cases} 1 & \text{an optimal solution } S^* \text{ is found ;} \\ 2 & \text{feasible solution } S \text{ is found but not proven as optimal;} \\ 3 & \text{no feasible solution is found (lack of memory);} \\ 4 & \text{there is no feasible solution of the problem at all.} \end{cases}$$

Here we explain some of the steps of algorithm TOL-ADVRP: At the root node we find solution S by solving AP problem. Then we calculate the total distance for every tour of S and check the feasibility of the solution. If it is feasible, then the optimal solution exists and the program stops. Otherwise, we repeat the following steps until the memory limit is reached:

- *Branching.* Choose the subproblem $b_i \in L$ with the smallest value of the objective function where i denoted to the iteration. In the case when more than one subproblem has the smallest value, the choice is made according to value of β. If $\beta = 0$ we choose last subproblem in
  the list *L*; otherwise choose one randomly to develop further. This second option will be used after in our multistart method.
- *Best first.* Find the ratio between the total distance $d(T_k)$ and the number of arcs in the chosen subproblem $t(T_k)$ for every infeasible tour. Choose the tour with the largest ratio to branch.
- *Tolerance (expanding search tree).* Calculate upper tolerances for all arcs in this tour as follows: Exclude in turn one arc from this tour by putting ∞ in the distance matrix. Find *AP* solution to those subproblems. Check feasibility for each new subproblem. If this subproblem produces a feasible solution *S*, and its value is smaller than UB then update the value of the upper bound ($UB = f(S)$). Find the difference between the value of the objective function before and after excluding the arc. This gives the value of upper tolerance (UT) to this arc. Note that the value of the counter *APcnt* is increased by the number of arcs in the chosen tour:

$$APcnt = APcnt + t(T_{k^*})$$

- *Check feasibility.* Check feasibility of the solution at every new subproblem which the program generates,
- *Update.* If a feasible solution is found and its value is smaller than current upper bound, then update the value of upper bound, update the list of active subproblems by removing the subproblems that have value greater than the value of upper bound and by adding the new expanded subproblems that has value smaller than UB as follows:

$$L \leftarrow \{r | f(S_r) < UB\} \backslash \cup \{b_j | j = 1, ..., i\}$$

  Note S_r is infeasible solutions.
- *Optimality conditions.* Check if $L = \emptyset$ and UB is updated then stop with the value of optimal solution $f(S^*) = UB$ ($ind = 1$). Otherwise, if $L = \emptyset$ and UB is not updated, stop with the message that no feasible solution exists ($ind = 4$),
- *Termination.* When there in no memory we get two possible outputs: If UB is updated, then S^* is returned as a feasible but not proved as optimal solution ($ind = 2$). Otherwise, a feasible solution has not been found, but that does not mean the feasible solution does not exist ($ind = 3$).

Proposition 2. *Algorithm* `TOL-ADVRP` *finds an optimal solution to* `ADVRP` *or proves that such a solution does not exist.*

Proof. It is enough to show that our `B&B` algorithm enumerates all possible `VRP` tours with *m* vehicles (assuming that $D_{\max} = \infty$ and there is no memory restriction). Really, our enumeration is based on eliminating arcs by giving them value ∞ (in step 8 in Algorithm 2.1). It is followed by solving lower bounding `AP` problem. `AP` provides a solution *S* as a set of cycles (Proposition 1). Clearly the set of all

Table 1 Original distance matrix for ADVRP with $n = 8$ and $m = 2$

	0	1	2	3	4	5	6	7
0	∞	2	11	10	8	7	6	5
1	6	∞	1	8	8	4	6	7
2	5	12	∞	11	8	12	3	11
3	11	9	10	∞	1	9	8	10
4	11	11	9	4	∞	2	10	9
5	12	8	5	2	11	∞	11	9
6	10	11	12	10	9	12	∞	3
7	7	10	10	10	6	3	1	∞

Table 2 New distance matrix

	0	1	2	3	4	5	6	7	8
0	∞	2	11	10	8	7	6	5	∞
1	6	∞	1	8	8	4	6	7	6
2	5	12	∞	11	8	12	3	11	5
3	11	9	10	∞	1	9	8	10	11
4	11	11	9	4	∞	2	10	9	11
5	12	8	5	2	11	∞	11	9	12
6	10	11	12	10	9	12	∞	3	10
7	7	10	10	10	6	3	1	∞	7
8	∞	2	11	10	8	7	6	5	∞

solutions generated with our method contains all possible cycles with m vehicles. On the other hand, the set of all feasible ADVRP tours are the subset of all this generated set. Therefore, our TOL-ADVRP enumerates all feasible tours. □

3.4 Illustrative Example

The example from Table 1 is taken from [2], where the number of customers is $n = 8$ and the number of vehicles is $m = 2$. The location of the first customer is considered as a depot.

In matrix D the first row represents the distances from the depot to all other customers. The first column represents the distances from each customer to the depot, and all other entries represent distances between the remaining customers. To solve this problem, we have to add $m - 1 = 2 - 1 = 1$ copy of a depot. Table 2 illustrates the new distance matrix after adding the last row and the last column according to the new distance function (see Sect. 2), where 0 and 8 represent two depots for this problem. We will consider 2 problems with this dataset with 2 different values of $D_{\max}$. First value of $D_{\max(1)}$ is ∞, which produce as output the longest tour LT in the optimal solution, then the second value of maximum distance allowed $D_{\max(2)}$ is chosen to be $0.90 \times LT$. In addition, we will denote by $\overline{f_b}(i,j)$ the value of AP at the subproblem b with $d(i,j) = \infty$;

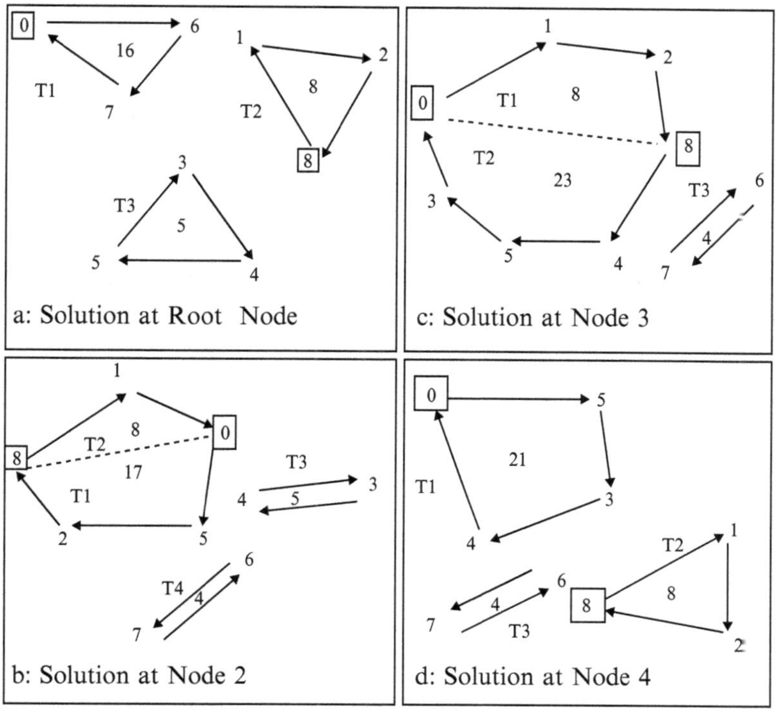

Fig. 1 Solutions at root node (node 1), node 2, node 3 and node 4

LB solution at the root node of B&B tree, obtained by solving AP, is given in Fig. 1a. Each depot label is written inside a squared box and the total distance is written inside each tour. The value of the objective function to the relaxed problem at the root node, obtained by solving AP, is $f_1 = 29$ (see Fig. 1a).

Problem 1 ($D_{\max(1)} = \infty$). We consider the problem with $D_{\max} = \infty$. we set $APcnt = 1, L = \{1\}, i = 1$, and $b_1 = 1$. Clearly, initial value of $UB = m \times D_{\max} = \infty$.

Iteration 1. It can be seen from Fig. 1a that we obtain solution S with three tours: one of them is infeasible, because it does not contain depot. Therefore this solution S is not feasible.

The program will check the total distance for all tours in the solution at the root node $d(T_m) = \{d(T_1), d(T_2), d(T_3)\} = \{16, 8, 5\}$. Since only the last tour T_3 is *infeasible*, there is one possible subtour for branching. The number of arcs in T_3 is equal to 3 ($t(T_3) = 3$) and its total distance is equal to 5 ($d(T_3) = 5$).

The value of upper tolerance for every arc $\{(5,3), (3,4), (4,5)\}$ in the chosen tour T_3 is calculated as follows:

1. Arc $(5,3)$: Exclude this arc from the solution by putting that its length is equal to ∞ instead of its original value $d(5,3) = 2$ in the new distance matrix (Table 2). Then find the solution of AP with $d(5,3) = \infty$. This solution is not feasible for

ADVRP (Fig. 1b) because there is at least one infeasible tour (in fact there are two infeasible tours T_3, T_4). The value of the optimal AP solution for the new distance matrix at node 2 in the search tree is $f_2 = \overline{f}_1(5,3) = 34$. The upper tolerance value (UT) for the arc $(5,3)$ in the optimal solution S is calculated as:

$$UT(5,3) = \overline{f}_1(5,3) - f_1 = 34 - 29 = 5$$

2. Arc $(3,4)$: First restore $d(5,3)$ into its previous value 2. By excluding an arc $(3,4)$ as before, we get $f_3 = \overline{f}_1(3,4) = 35$. The solution at node 3 is also not feasible for ADVRP since it contains one infeasible tour T_3 (Fig. 1c). The value of the upper tolerance value for arc $(3,4)$ is:

$$UT(3,4) = \overline{f}_1(3,4) - f_1 = 35 - 29 = 6$$

3. Arc $(4,5)$: As before restore $d(3,4)$ into its previous value 1. Excluding an arc $(4,5)$ produces $f_4 = \overline{f}_1(4,5) = 33$ (Fig. 1d). The solution at node 4 is not a feasible solution because it contains one infeasible tour T_3. The value of upper tolerance for arc $(4,5)$ is:

$$UT(4,5) = \overline{f}_1(4,5) - f_1 = 33 - 29 = 4$$

The value of $APcnt = 1 + 3 = 4$. No need to update UB since no feasible solution is found, while we need to update L. The list of active subproblems $L = \{2,3,4\}$.

Iteration 2. The smallest upper tolerance is at node $b_2 = 4$. Therefore, the arc $(4,5)$ is excluded and $LB = f_4 = 33$. We start now from subproblem 4, which gives an infeasible solution (Fig. 1d). Among 3 tours in that solution S only one is infeasible. It has two arcs. Thus, in this case, the two new subproblems are generated by the program:

$$f_5 = \overline{f}_4(7,6) = 34 \text{ (feasible solution : d(T1) = 26, d(T2) = 8)}$$
$$f_6 = \overline{f}_4(6,7) = 37 \text{ (infeasible solution)}.$$

Update the value of upper bound UB ($UB = 34$) and the list of active subproblems ($L = \emptyset$) and the counter $APcnt = 6$. So the optimal solution is found at node 5 in the search tree (Fig. 2b). In this small example, the total number of subproblems generated in the search tree is 6.

Problem 2 ($\mathbf{D_{max(2)}} = 23$). The longest tour in the optimal solution is 26, we use this value to get new value to $D_{\max}$ where $D_{\max} = 0.9 \times 26 = 23.4$. We will consider only the integral part of this number so that the new value of maximum distance allowed is $D_{\max} = 23$. We update the value of $D_{\max}$, and we run the same example.

Iteration 1. It will be the same as iteration 1 in problem 1, i.e., $L = \{2,3,4\}$.

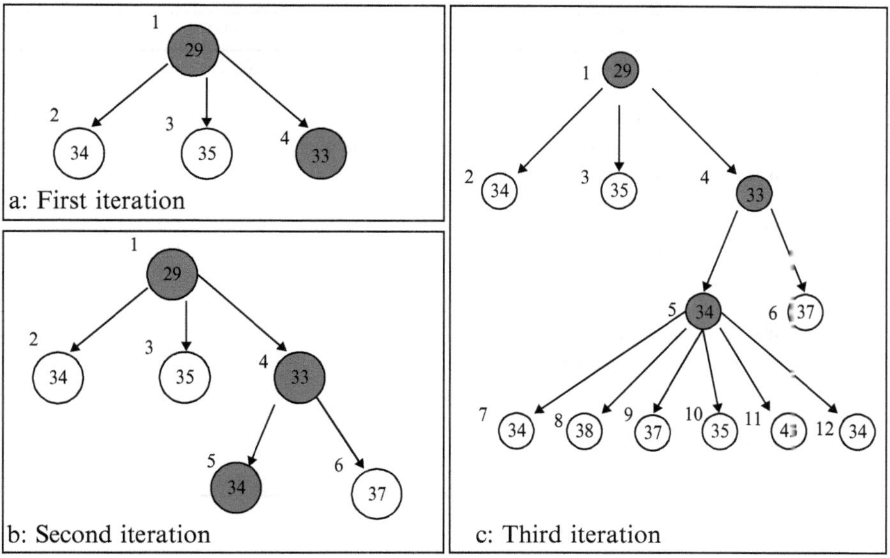

Fig. 2 TOL-ADVRP tree

Iteration 2. In this iteration we have $L = \{2,3,5,6\}$. Moreover we have two nodes (node 2 and node 5) that have the same smallest value 34 (Fig. 2b). Therefore, according to tie-breaking rule for TOL-ADVRP, we have to choose the last node to branch next ($b_3 = 5$). When we branch at node 5, we find first feasible solution at node 9, so we update $UB = 37, APcnt = 12, L = \{2,3,7,10,12\}$.

Iteration 3. In this iteration we have 3 nodes (2, 7 and 12) that have the same smallest value 34 (Fig. 2c), we have to choose the last node to branch next i.e., $b_4 = 12$, etc.

At the end, after generating 58 subproblems ($APcnt = 58$) we get optimal solution value 37.

4 Multistart Method

As mention earlier, motivation for developing multistart B&B method for solving ADVRP is based on the fact that TOL-ADVRP is very fast but requires a lot of computer memory. It usually stops after a few seconds reaching the maximum number of nodes visited in the search tree. The main idea of (MSBB-ADVRP) is to introduce random selection of the next subproblem among those with the same (smallest) objective function value. This random choice may cause the generation of smaller search tree. Therefore, if we reach the maximum number of subproblems allowed, we restart exact B&B method hoping that in the next attempt we will get

Algorithm 2.2: Algorithm of Multistart B&B (MSBB-ADVRP)

ProcedureMSBB-ADVRP $(n, m, D_{max}, D, Maxnodes, \beta, ntrail, S_{best})$;
$f_{best} \leftarrow \infty$, $\beta \leftarrow 1$;
for $i = 1$ **to** $ntrail$ **do**
    TOL-ADVRP$(n, m, D_{max}, D, Maxnodes, \beta, S^*, ind)$;
    if $(ind = 1)$ **then**
        $S_{best} = S^*$, stop ;
    end
    if $(ind = 4)$ **then**
        stop ;
    end
    if $(1 < ind < 4)$ **then**
        if $(f(S^*) < f_{best})$ **then**
            $S_{best} = S^*$, $f_{best} = f(S^*)$;
        end
    end
end

an optimal solution. For that reason we rerun TOL-ADVRP (with $\beta = 1$) many times until an optimal solution is found or infeasibility is proven. Therefore, the (MSBB-ADVRP) may be summarized as follows (see Algorithm 2.2):

1. Re-run each instance given number of times (*ntrail*—a parameter); stop when an optimal solution is found ($ind = 1$) or infeasibility of the problem instance is proved ($ind = 4$). This will increase the chance to find an optimal solution or at least improve the value of the best feasible solution found so far
2. Choose randomly one subproblem from the list of active subproblems (L) as follows:

   - Generate uniform random number $\alpha \in [0, 1]$
   - Find the number of nodes in the list (ns) which has value equal to the smallest objective function value in this list
   - Find $k \in [1, ns]$ as $k = 1 + ns * \alpha$
   - Branch on the node corresponding to kth position in the list ns

4.1 Algorithm

4.2 Illustrative Example

We now present our MSBB-ADVRP on the same example from the previous sect. We do not consider Problem 1, since there were no ties.

Problem 2. We run the example with $D_{\max(2)} = 23$. In the first iteration both programs do the same because we have only one smallest value in L (see Fig. 2a).

At the second iteration two nodes 2 and 5 have the same smallest value of 34 (Fig. 2b). So, there are two options. Suppose we randomly choose node 5. Then we get the solution as given at Fig. 2c. In the third iteration we have the number of subproblems with the same value is equal to 3 ($ns = 3$). So we need to choose one k among three nodes: 2, 7 or 12 where $k = 1 + 3\alpha$. Assume that $\alpha = 0.4$, then $k = 2$ and the second node (7) is chosen for branching. This step is repeated at each iteration until the optimal solution is found.

5 An Efficient Implementation: Data Structure

The most important task in implementation of TOL-ADVRP is to keep track of excluded arcs (a,b) during complete enumeration within B&B tree. Left-hand side arcs (a) are stored in the first matrix A and the right-hand side arcs (b) are stored in the second matrix B. Those matrices are expanded during the execution of the code. Each row of A and B represents a node in the search tree. Thus, we always start with $A = [-1]$ and $B = [-1]$ since there are no excluded arcs at the root node. Note that we use symbol (-1) to denote a dummy vertex. Each time some arc is excluded (its value set to ∞ and AP solved again), new row in both matrices are added, containing end points of excluded arc. In addition, each iteration brings a new column, where rows from previous iteration are filled with dummy vertices. Of course based on best-first-search criterion and tie-breaking rule, new rows of A and B (in new iteration) will keep track of excluded parent vertices from previous iterations. We will now present our data structure on the same example from the previous sect. At the root node we have $A = [-1]$, $B = [-1]$ (see Fig. 3).

At the first iteration we have 3 new nodes in the search tree, or 3 possible arcs to be excluded (see Fig. 2a). So we add 3 more rows to our matrices A and B. Thus, we have only one column and 4 rows in both matrices. At node 2 we exclude arc $(5,3)$. So we put 5 in the row 2 of A, and 3 in the row 2 of B. The same is done for other two nodes (see Fig. 3: First iteration). Node 4 is chosen to branch since the AB solution, after excluding arc (4, 5), was the smallest. So we copy all the numbers from both matrices of node 4 and put them as initial values to all sons of node 4.

The sons of each parent node have identical rows except the last value. The number of values in each row which is not equal (-1) represents the number of excluded arcs in this node. We use temporary vector (V) to save the values of excluded arcs in the node before we solve AP.

In the second iteration we have 2 more nodes. We add two rows to both matrices (exclude two arcs from some nodes) and one column. According to our example we choose node 4 to branch. The new two nodes will copy the information from the row of node 4. Then it will add the new arc that was excluded in each case (see Fig. 3: Second iteration). For the root node all the row contains (-1), and for nodes (2,3,4) we put (-1) in the second column because we have excluded only one arc for them. Each time we increase the dimension of both matrix we have to put (-1) for all the previous nodes in the new columns.

At root node:

A:

→	1	−1

B:

1	−1

First iteration:

A:

	1	−1
	2	5
	3	3
→	4	4

V= [d (4,5)

B:

1	−1
2	3
3	4
4	5

Second iteration:

A:

	1	−1	−1
	2	5	−1
	3	3	−1
	4	4	−1
→	5	4	7
	6	4	6

V=[d (4, 5) d (7, 6)]

B:

1	−1	−1
2	3	−1
3	4	−1
4	5	−1
5	5	6
6	5	7

Fig. 3 First two iterations

Third iteration:

A:

1	−1	−1	−1
2	5	−1	−1
3	3	−1	−1
4	4	−1	−1
5	4	7	−1
6	4	6	−1
7	4	7	0
8	4	7	6
9	4	7	7
10	4	7	5
11	4	7	3
12	4	7	4

B:

1	−1	−1	−1
2	3	−1	−1
3	4	−1	−1
4	5	−1	−1
5	5	6	−1
6	5	7	−1
7	5	6	6
8	5	6	7
9	5	6	5
10	5	6	3
11	5	6	4
12	5	6	0

Fig. 4 Third iteration

In the third iteration of branching we choose to branch on node 5. This will produce 6 new nodes. So we need to add 6 rows and one more column to both matrices (see Fig. 4: Third iteration).

6 Computational Results

Computers: All experiments were implemented under windows XP and on intel(R) Core(TM)2 CPU 6600@2.40GHz, with 3.24 GB of RAM. The code is written in C++ language. Some parts of the code are taken from.

Test Instances. A full asymmetric distance matrices are generated at random using uniform distribution to generate integer numbers between 1 and 100. The generator for random test instances needs the following input data:

n—the size of distance matrix
γ—the parameter that controls the degree of symmetry in the distance matrix, $\gamma \in [0,1]$: 0 means completely random and asymmetric; 1 means completely symmetric; 0.5 means 50% symmetric, etc.
seed—the random number: when $n \leq 200$, four different seeds were chosen to generate four different distance matrices for each combination of (n,m). However, only one distance matrix is generated in case: $200 < n \leq 1{,}000$

In addition, the shortest distances between each two customers are calculated to get input matrix C. Test instances are divided into two groups: small size (with $n = 40,60,...,200$) and large size (with $n = 240,280,\ldots,1{,}000$). For each n belonging to the small set, two different types of instances are generated, based on the different number of vehicles: $m_1 = n/20$ and $m_2 = n/10$. For instances belonging to the large set, we use only m_1. In addition, for each distance matrix we consider 3 problems with 3 different values of $D_{\max}$ (see Sect. 3.4). First value of $D_{\max}$ is ∞, then we use this formulae to obtain new value of $D_{\max}$:

$$D_{\max(i)} = 0.90 \times LT(i-1),$$

where $i \in \{2,3\}$, and $LT(i-1)$ is the longest tour in the optimal solution when the value of maximum distance allowed is $D_{\max(i-1)}$. Thus, the total number of instances is 257. All test instances used in this chapter can be found on the web site www.mi.sanu.ac.rs/~nenad/advrp/ as well as the code for generator coded in C++.

Methods Compared. In this chapter we compare three methods to find an optimal solution to ADVRP: MSBB-ADVRP, TOL-ADVRP, CPLEX-ADVRP. In all our experiments reported below, we run MSBB-ADVRP only two times. We note that increasing the number of restarts might improve chances to find an optimal solution, but with the cost of larger CPU time. In CPLEX-ADVRP the process will continue until an optimal solution is found or the time limit is reached. We choose the time limit to be 10,800 s (3 h) for all test instances.

Comparison. Tables 3–5 contain summary results to all 257 test instances from $n = 40$ up to $n = 1{,}000$ customers with $D_{\max(1)} = \infty$, $D_{\max(2)} = 0.90 \times LT(1)$, and $D_{\max(3)} = 0.90 \times LT(2)$, respectively. Detailed results for all three methods may be found on our web site www.mi.sanu.ac.rs/~nenad/advrp/. The rows in all tables give the following characteristics:

Table 3 Results for 92 instances with $D_{\max(1)} = \infty$

	72 small test inst.			20 larg test inst.			Total number		
	TOL	CPLEX	MSBB	TOL	CPLEX	MSBB	TOL	CPLEX	MSBB
♯ Opt ($ind = 1$)	72	72	72	20	2	20	92	74	92
♯ Feas ($ind = 2$)	0	0	0	0	0	0	0	0	0
♯ No Mem ($ind = 3$)	0	0	0	0	18	0	0	18	0
♯ No Feas ($ind = 4$)	0	0	0	0	0	0	0	0	0
Total time	0.72	21896.33	0.71	3.89	3766.31	3.6	4.61	25662.64	4.31
Average time	0.01	304.12	0.01	0.19	1883.16	0.18	0.05	346.79	0.05
% of solved	100	100	100	100	10	100	100	80.43	100

Table 4 Results for 92 instances with $D_{\max(2)} = 0.90 \times LT(1)$

	72 small test inst.			20 larg test inst.			Total number		
	TOL	CPLEX	MSBB	TOL	CPLEX	MSBB	TOL	CPLEX	MSBB
♯ Opt ($ind = 1$)	45	70	53	19	2	20	64	72	73
♯ Feas ($ind = 2$)	19(11)	0	16(14)	0	2(0)	0	19(11)	2(0)	16(14)
♯ No Mem ($ind = 3$)	6	0	1	1	16	0	7	16	1
♯ No Feas ($ind = 4$)	2	2	2	0	0	0	2	2	2
Total time	13.9	40,972	141.65	945.5	12,671	120.49	959.53	53,644	262.14
Average time	0.30	569.06	2.58	49.76	6335.93	6.02	14.54	724.93	3.50
% of solved	65.28	100.00	76.39	95.00	10.00	100.00	69.57	78.26	79.35

Table 5 Results for 73 instances with $D_{\max(3)} = 0.90\ of\ LT(2)$

	53 small test inst.			20 larg test inst.			Total number		
	TOL	CPLEX	MSBB	TOL	CPLEX	MSBB	TOL	CPLEX	MSBB
♯ Opt ($ind = 1$)	26	51	36	14	1	19	40	52	55
♯ Feas ($ind = 2$)	16(5)	2(0)	14(13)	1(0)	0	0	17(5)	2(0)	14(13)
♯ No Mem ($ind = 3$)	11	0	3	5	19	1	16	19	4
♯ No Feas ($ind = 4$)	0	0	0	0	0	0	0	0	0
Total time	42.07	50,869	206.17	1260.69	4,032	1620.3	1302.76	54,902	1826.47
Average time	1.62	997.45	5.73	90.05	4032.28	85.28	32.56	1055.80	33.20
% of solved	49.06	96.23	67.92	70.00	5.00	95.00	54.79	71.23	75.34

1. ♯ Opt ($ind = 1$)—how many times each program finds an optimal solution
2. ♯ Feas ($ind = 2$)—how many times feasible (but not optimal) solutions have been found. It presented as $a(b)$ where a (or b) is the number of times the feasible solutions (or the same value as optimal) has been found
3. ♯ No Mem ($ind = 3$)—how many times feasible solutions are not found because of lack of memory
4. ♯ No Feas ($ind = 4$)—how many instances with proven infeasibility are detected
5. Total time—the total time spent only for instances where the optimal solutions are found
6. Average time—the average time for instances which an optimal solution was found, where (Average time = Total time/♯ Opt)

The numerical analysis identifies:

1. The most effective method on average is our Multistart Branch and Bound for ADVRP (`MSBB-ADVRP`). For 92 instances in the first stage (with $D_{max(1)} = \infty$), the rate of success is 100% for `TOL-ADVRP`, 80% for `CPLEX-ADVRP` and 100% for `MSBB-ADVRP`. For the second stage (with $D_{max(2)} = 0.90 \times LT(1)$) the rate of success is 69%, 78% and 79% for `TOL-ADVRP`, `CPLEX-ADVRP` and `MSBB-ADVRP`, respectively. Finally, in the third stage (with $D_{max(3)} = 0.90 \times LT(2)$) the rate of success for the three programs `TOL-ADVRP`, `CPLEX-ADVRP` and `MSBB-ADVRP` is 55%, 71% and 75%, respectively.
2. Regarding efficiency, it can be seen that `TOL-ADVRP` and `MSBB-ADVRP` are much faster than `CPLEX-ADVRP`. The `MSBB-ADVRP` is more efficient than `TOL-ADVRP` for solving instances in the first two stages (total time for `TOL-ADVRP` in the first two stages are 4.61 and 959.53 s, and for `MSBB-ADVRP` are 4.31 and 262.14 s), while the opposite holds for the third stage (1302.76 s for `TOL-ADVRP`, and 1826.47 s for `MSBB-ADVRP`). However in that stage the number of instances solved exactly by `MSBB-ADVRP` and `TOL-ADVRP` is 55 and 40, respectively;
3. When compare small and large test instances, it can be concluded that the `CPLEX` is most effective for small instances in the second stage 100% of test instances are solved (Table 4) and in the third stage 96.23% of test instances are solved (Table 5). However, for large test instances, `MSBB-ADVRP` is the most effective (100% for the first and the second stage, 95% for the third stage);
4. Regarding average time for small instances, the most efficient method is `TOL-ADVRP` (0.01, 0.30 and 1.62 s in the first, the second and the third stage), while for large instances `MSBB-ADVRP` is the most efficient: it takes 0.18, 6.02, and 85.28 s in the first, the second and the third stage, respectively. However, the most efficient on average is `MSBB-ADVRP`.

7 Conclusions

We consider ADVRP and suggest exact algorithms for solving it. They are based on Assignment problem relaxation, as first time proposed by [20]. In order to rebuild feasibility, we branch by using tolerance criterion. We found that our simple method is fast but it has the memory consumption problem. Therefore we introduce the new method based on randomness in choosing the next node in the branch and bound tree.

Computational experiments show that with our multistart approach we are able to solve at least 75% of instances in all stages with up to 1,000 customers. It appears that MSBB-ADVRP on average needs 0.04 s for the first stage, 3.59 s for the second stage and 33.20 s for the third stage, while CPLEX needs on average 346.79, 724.93, 1055.81 s for the first, the second and the third stage, respectively. In summary, the results of experiments emphasize that using MSBB-ADVRP provides good solutions in reasonable computational time. Moreover, as far as we know, we are able to exactly solve problems with larger dimension than previous methods from the literature. For example the largest problem solved by CPLEX has $n = 360$ customers, while we are able to solve exactly with $n = 1{,}000$ customers. It is interesting to note that the dimension of problems solved by our methods depends on available memory of the computer used. Thus, our approach may be used in the future using new computers having larger memory.

We are working currently to improve the running computational times of the algorithm by trying to develop a good heuristic such as Variable Neighborhood Search [11,25], and to use it as initial upper bound. Another possibility is to improve lower bounds that we do not explore in this chapter by adding more constraints to the assignment problem or to relax some of them using Lagrangian multipliers. Such an approach does not use advantage provided by fast Hungarian method and could be research topic for the future work. Although we implement our B&B-based method on DVRP problem, the method is quite general and may be adapted for solving other VRP variants.

References

1. Almoustafa, S., Goldengorin,B., Tso, M., Mladenović, N.: Two new exact methods for asymmetric distance-constrained vehicle routing problem. Proceedings of SYM-OP-IS. Belgrade, pp. 297–300 (2009)
2. Balas, E., Toth, P.: Branch and bound methods. In: Lawer, et al. (eds.) The Traveling Salesman Problem, pp. 361–401. Wiley, Chichester (1985)
3. Baldacci, R., Mingozzi, A.: An unified exact method for solving different classes of vehicle routing problems. Math. Program. Ser. A **120**(2), 347–380 (2009)
4. Baldacci, R., Toth, P., Vigo, D.: Recent advances in vehicle routing exact algorithms. 4OR **5**(4), 269–298 (2007)

5. Baldacci, R., Mingozzi, A., Roberti, R.: Recent exact algorithms for solving the vehicle routing problem under capacity and time window constraints (invited review). Eur. J. Oper. Res. doi:10.1016/j.ejor.2011.07.037, **218**(1), 1–6 (2011, in press)
6. Christofides, N., Mingozzi, A., Toth, P.: State space relaxation procedures for the computation of bounds to routing problems. Networks **11**(2), 145–164 (1981)
7. Clarke, G., Wright, J. V.: Scheduling of vehicles from a central depot to a number of delivery points. Oper. Res. **12**(4), 568–581 (1964)
8. Dantzig, G.B., Fulkerson, D.R., Johnson, S.M.: Solution of a large-scale traveling salesman problem. Oper. Res. **2**, 393–410 (1954)
9. Goldengorin, B., Jager, G., Molitor, P.: Tolerances applied in combinatorial optimization. J. Comp. Sci. **2**(9), 716–734 (2006)
10. Haimovich, M., Rinnooy Kan, A.H.G., Stougie, L.: Analysis of heuristic routing problems. In: Golden, et al. (eds.) Vehicle Routing: Methods and Studies, pp. 47–61. North Holland, Amsterdam (1988)
11. Hansen, P., Mladenović, N., Moreno Pé, J.A.: Variable neighbourhood search: methods and applications. Ann. Oper. Res. **175**(1), 367–407 (2010)
12. Jonker, R., Volgenant, A.: Improving the hungarian assignment algorithm. Oper. Res. Lett. **5**(4), 171–175 (1986)
13. Kara, I.: Two indexed polynomial size formulations for vehicle routing problems. Technical Report (2008/01). Baskent University, Ankara/Turkey (2008)
14. Koltai, T., Terlaky, T.: The difference between the managerial and mathematical interpretation of sensitivity analysis results in linear programming. Int. J. Prod. Econ. **65**(3), 257–274 (2000)
15. Kuhn, H.W.: The Hungarian method for the assignment problem. Naval Res. Logist. Q. **2**, 83–97 (1955)
16. Laporte, G.: The vehicle routing problem: An overview of exact and approximate algorithms. Eur. J. Oper. Res. **59**(3), 345–358 (1992)
17. Laporte, G.: What you should know about the vehicle routing problem. Naval Res. Logist. **54**(8), 811–819 (2007)
18. Laporte, G., Nobert, Y.: Exact algorithms for the vehicle routing problem. Ann. Discrete Math. **31**, 147–184 (1987)
19. Laporte, G., Nobert, Y., Desrochers, M.: Optimal routing under capacity and distance restractions. Oper. Res. **33**(5), 1050–1073 (1985)
20. Laporte, G., Nobert, Y., Taillefer, S.: A branch and bound algorithm for the asymmetrical distance-constrained vehicle routing problem. Math. Model. **9**(12), 857–868 (1987)
21. Lawer, E.L., Lenstra, J.K., Rinnooy Kan, A.H.G., Shmoys, D.B.: The Traveling Salesman Problem: A Guided Tour of Combinatorial Optimization. Wiley-Interscience series in Discrete Mathematics. Chichester, Wiley (1985)
22. Lenstra, J.K., Rinnooy Kan, A.H.G.: Some simple applications of the traveling salesman problem. Oper. Res. Q. **26**(4), 717–734 (1975)
23. Letchford. A. N., Salazar-Gonźalez, J.J.: Projection results for vehicle routing. Math. Program. Ser. B. **105**, 251–274 (2006)
24. Lin, C., Wen, U.: Sensitivity analysis of the optimal assignment. Discrete Optim. **149**(1), 35–46 (2003)
25. Mladenović, N., Hansen, P.: Variable neighbourhood search. Comp. Oper. Res **24**(11), 1097–1100 (1997)
26. Nemhauser, G.L., Wolsey, L.A.: Integer and combinatorial optimization. Discrete Math. Optim. Wiley, New York (1988)
27. Paschos, V.Th.: An overview on polynomial approximation of NP-hard problems. Yugoslav J. Oper. Res. **19**(1), 3–40 (2009)

28. Pessoa, A., Poggi de Aragão, M., Uchoa, E.: Robust branch-cut-and-price algorithms for vehicle routing problems. In: Golden, B., et al. (eds.) The Vehicle Routing Problem Latest Advances and New Challenges. Operations Research/Computer Science Interfaces Series, Springer, New York, vol. **43**, Part II, pp. 297–325 (2008)
29. Toth, P., Vigo, D.: The Vehicle Routing Problem. SIAM Monographs on Discrete Mathematics and Applications. Society for Industrial and Applied Mathematics, Philadelphia (2002)
30. Turkensteen, M., Ghosh, D., Goldengorin, B., Sierksma, G.: Tolerance-based branch and bound algorithms for the ATSP. Eur. J. Oper. Res. **189**, 775–788 (2008)
31. Volgenant, A.: An addendum on sensitivity analysis of the optimal assignment. Eur. J. Oper. Res. **169**(1), 338–339 (2006)

On a Relationship Between Graph Realizability and Distance Matrix Completion

Leo Liberti and Carlile Lavor

Abstract We consider a certain subclass of Henneberg-type edge-weighted graphs which is related to protein structure, and discuss an algorithmic relationship between the DISTANCE GEOMETRY PROBLEM and the EUCLIDEAN DISTANCE MATRIX COMPLETION PROBLEM.

1 Introduction

The structure of proteins is strongly related to its function. Efforts for finding the three-dimensional structure of proteins include minimization of a potential energy function and exploitation of known chemical properties such as inter-atomic distances [29]. Such distances may be known because they refer to covalent bonds and angles, or because they can be found using Nuclear Magnetic Resonance (NMR) [4]. In this paper we focus on finding the protein structure using distance information only.

2 The Distance Geometry Problem

We employ an abstract generalized model of this problem, whereby we look for the realization in $\mathbb{R}^K$ of a weighted simple undirected graph $G = (V, E, d)$, where we assume, to avoid the trivial case, that $n = |V| > K$. In the case of proteins, $K = 3$,

L. Liberti (✉)
LIX, Ecole Polytechnique, 91128 Palaiseau, France
e-mail: liberti@lix.polytechnique.fr

C. Lavor
Department of Applied Mathematics (IMECC-UNICAMP), State University of Campinas, 13081-970, Campinas - SP, Brazil
e-mail: clavor@ime.unicamp.br

A. Migdalas et al. (eds.), *Optimization Theory, Decision Making, and Operations Research Applications*, Springer Proceedings in Mathematics & Statistics 31,
DOI 10.1007/978-1-4614-5134-1_3,

V represents the set of atoms, E represents the set of atom pairs for which a distance is available, and $d: E \to \mathbb{R}_+$ is the edge weight function encoding the distances. Given a positive integer K and a weighted simple undirected graph $G = (V, E, d)$, the DISTANCE GEOMETRY PROBLEM (DGP) asks whether there exists a realization $x: V \to \mathbb{R}^K$ such that:

$$\forall \{u,v\} \in E \quad \|x_u - x_v\|^2 = d_{uv}^2, \tag{1}$$

where the norm is assumed to be the Euclidean norm. In order to fix an orthogonal frame of reference and to avoid translations and rotations, we assume that a subset $V_0 = \{v_1, \ldots, v_K\} \subseteq V$ and a partial realization $x': V_0 \to \mathbb{R}^K$ are also given as part of the input. The DGP is also called the *graph realization problem*. Realizations satisfying (1) are called *valid*. Once a valid realization is found, distances between *all* pairs of vertices (not just those in E) can be determined. Formally, this extends $d: E \to \mathbb{R}_+$ to a function $\bar{d}: V \times V \to \mathbb{R}_+$. The values of the function $\bar{d}$ can be arranged into a square *Euclidean distance matrix* on the point set $\mathscr{X} = \{x_v \mid v \in V\} \subseteq \mathbb{R}^K$. The pair $(\mathscr{X}, \bar{d})$ is known as a *distance space* [1].

3 The Euclidean Distance Matrix Completion Problem

In the EUCLIDEAN DISTANCE MATRIX COMPLETION PROBLEM (EDMCP) [8], the input is a partial square symmetric matrix A (i.e. a symmetric matrix where certain entries are missing) and the output is a pair $(\bar{A}, K)$ where $\bar{A}$ is a symmetric completion of A, and $K \in \mathbb{N}$ such that: (a) $\bar{A}$ is a Euclidean distance matrix in $\mathbb{R}^K$ and (b) K is minimum possible. We consider here a variant of the EDMCP, which we call EDMCP$_K$, where K is actually given as part of the input and the output certificate for YES instances only consists of the completion matrix $\bar{A}$ of the partial matrix A as a Euclidean distance matrix ($\bar{A}$ is also called a *valid completion* of A). It is easy to see that the EDMCP$_K$ is strongly related to the DGP: if x is a valid realization of G, then the partial distance matrix can be completed in polynomial time, and if $\bar{A}$ is a valid completion of A, then the corresponding DGP graph is a clique, whose realization in $\mathbb{R}^K$ can be found in polynomial time [3].

This mapping in the output parallels a mapping in the input data. A partial square symmetric matrix $A = (a_{uv})$ with missing components indexed by the set $\bar{E}$ of unordered index pairs $\{u,v\}$ encodes the weighted simple undirected graph $G = (V, E, a)$ where V is the set of row/column indices, E is the complement of $\bar{E}$ with respect to the set of all unordered pairs of V, and the edge weight a maps $\{u,v\}$ to a_{uv}. Conversely, a weighted simple undirected graph $G = (V, E, a)$ can be encoded in a partial square symmetric matrix A where the $\{u,v\}$th component is a_{uv} for all $\{u,v\} \in E$ and the other components are missing. We formalize this correspondence by setting $\mathscr{M}(G) = A$ for a graph G and its corresponding partial matrix A, and $\mathscr{G}(A) = G$ for a partial matrix A and its corresponding graph G. It is trivial to see that $\mathscr{M}$ and $\mathscr{G}$ are inverse operators.

4 Rigidity and Henneberg type graphs

The DGP refers to a field of study which is known as Distance Geometry (DG). DG was formally started in the 1930s, when Menger found how to decide whether a given square matrix encodes a distance matrix using Cayley determinants [19]. Blumenthal then extended Menger's findings to a well-developed theory [1], and re-christened Cayley determinants "Cayley–Menger determinants." The study of realizations of graphs in the plane and in space, however, dates much further back. The ancient Greeks were concerned with finding all polyhedra in space, for example. Statics, which is necessary to ensure that buildings will not collapse under the action of external forces, has existed ever since man got tired of being rained on and decided to build himself a roof. Realizations of graphs in space from the point of view of statics are known as "bar-and-joint frameworks." Several important results on the rigidity of such frameworks date from the end of the nineteenth century [2, 28]. Henneberg [6] was the first to formalize an iterative procedure for verifying whether such frameworks are rigid. In particular, one of his two "steps" (known as *Henneberg type I* step [7, 30]) can be paraphrased (and generalized) as follows: if there is an order on V such that the first K vertices have a known realization, and such that every subsequent vertex is adjacent to at least K predecessors, then the graph almost certainly has a rigid realization in $\mathbb{R}^K$. This idea was already present in the works of Saviotti [27], as testified by the two-dimensional case shown in Fig. 1.

The set of Henneberg type I graphs gives rise to a subset of DGP instances known as the DISCRETIZABLE DGP (DDGP) [21].

In the above paragraph we used the term "rigid framework" and "almost certainly" intuitively. These can be formally defined as follows. A framework is a pair (G,x) where G is a graph and x is a valid realization; a flexing of a framework is a continuous map p from $[0,1]$ to the set of all realizations of V such that $p(0)=x$, $p(t)$ satisfies (1) for all $t \in [0,1]$, and $p(t)$ is not an isometry of x for all $t \in (0,1]$. A framework is rigid if it has no flexing. As for "almost certainly rigid," this means that the set of realizations which are not rigid has Lebesgue measure 0 in the set of all possible realizations.

If the Henneberg type I order is not explicitly given, it may not be immediately obvious how to find one. The problem of finding a Henneberg type I order is defined in [9] as the DISCRETIZATION VERTEX ORDER PROBLEM (DVOP). There is an exponential algorithm $O(n^{K+3})$ for solving the DVOP, which is polynomial for fixed K. Implementations for $K=3$ are very fast and can successfully be used as a preprocessing step to solving the DDGP.

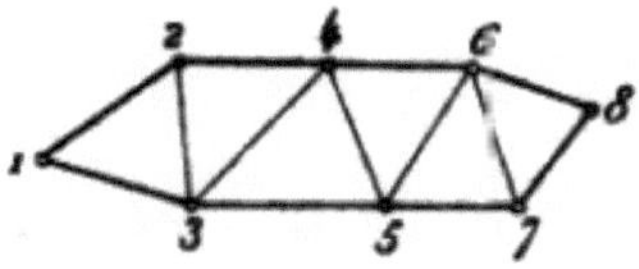

Fig. 1 Figure 30 in [27]

5 Branch-and-Prune

For almost all edge weightings, Henneberg type I graphs can have finitely many different valid realizations whose corresponding distance spaces are incongruent. With a slight abuse of notation, we call two different valid realizations with incongruent distance spaces *incongruent realizations*. If vertex v has *exactly* K adjacent predecessors, then x_v is at the intersection P of K spheres in $\mathbb{R}^K$. The cardinality of P is in $\{0,1,2\}$ as long as the position of the K adjacent predecessors of v affinely spans a $(K-1)$-dimensional subspace of $\mathbb{R}^K$. The case $|P|=0$ occurs when the edge weighting is such that G has no realization in $\mathbb{R}^K$. The case $|P|=1$ only occurs when the subgraph induced by v and its K adjacent predecessors defines a flat simplex in $\mathbb{R}^K$. Since the set of flat simplices have Lebesgue measure 0 in the set of all simplices, this is a case which can be ignored almost all the time. The remaining case is $|P|=2$, shown for $K=3$ in Fig. 2.

Thus, one can find all incongruent realizations of a Henneberg type I graph G by the following method:

(a) place the first K vertices arbitrarily (this essentially fixes the reference system up to reflection);
(b) place the vth vertex in the Henneberg type I order in one of the points in P;
(c) for each position x_v for v in P, recursively call Step (b) with v replaced by $v+1$.

In the worst case (i.e. whenever $|P|$ is always 2 and there are no other edges but those that define the Henneberg type I step), this gives rise to a full binary search tree after level K, which amounts to 2^{n-K} different realizations, 2^{n-K-1} of which are incongruent, the other 2^{n-K-1} being their reflection through the first K vertices [11, Thm. 2]. We let X be the set of all realizations found by this method.

We remark that the recursive call at Step (c) may occur fewer than twice whenever vertex v has more than K adjacent predecessors, as the intersection of more than K spheres in $\mathbb{R}^K$ almost always has either 0 or 1 point. Thus, vertices with more than K adjacent predecessors are used to "prune out" certain

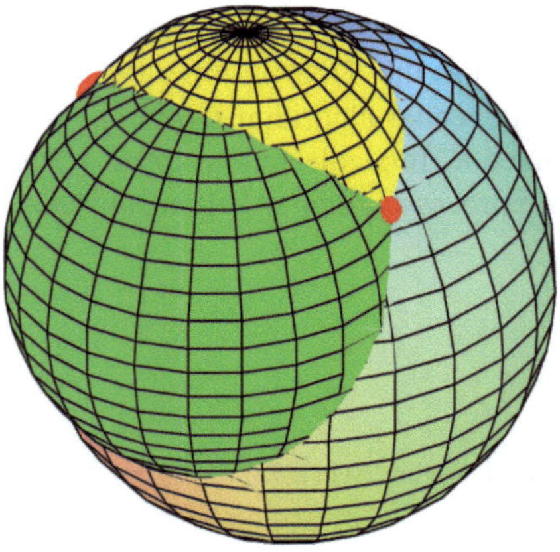

Fig. 2 General case for the intersection P of three spheres in $\mathbb{R}^3$

Algorithm 3.1: BP($v, \bar{x}, X$)

Require: A vertex $v \in V \smallsetminus [K]$, a partial realization $\bar{x} = (x_1, \ldots, x_{v-1})$, a set X.
$P = \bigcap_{\substack{u \in N(v) \\ u < v}} S^{\bar{x}}_{uv}$
for $p \in P$ **do**
  $x \leftarrow (\bar{x}, p)$
  if $\rho(v) = n$ **then**
    $X \leftarrow X \cup \{x\}$
  else
    BP($v+1, x, X$)
  end if
end for

branches of the binary search tree. This is why the corresponding algorithm is called *Branch-and-Prune* (BP) [10, 17]. The BP algorithm was originally only defined for immediate predecessors [11], but was henceforth extended to work in several different situations: for Henneberg type I graphs [21], for certain types of interval-weighted graphs related to proteins [12, 15, 20], and for the purpose of overcoming a technical limitation of NMR machinery, which generally only provides reliable distance measures for pairs of hydrogen atoms [13, 14, 16, 22]. A publically available BP implementation is described in [25]. The current computational state-of-the-art for the BP algorithm is attained with a parallel BP implementation [23, 24], which can realize a protein backbone of 10^4 atoms in $\mathbb{R}^3$ in just over 10s of CPU time on a cluster of 8 nodes.

Step (b) of the BP algorithm is formalized in Algorithm 3.1. It takes as input a vertex v of rank $\rho(v) > K$, a partial realization $\bar{x}$ on the predecessors of v and a set X which will contain all the valid realizations at the end of the execution. We identify for convenience V with the set $[n] = \{1, \ldots, n\}$ of vertex ranks, we denote the set of vertices adjacent to v by $N(v)$, and we use $S^{\bar{x}}_{uv}$ to denote the sphere centered at $\bar{x}_u$ with radius d_{uv}. The BP algorithm starts with the call BP($K+1, x', \emptyset$), where x' is the (given) realization of the first K vertices.

It was shown in [18, Lemma 3.4] that the BP algorithm finds all incongruent solutions of the DDGP, and in [18, Prop. 3.5] that for almost all instances, no two distinct search tree nodes at a given level v will be such that one node has two subnodes and the other node only one.

5.1 *Partial reflections*

What do incongruent realizations of G look like? Partition the edges of G in those edges which ensure the existence of a Henneberg type I order (which we call *discretization edges*) and all the other edges (which we call *pruning edges*). Let $\bar{X}$ be the set of all realizations of the subgraph of G defined by the discretization edges. Then, by definition, $|\bar{X}| = 2^{n-K}$. A *partial reflection* of $x \in \bar{X}$ with respect to a vertex $v > K$ is a map $\pi : (V \smallsetminus [K]) \times \bar{X} \to \bar{X}$ such that $\pi_v = I^v \times R_{v,x}^{n-v}$, where $R_{v,x}$ is the reflection operator through the hyperplane defined by $x_{v-K}, \ldots, x_{v-1}$; in other words, $\pi_v(x) = (x_1, \ldots, x_{v-1}, R_{v,x}(x_v), \ldots, R_{v,x}(x_n))$. As remarked in [5, Sect. 2.1], partial reflections are also maps $X \to X$ (a proof of this is found in [18, Thm. 4.9]). A strong converse is also true for DDGP instances where each vertex is adjacent to at least K *immediate* predecessors (such instances are collectively known as the DISCRETIZABLE MOLECULAR DGP in general dimensions, or ^{K}DMDGP, see [11] for the case with $K = 3$), namely, that for any distinct $x, y \in X$ there is a composition ρ of partial reflections such that $y = \rho(x)$ [18, Thm. 5.4].

6 BP in distance space

As remarked in [26], the completion in $\mathbb{R}^3$ of a distance (sub)matrix with the following structure:

$$\begin{pmatrix} 0 & d_{12} & d_{13} & d_{14} & \boxed{\delta} \\ d_{21} & 0 & d_{23} & d_{24} & d_{25} \\ d_{31} & d_{32} & 0 & d_{34} & d_{35} \\ d_{41} & d_{42} & d_{43} & 0 & d_{45} \\ \boxed{\delta} & d_{52} & d_{53} & d_{54} & 0 \end{pmatrix} \tag{2}$$

can be carried out in constant time by solving a quadratic system in the unknown δ derived from setting the Cayley–Menger determinant [9, (9)] of the distance space $(\mathscr{X}, d)$ to zero, where $\mathscr{X} = \{x_1, \ldots, x_5\}$ and d is given by (2). This is because, for general K, the Cayley–Menger determinant is proportional to the K-volume of the simplex on $K+1$ points, which is the (unique, up to rotations and translations) realization of the weighted 5-clique defined by a full distance matrix. Since a simplex on 5 points embedded in $\mathbb{R}^3$ necessarily has 4-volume equal to zero, it suffices to set the Cayley–Menger determinant of (2) to zero to obtain a quadratic equation in δ. We denote the pair $\{u, v\}$ indexing the unknown distance δ by $\mathsf{U}(D)$, the Cayley–Menger determinant of a matrix D by $\mathsf{CM}(D)$, and the corresponding quadratic equation in δ by $\mathsf{CM}(D, \delta) = 0$. This equation has real solutions only if (2) is a Euclidean distance matrix. Furthermore, if it has real solutions at all, it almost always has two distinct solutions δ^1, δ^2. These are two valid values for the missing distance d_{15}. This observation trivially extends to general K, where we consider a $K+2$ point simplex realization of a weighted near-clique on $K+2$ vertices with one missing edge.

6.1 *The main idea*

We consider a Henneberg type I graph G and a partial embedding $\bar{x}$ for the subgraph $G[[K]]$ of G induced by the set $[K]$ of the first K vertices. The DDGP order on V

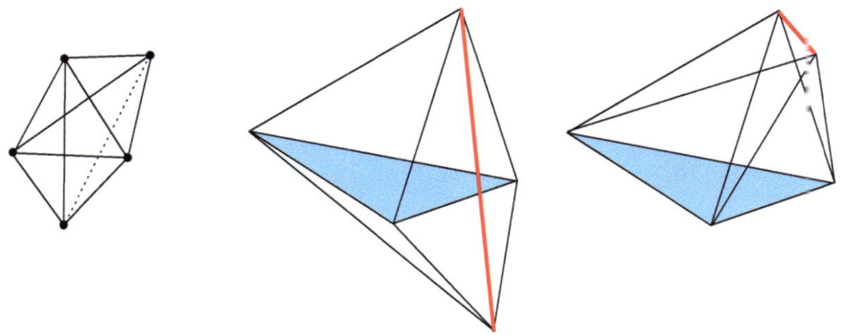

Fig. 3 On the *left*: a near clique on 5 vertices with one missing edge (*dotted line*). On the *right*: its two possible realizations in $\mathbb{R}^3$ for a given feasible edge weighting (distance values for the missing edge shown in *red*)

guarantees that the vertex of rank $K+1$ has K adjacent predecessors, hence it is adjacent to all the vertices of rank $v \in [K]$. Thus, $G[[K+1]]$ is a full $(K+1)$-clique. Consider now vertex v_{K+2}: the Henneberg type I order guarantees that v_{K+2} has at least K adjacent predecessors. If it has $K+1$, then $G'[[K+2]]$ is the full $(K+2)$-clique. Otherwise $G'[[K+2]]$ is a near-clique on $K+2$ vertices with one missing edge (say $\{u, K+2\}$ for some $u \in [K+1]$). We can therefore use the Cayley–Menger determinant to compute two possible values for $d_{u,K+2}$, as discussed above. Because the Henneberg type I order always guarantees at least K adjacent predecessors, this procedure can be generalized to vertices of any rank v in $V \setminus [K]$, and so it defines a recursive algorithm which branches whenever a distance can be assigned two different values, simply continues to the next rank whenever the subgraph induced by the current $K+2$ vertices is a full clique, and prunes all branches whenever the partial distance matrix defined on the current $K+2$ vertices has no Euclidean completion.

In general, this procedure holds for realizations in $\mathbb{R}^K$ whenever there is a vertex order such that each next vertex v is adjacent to K predecessors: thus we can define a subgraph containing v and $K+1$ predecessors consisting of two $(K+1)$ cliques whose intersection is a K-clique (i.e., a near-clique with one missing edge). There are in general two possible realizations in $\mathbb{R}^K$ for such subgraphs, as shown in Fig. 3.

6.2 *Formalization and properties*

Algorithm 3.2 formalizes such a recursive algorithm. It takes as input a vertex v of rank greater than $K+1$, a partial matrix A and a set $\mathscr{A}$ which will eventually contain all the possible completion of the partial matrix given as the problem input. For a given partial matrix A, a vertex v of $\mathscr{G}(A)$ and an integer $\ell \leq K$, let A_v^ℓ be the $\ell \times \ell$ symmetric submatrix of A including row and column v that has fewest missing components. Whenever A_v^{K+2} has no missing elements, the equation

Algorithm 3.2: dBP(v, A, $\mathscr{A}$)

Require: A vertex $v \in V \smallsetminus [K+1]$, a partial matrix A, a set $\mathscr{A}$.
1: $P = \{\delta \mid \mathsf{CM}(A_v^{K+2}, \delta) = 0\}$
2: **for** $\delta \in P$ **do**
3: $\quad \{u,v\} \leftarrow \mathsf{U}(A_v^{K+2})$
4: $\quad d_{uv} \leftarrow \delta$
5: $\quad$ **if** A is complete **then**
6: $\qquad \mathscr{A} \leftarrow \mathscr{A} \cup \{A\}$
7: $\quad$ **else**
8: $\qquad$ dBP($v+1$, A, $\mathscr{A}$)
9: $\quad$ **end if**
10: **end for**

$\mathsf{CM}(A_v^{K+2}, \delta) = 0$ is either a tautology if A_v^{K+2} is a Euclidean distance matrix, or unsatisfiable in $\mathbb{R}$ otherwise. In the first case, we define it to have $\delta = d_{uv}$ as a solution, where u is the smallest row/column index of A_v^{K+2}. In the second case, we define it to have no solutions.

Lemma 1. *In Step 1 of Algorithm 3.2, A_v^{K+2} always has at most one missing distance involving v.*

Proof. At level v of Algorithm 3.2, all distances for $\{u,w\}$ for $u,w < v$ are known by the induction hypothesis. The induction starts because either $d_{1,K+1}$ is part of the input partial matrix, or, if not, by calling BP($K+1, \bar{x}, \varnothing$) just for level $K+1$, without recursion: then the distance $d_{1,K+1}$ can be computed. By the Henneberg type I order, v is adjacent to at least K predecessors, so that the densest $(K+2) \times (K+2)$ symmetric submatrix of A involving row and column v must be such that all other rows/columns are indexed by as many adjacent predecessors of v as possible. Since there are at most $K+1$ such adjacent predecessors, there is at most one missing distance in A_v^{K+2}, and it involves v. If A can be completed to a Euclidean distance matrix, then the missing distance is assigned a feasible value in Step 4. This completes the induction step.

Corollary 1. *In Step 3 of Algorithm 3.2, U is well defined.*

Theorem 1. *At the end of Algorithm 3.2, $\mathscr{A}$ contains all possible completions of the input partial matrix.*

Proof. By contradiction, if not then there must be a recursive call when there is a $\gamma \in \mathbb{R}_+$ such that $d_{uv} = \gamma$ yields a partial matrix which can be completed to a Euclidean distance matrix, but $\gamma \notin P$. But by Lemma 1 this would mean that the quadratic equation $\mathsf{CM}(A_v^{K+2}, \delta) = 0$ in δ has more than two solutions, which is impossible.

6.3 A dual Branch-and-Prune

The resemblance of Algorithms 3.1 and 3.2 is such that it is very easy to assign dual meanings to the original (otherwise known as *primal*) BP algorithms. As was made clear in Sect. 3, weighted graphs and partial symmetric matrices are dual to each other through the inverse mappings $\mathscr{M}$ and $\mathscr{G}$. Whereas in the primal BP we decide realizations of the graph, in the dual BP we decide the completions of partial matrices, so realizations and distance matrix completions are dual to each other. The primal BP decides on points $x_v \in \mathbb{R}^K$ to assign to the next vertex v, whereas the dual BP decides on distances δ to assign to the next missing distance incident to v and to a predecessor of v; there are at most two choices of x_v as there are at most two choices for δ; only one choice of x_v is available whenever v is adjacent to strictly more than K predecessor, and the same happens for δ; finally, no choices for x_v are available in case the current partial realization cannot be extended to a full realization of the graph, as well as no choices for δ are available in case the current partial matrix cannot be completed to a Euclidean distance matrix. This means that weighted edges and points in Euclidean space are dual to each other. The same vertex order can be used by both the primal and the dual BP (so the order is self-dual).

There is one clear difference between primal and dual BP: namely, that the dual BP needs an initial $(K+1)$-clique, whereas the primal BP only needs an initial K-clique. This difference also has a dual interpretation: a complete Euclidean distance matrix corresponds to two (rather than one) realizations, one the reflection of the other through the hyperplane defined by the first K points.

References

1. Blumenthal, L.: Theory and Applications of Distance Geometry. Oxford University Press, Oxford (1953)
2. Cremona, L.: Le figure reciproche nella statica grafica. G. Bernardoni, Milano (1872)
3. Dong, Q., Wu, Z.: A linear-time algorithm for solving the molecular distance geometry problem with exact inter-atomic distances. J. Global Optim. **22**, 365–375 (2002)
4. Gunther, H.: NMR Spectroscopy: Basic Principles, Concepts, and Applications in Chemistry. Wiley, New York (1995)
5. Hendrickson, B.: Conditions for unique graph realizations. SIAM J. Comput. **21**(1), 65–84 (1992)
6. Henneberg, L.: Die Graphische Statik der starren Systeme. Teubner, Leipzig (1911)
7. John, A.L.S.: Geometric constraint systems with applications in cad and biology. Ph.D. thesis, University of Massachusetts at Amherst (2008)
8. Laurent, M.: Matrix completion problems. In: Floudas, C., Pardalos, P. (eds.) Encyclopedia of Optimization, 2nd edn., pp. 1967–1975 Springer, New York (2009)
9. Lavor, C., Lee, J., John, A.L.S., Liberti, L., Mucherino, A., Sviridenko, M.: Discretization orders for distance geometry problems. Optim. Lett. doi:10.1007/s11590-011-0302-6, 6:783–796 (2012)
10. Lavor, C., Liberti, L., Maculan, N.: The discretizable molecular distance geometry problem. Tech. Rep. q-bio/0608012, arXiv (2006)

11. Lavor, C., Liberti, L., Maculan, N., Mucherino, A.: The discretizable molecular distance geometry problem. Comput. Optim. Appl. doi:10.1007/s10589-011-9402-6, 52:115–146 (2012)
12. Lavor, C., Liberti, L., Mucherino, A.: The *i*Branch-and-Prune algorithm for the discretizable molecular distance geometry problem with interval data. J. Global Optim. (accepted) doi: 10.1007/s10898-011-9799-6
13. Lavor, C., Mucherino, A., Liberti, L., Maculan, N.: An artificial backbone of hydrogens for finding the conformation of protein molecules. In: Proceedings of the Computational Structural Bioinformatics Workshop, pp. 152–155. IEEE, Washington, DC (2009)
14. Lavor, C., Mucherino, A., Liberti, L., Maculan, N.: Computing artificial backbones of hydrogen atoms in order to discover protein backbones. In: Proceedings of the International Multiconference on Computer Science and Information Technology, pp. 751–756. IEEE, Mragowo (2009)
15. Lavor, C., Mucherino, A., Liberti, L., Maculan, N.: Discrete approaches for solving molecular distance geometry problems using NMR data. Int. J. Comput. Biosci. **1**, 88–94 (2010)
16. Lavor, C., Mucherino, A., Liberti, L., Maculan, N.: On the computation of protein backbones by using artificial backbones of hydrogens. J. Global Optim. **50**, 329–344 (2011)
17. Liberti, L., Lavor, C., Maculan, N.: A branch-and-prune algorithm for the molecular distance geometry problem. Int. Trans. Oper. Res. **15**, 1–17 (2008)
18. Liberti, L., Masson, B., Lee, J., Lavor, C., Mucherino, A.: On the number of solutions of the discretizable molecular distance geometry problem. In: Combinatorial Optimization, Constraints and Applications (COCOA11). LNCS, vol. 6831, pp. 322–342. Springer, New York (2011)
19. Menger, K.: Untersuchungen über allgemeine metrik. Mathematische Annalen **103**, 466–501 (1930). doi:10.1007/BF01455705
20. Mucherino, A., Lavor, C.: The branch and prune algorithm for the molecular distance geometry problem with inexact distances. In: Proceedings of the International Conference on Computational Biology, vol. 58. World Academy of Science, Engineering and Technology, pp. 349–353 (2009)
21. Mucherino, A., Lavor, C., Liberti, L.: The discretizable distance geometry problem. Optim. Lett. (accepted) doi: 10.1007/s11590-011-0358-3
22. Mucherino, A., Lavor, C., Liberti, L., Maculan, N.: On the definition of artificial backbones for the discretizable molecular distance geometry problem. Mathematica Balkanica **23**, 289–302 (2009)
23. Mucherino, A., Lavor, C., Liberti, L., Talbi, E.G.: On suitable parallel implementations of the branch & prune algorithm for distance geometry. In: Proceedings of the Grid5000 Spring School. Lille, France (2010)
24. Mucherino, A., Lavor, C., Liberti, L., Talbi, E.G.: A parallel version of the branch & prune algorithm for the molecular distance geometry problem. In: Proceedings of the ACS/IEEE International Conference on Computer Systems and Applications (AICCSA10). IEEE, Hammamet, Tunisia (2010)
25. Mucherino, A., Liberti, L., Lavor, C.: `MD-jeep`: an implementation of a branch-and-prune algorithm for distance geometry problems. In: Fukuda, K., van der Hoeven, J., Joswig, M., Takayama, N. (eds.) Mathematical Software. LNCS, vol. 6327, pp. 186–197. Springer, New York (2010)
26. Porta, J., Ros, L., Thomas, F.: Inverse kinematics by distance matrix completion. In: Proceedings of the 12th International Workshop on Computational Kinematics, pp. 1–9 (2005)
27. Saviotti, C.: Nouvelles méthodes pour le calcul des travures réticulaires. In: Appendix to L. Cremona, "Les figures réciproques en statique graphique", pp. 37–100. Gauthier-Villars, Paris (1885)
28. Saviotti, C.: La statica grafica: Lezioni. U. Hoepli, Milano (1888)
29. Schlick, T.: Molecular modelling and simulation: an interdisciplinary guide. Springer, New York (2002)
30. Tay, T.S., Whiteley, W.: Generating isostatic frameworks. Struct. Topology **11**, 21–69 (1985)

Effect Oriented Planning of Joint Attacks

Nils-Hassan Quttineh, Torbjörn Larsson, Kristian Lundberg, and Kaj Holmberg

Abstract We consider tactical planning of a military operation on a large target scene where a number of specific targets of interest are positioned, using a given number of resources which can be, for example, fighter aircraft, unmanned aerial vehicles, or missiles. The targets could be radar stations or other surveillance equipment, with or without defensive capabilities, which the attacker wishes to destroy. Further, some of the targets are defended, by, for example, Surface-to-Air Missile units, and this defense capability can be used to protect also other targets. The attacker has knowledge about the positions of all the targets and also a reward associated with each target. We consider the problem of the attacker, who has the objective to maximize the expected outcome of a joint attack against the enemy.

The decisions that can be taken by the attacker concern the allocation of the resources to the targets and what tactics to use against each target. We present a mathematical model for the attacker's problem. The model is similar to a generalized assignment problem, but with a complex objective function that makes it intractable for large problem instances. We present approximate models that can be used to provide upper and lower bounds on the optimal value, and also provide heuristic solution approaches that are able to successfully provide near-optimal solutions to a number of scenarios.

1 Introduction

Effect-Based Operations (EBO) is a military concept which emerged during the 1991 Gulf war for the planning and conduct of operations combining military and non-military methods to achieve a particular effect. The doctrine was developed

N.-H. Quttineh (✉) • T. Larsson • K. Lundberg • K. Holmberg
Department of Mathematics, Linköping University, Sweden
e-mail: nils-hassan.quttineh@liu.se; torbjorn.larsson@liu.se; kristian.lundberg@liu.se; kaj.holmberg@liu.se

A. Migdalas et al. (eds.), *Optimization Theory, Decision Making, and Operations Research Applications*, Springer Proceedings in Mathematics & Statistics 31,
DOI 10.1007/978-1-4614-5134-1_4,

to take advantage of advancements in weaponry and tactics, from an emerging understanding that attacking a second-order target may have first order consequences for a variety of objectives. The Commander's intent can be satisfied with a minimum of collateral damage or risk to own forces, but EBO planning is complex and hard since it embraces political factors as well as economic.

Despite its complexity, this is not an impossible task. We have been dealing with these challenges on an ad hoc basis throughout history, but we can now use modern technologies and process thinking to provide all ingredients of successful effect-based operations.

A network-centric system is a system-of-systems concept where a number of actors are attached to each other in a network sharing information in an adaptable and interoperable manner. Obviously networking enables an enormous rise in accessible information and the intrinsic challenge is the development of systems and functions to shape this information into guidance and control of a variety of operations with multiple objectives. For example, Yost and Washburn [6] present an optimization methodology for finding a correct balance between weapons and attack damage assessment sensors.

The above-mentioned pinpoints the trend in military operational planning, also at the Swedish military arena. In our case we can use this paradigm shift to put functional and algorithmic requirements on planning of air-to-ground missions. This leads to adaptation to new doctrines of command and control and to a tool that contains the most of planning experience implemented by planning specialist personnel in cooperation with algorithm experts. Mission performance can be driven to its limits with a model-based planning, which simultaneously keeps control of both objective and system performance, which is probably the most cost effective way to gain performance.

1.1 Network Centric Framework

In a network centric framework, a resource is not an entity tightly coupled to a sluggish hierarchical organization but a resource with own intelligence to offer specific effects to a variety of effect customers. Our work does not embrace the full meaning of EBO but is guided by quantifying and responding to effect requests and hence becoming a true entity of a network centric system. In order to understand the paradigm shift in EBO planning or network centric planning, Fig. 1 shows the principles of future effect-based operations.

Initially an effect must be achieved in order to answer what to do. Thereafter possible systems are considered and how these systems could manage to do it. The last issue of the effect chain is to decide the resource allocation. As can be noticed, resource owners are considered in the later planning stages, which is quite a change from traditional planning.

Obviously there are two dimensions in the effect chain, the mission-conduction and the resource owner dimensions. The resource owner dimension keeps and

Fig. 1 The effect chain including an EBO principle of a split up of the planning process into stages from the target to allocation of individual platforms

conducts resource supply chains as well as allocation schemes and schedules. The mission-conduction states individual missions and how they shall be implemented.

In order to fulfill requirements on future EBO planning systems, effort must be put on scalable model-based algorithms which promote an easy workflow and a high speed planning performance. Each scenario shall be individually stated by the set of input data, but planning shall always be performed via implemented tactics and knowledge of actual resource performance and mission pattern.

1.2 Mission Planning

An air-to-ground mission planning system is modular and contains a planning system and weapon systems, hosted by a variety of carriers such as unmanned aerial vehicles or fighter aircraft. In order to perform effect oriented planning in line with Fig. 1 we transform the planning process according to Fig. 2, where each platform is separated into carrier and weapon performance and tactics producing a certain effect which can be matched with the effect customers needs.

Initially we maximize system effect in the target area by optimally allocating the number of weapons to suppress enemy defense and destroy vital targets. A target area can consist of different ground-based targets and sheltering air defense units. Each target has a specific value which indicates its importance. The effect-oriented weapon allocation of the target area is followed by a search for appropriate platforms, where platform location and scheduling parameters are considered.

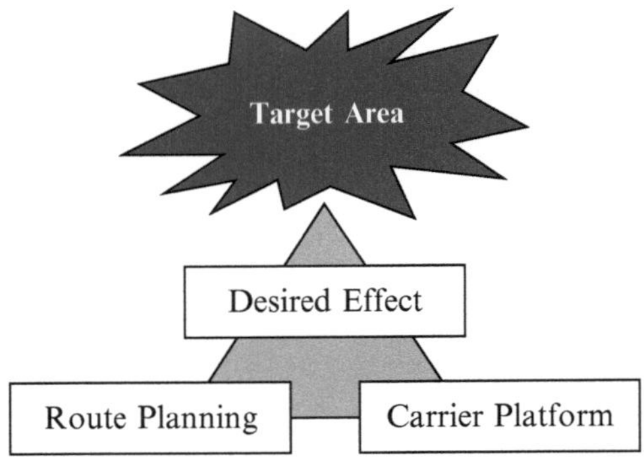

Fig. 2 A resource, a fighter with weapon system, has a relationship between route planning, type of weapon and a setup of tactics which forms the final effect

Each platform must further have a route to the firing position, including tactical features such as hiding and a limited exposure of radar cross section during the flight phase.

These planning aspects are coupled, but with an acceptable loss of generality the effect planning task can be separated from the platform in order to start an overall planning process. Our work addresses a model-based approach to rapidly calculate weapon allocation to optimize system effect in an hostile ground-based target area. Early work on a similar problem was done by Miercort and Soland in [4], but they consider a less complicated model without intricate dependencies. In a recent paper by Kwon et al. [3], a new weapon-target allocation problem is presented together with a branch-and-price algorithm for solving it. In contrast, Kaminer and Ben-Asher present a model in [2] for maximizing the effectiveness of a defense.

1.3 Paper Overview

In Sect. 2 we describe the problem at hand, which is basically a weapon-targeting problem, together with some basic concepts that will be used throughout the paper. Section 3 gives a generic mathematical model for the problem. It is straightforward with only simple linear constraints, but comes with a difficult objective function. This section also gives optimistic and pessimistic models that can be used to find upper and lower bounds on the optimal objective value.

In order to use the generic model and solve realistic scenarios, it is necessary to specify how to evaluate a given situation, and especially how the defenders act in different situations. One possible way to do this is presented in Sect. 4.

Section 5 looks into different heuristic approaches, who cannot guarantee optimality but find high quality solutions for larger scenarios within the reasonable time frames. Section 6 contains results for these heuristics. Finally, in Sect. 7, we present some remarks and conclusions together with suggestions on future work. This paper is based on material that can be found in [5].

2 The Joint Attack Problem

Imagine a large open area, like a desert, where a number of enemy targets are positioned. These can be radar stations or other surveillance equipment, which the attacker wishes to destroy. The targets are, however, guarded by defenders, like Surface-to-Air Missile (SAM) units. The defenders are also considered to be potential targets for the attack, since the destruction of defenders can improve upon the overall outcome of the attack.

The positions of all targets, both those with and without defense, are known. The set of targets is denoted S, and the subset $\bar{S}$ denotes the targets with defensive capabilities, which are defined by radii of defense and armament. Each target $s \in S$ is given a specified reward r_s, where important targets have higher values.

The attacker's problem is to maximize the expected outcome of a simultaneous attack against the enemy, using at most R identical resources, like aircraft or unmanned aerial vehicles. Each target should be assigned an attack plan which specifies the number of resources to be used against it, and also from which directions.

As illustrated in Fig. 3, some targets do not have a defensive system of their own, but depends on the defense of others. Also, the radius of defense for different defenders might overlap. A defender will always protect itself primarily, and then engage resources passing by inside its radius of defense towards other targets.

2.1 Tactics and Angles of Attack

If a target s is attacked, it is done so by a tactic t chosen from a set of tactics, T. In real life there are numerous possible tactics for an attack, but we limit ourselves to tactics using at most 3 resources, as described graphically in Fig. 4. The idea

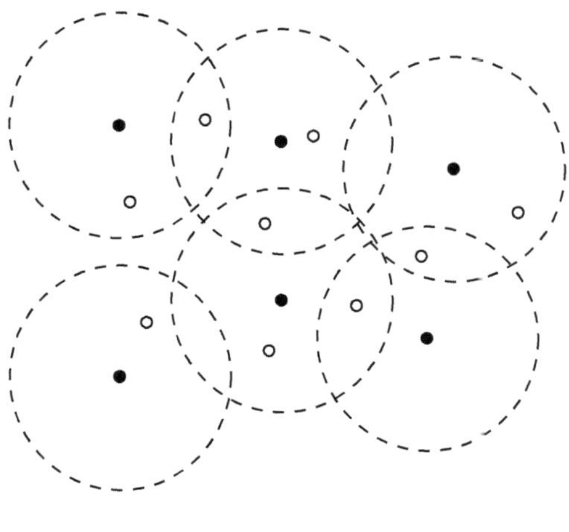

Fig. 3 A possible attack scenario. Some targets, here shown in *black*, are air defense units. The other targets are radar stations or similar surveillance units who are valuable to destroy

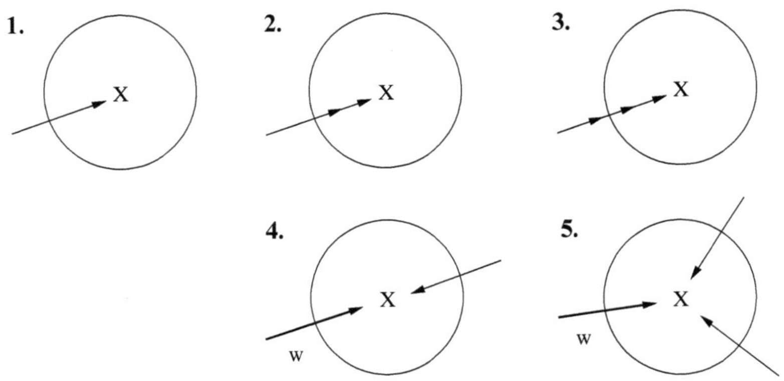

Fig. 4 A graphical description of the 5 tactics considered

behind these tactics is to overload the defensive system of a single defender. This is done either by sending multiple resources from one direction (see tactics 1–3), or by attacking from multiple and evenly spread directions (see tactics 4–5).

Each tactic t has its own features, such as the number of resources needed, n_t, and the number of attacking directions involved, denoted V_t. The number of resources that is launched from each of the angles V_t is denoted by m_t. Each tactic gives rise to a probability of success, for each of the n_t resources, against a single target s. This probability is denoted p_{st} and might vary between the targets, depending on their respective defensive capabilities.

We consider a coarse angle discretization (every 30°), defining a set V of angles. Each tactic $t \in T$ is associated with a reference angle of attack, w, which defines from which direction the attack is launched. For tactics which involve more than one angle of attack (i.e. $V_t > 1$), multiple angles w might give rise to exactly the same attack, since we consider evenly spread angles. To avoid such symmetries, we introduce a subset W_{st} which contains all reference angles w to be used together with tactic t against target s.

For tactics involving multiple angles, we define

$$w_j = w + (j-1) \cdot \frac{2\pi}{V_t}, \quad j = 1, \ldots, V_t.$$

We also introduce the concept of an engagement path (s,v), which is the line emanating from target s at angle v. In total, there are $|S| \cdot |V|$ different engagement paths. For a certain tactic and angle, though, only a few of these paths will be used. If there is at least one resource on the path, we call it an active path.

In the following, a reference angle of attack is always denoted w and defined by the set W_{st}, whereas an angle v refers to an individual angle in V used for general discussions involving engagement paths (s,v).

2.2 The Objective

The essence of the attacker's problem is to decide for each target s which tactic t that shall be used (if any) and specify a reference angle of attack w. We therefore introduce the binary variable

$$z_{stw} = \begin{cases} 1 \text{ if target } s \text{ is attacked using tactic } t \text{ from angle } w, \\ 0 \text{ otherwise.} \end{cases}$$

These decisions, at most one for each target s, are defined as an attack plan $\mathbf{z}$. Let $p^{\text{kill}}_{stw}(\mathbf{z})$ be the probability of successfully incapacitating target s when attacked by tactic t from reference angle w. As will be clear from the upcoming analysis, this probability depends on the overall attack plan $\mathbf{z}$, which is a complicating fact.

The probability for a resource to survive the defense of a defender $i \in \bar{S}$ which it passes by on its way towards the target s on path (s,v) is denoted $p_{isv}(\mathbf{z})$, and it depends on what tactics are used against the other targets. Whenever an engagement path (s,v) does not intersect the area of defense for target i, $p_{isv}(\mathbf{z}) = 1$ holds.

The success of an attack against a certain target depends on the following.

1. The number of resources used against the target ($n_t = V_t \cdot m_t$).
2. The target's ability to defend itself against incoming resources (p_{st}).
3. The probability of successfully surviving the defense of every other target which the resource pass by on its way towards the target (p_{isw_j}).

For a given target s, tactic t and angle of attack w, the probability of successfully eliminating target s is

$$p^{\text{kill}}_{stw}(\mathbf{z}) = 1 - \prod_{j=1}^{V_t} \left[1 - p_{st} \prod_{i \in \bar{S} \setminus \{s\}} p_{isw_j}(\mathbf{z})\right]^{m_t}. \tag{1}$$

The probability of success for a tactic t and angle w against a target s generally depends on which tactics are applied against every other target, that is, the whole attack plan, which means that the probabilities p_{isw_j} are related to each other. This dependence is the core difficulty of the attacker's problem.

The objective is to maximize the expected total reward of the attack, found by multiplying the probability of success of an attack against a target with its reward. Since we want to optimize the total reward of the attack, these expected values should be added. The objective then becomes

$$\max \sum_{s \in S} \left[\sum_{t \in T} \sum_{w \in W_{st}} p^{\text{kill}}_{stw}(\mathbf{z}) \cdot z_{stw}\right] \cdot r_s.$$

For each target $s \in S$, at most one of the decision variables z_{stw}, $t \in T$, $w \in W_{st}$, takes the value one, since it is attacked at most once.

3 Mathematical Models

We here give a generic model for the joint attack problem and two approximate models that can be used to find upper and lower bounds on the optimal value.

3.1 A Generic Model

As stressed above, the probability $p_{isv}(\mathbf{z})$ depends in general on the whole attack plan $\mathbf{z}$, but in the generic model we make no assumptions on the exact nature of this dependence.

$$\max \sum_{s\in S}\left[\sum_{t\in T}\sum_{w\in W_{st}} p^{\text{kill}}_{stw}(\mathbf{z})\cdot z_{stw}\right]\cdot r_s \qquad [GENERIC]$$

$$\begin{aligned} s.t. \quad & \sum_{s}\sum_{t}\sum_{w\in W_{st}} n_t\cdot z_{stw} \le R && (i)\\ & \sum_{t}\sum_{w\in W_{st}} z_{stw} \le 1, \quad s\in S && (ii)\\ & z_{stw} \in \{0,1\}, \quad s\in S, t\in T, w\in W_{st} \end{aligned}$$

It is not necessary to attack all targets. Depending on the rewards specified for the targets, it might be optimal not to do so. Constraint (i) states that we cannot use more resources than we have. Constraint (ii) makes sure that each target is attacked at most once. Both constraints are linear, but the objective is in general nonlinear, non-convex and non-separable.

3.2 Optimistic Model

It is possible to construct two auxiliary problems that provide upper and lower bounds, respectively, on the optimal value of the generic problem. We analyze the expression for $p^{\text{kill}}_{stw}(\mathbf{z})$, under two specific assumptions.

Assume that no target will shoot against resources passing by towards other targets, but just against resources targeting themselves. This means that $p_{isv}(\mathbf{z}) = 1$ would hold for all targets $s \in \bar{S}$, and that $p^{\text{kill}}_{stw}(\mathbf{z})$ would collapse into the quantity

$$P_{st} = 1 - \prod_{j=1}^{V_t}\left[1 - p_{st}\prod_{i\in\bar{S}\setminus\{s\}} 1\right]^{m_t} = 1-(1-p_{st})^{n_t}.$$

Now the probabilities of success no longer depend on the overall attack plan $\mathbf{z}$. Further, since this expression does not depend on the angle w anymore, we only have to decide which tactic t to use against each target s, if any tactic at all.

We then obtain the optimistic model

$$
\begin{array}{llll}
\max & \sum_s \sum_t r_s \cdot P_{st} \cdot z_{st} & & [OPTIMISTIC] \\
s.t. & \sum_s \sum_t n_t \cdot z_{st} \le R & & (i) \\
& \sum_t z_{st} \quad \le 1, & s \in S & (ii) \\
& z_{st} \quad \in \{0,1\}, & s \in S, t \in T. &
\end{array}
$$

Solutions to the optimistic model give upper bounds to the original problem, since the values of all coefficients in the objective function are systematically increased. Even more, this is a valid upper bound for all choices of discretization V.

The solution found is also a feasible solution in the original problem, if complemented with an arbitrary reference angle of attack for each tactic used. This means that we can easily calculate a true objective value and also get a lower bound. This bound is only valid for the considered discretization V though.

3.3 Pessimistic Model

In contrast to the assumption made above, we now assume that each target will shoot against all resources passing by, on their paths towards other targets, and with its full defensive capability. Denote by $\tilde{p}_{isv}$ the resulting probability of surviving the defense from another target. This probability is clearly a pessimistic estimate of the true probability of surviving the defense from this target.

If $p_{isv}(\mathbf{z}) = \tilde{p}_{isv}$ would always hold, then $p^{\text{kill}}_{stw}(\mathbf{z})$ would become the quantity

$$
P_{stw} = 1 - \prod_{j=1}^{V_t} \left[1 - p_{st} \prod_{i \in \bar{S} \setminus \{s\}} \tilde{p}_{isw_j} \right]^{m_t},
$$

and we then obtain the pessimistic model

$$
\begin{array}{llll}
\max & \sum_s \sum_t \sum_{w \in W_{st}} r_s \cdot P_{stw} \cdot z_{stw} & & [PESSIMISTIC] \\
s.t. & \sum_s \sum_t \sum_{w \in W_{st}} n_t \cdot z_{stw} \le R & & (i) \\
& \sum_t \sum_{w \in W_{st}} z_{stw} \quad \le 1, & s \in S & (ii) \\
& z_{stw} \quad \in \{0,1\}, & s \in S, t \in T, w \in W_{st}. &
\end{array}
$$

The values of $\tilde{p}_{isv}$ might of course be too pessimistic, and hence the solution could provide poor lower bounds on the optimal value of the generic model.

Hopefully, though, the structure of the solution (the attack plan $\mathbf{z}$) is close to the optimal one, and by evaluating the true objective one can find a better pessimistic bound.

4 Simulation Details

In order to fully specify the generic model presented in Sect. 3.1, one needs to describe how the probability $p_{isv}(\mathbf{z})$ depends on the attack plan $\mathbf{z}$. It is obviously a hard task to model a real-life situation. We will here give the assumptions used in our simulation study.

We will analyze the different factors that affect $p_{isv}(\mathbf{z})$, that is, the probability for a resource to survive the defense from another target as it passes by toward its own target, and how it depends on $\mathbf{z}$. To do this, we look into the details of the defensive systems of the targets and define their rules of engagement.

4.1 Specifications of the Defensive System

Since we consider the problem of the attacker, we need to specify a set of deterministic engagement rules for the defenders. Each target with defensive capability is assumed to have a specified number of defensive channels, such as cannons or anti-missile systems. It will primarily defend itself, and any residual defensive channels will be used to defend the other targets, by engaging resources passing by inside its radius of defense. We make the following assumptions for each defender $i \in \bar{S}$.

1. The defender will primarily defend itself.
2. If there are $D_i > 0$ residual defensive channels, then they will be evenly allocated against the active engagement paths that pass by the target.
3. At most F_i channels might be used against a single engagement path.
4. At most G_i different engagement paths might be engaged.
5. All defensive channels should be used if there is something to shoot at.
6. If there are more active paths than defensive channels, one defensive channel is allocated to each path as long as possible with respect to a ranking defined by the distances to the target.

Given an attack plan $\mathbf{z}$, we let auxiliary variables $u_{isv}(\mathbf{z})$ describe how many defensive channels that should be allocated against the resources on each active path (s,v) passing by. The values of these variables will comply with the above rules.

Specifically, the number of resources on each path, denoted N_{sv}, affects the probability of success for each of these resources. We define $K = \max_{t \in T}\{n_t \; : \; V_t = 1\}$ to be the maximum number of resources travelling on a single engagement path. Hence, N_{sv} is in the range $k = 0, \ldots, K$.

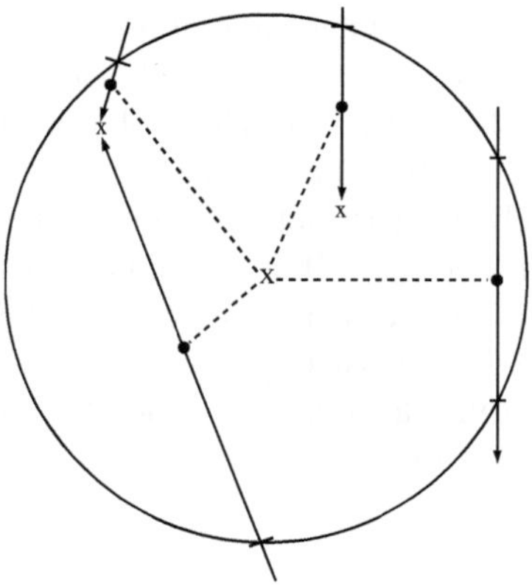

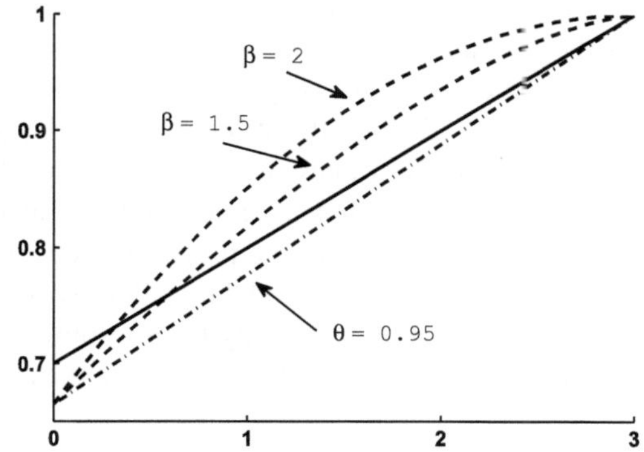

Fig. 5 To the *left*, an illustration of how the distance between a target and the active engagement path is measured. To the *right*, an example of how the design parameters β_{ik} and θ_{ik} affect the probability p_{isv}^k

We further define the parameter d_{isv} to be the orthogonal distance between a target $i \in \bar{S}$ and the engagement path (s, v). For other targets with positions inside the area of defense of target i, the distance to the mid-point of this path is used. This is illustrated in Fig. 5. Each active path is given a rank number, where the path closest to target i gets the highest rank, the second closest path gets the second rank, and so on. Closest path refers to the smallest distance d_{isv} and is thus relative to the target i. This ranking will be used when the defenders cannot engage all paths passing by, but need to prioritize.

4.2 *Specification of the Objective*

The probability for a resource to survive as it passes by target $i \in \bar{S}$ towards target $s \in S$ on path (s, v) is a function of the distance d_{isv} and the number of resources N_{sv} on the path, which are both a direct consequence of the attack plan $\mathbf{z}$. The obvious way to model this dependence would be to demand values for all such combinations as input data, but this is practically impossible. We instead introduce an analytic expression, based on both d_{isv} and N_{sv}.

Let p_{isv}^k be the probability for a resource to successfully pass by one defensive channel of target i. These probabilities are derived from the values of p_{st}, for tactics $t \in T$ where all $k = n_t$ resources are sent from the same angle ($V_t = 1$). Since this is only relevant for targets in $\bar{S}$, we denote this p_{ik} for all $i \in \bar{S}$ and $k = 1, \ldots, K$.

$$p_{isv}^k = 1 - \left(1 - \frac{d_{isv}}{\rho_i}\right)^{\beta_{ik}} \cdot (1 - \theta_{ik} \cdot p_{ik})$$

Here, ρ_i is the radius of defense, while β_{ik} and θ_{ik} are design parameters that model the defensive capacities of target i against different numbers of resources k.

The rightmost plot in Fig. 5 shows the probability p_{isv}^k on the y-axis as a function of the distance d_{isv} on the x-axis. Here, the probability $p_{ik} = 0.7$ is used, and the solid line corresponds to parameter values $\beta_{ik} = 1$ and $\theta_{ik} = 1$. The dash-dotted line is obtained when the value of θ_{ik} is changed to 0.95. The two dashed curves correspond to the values of 1.5 and 2, respectively, for parameter β_{ik}. In all, this expression for p_{isv}^k shows a reasonable behaviour. For $d_{isv} = 0$, its value becomes $\theta_{ik} \cdot p_{ik}$ and for $d_{isv} = \rho_i$ the probability becomes 1. For distances in-between, the parameter β_{ik} is used to model the effectiveness of the defensive system of target i.

Now finally, the probability for a resource to survive as it passes by target $i \in \bar{S}$ towards target $s \in S$ on path (s,v), given the attack plan $\mathbf{z}$, is

$$p_{isv}(\mathbf{z}) = \prod_{k=1}^{K} \left(p_{isv}^k\right)^{u_{isv}^k}.$$

Here, the auxiliary variable u_{isv}^k equals $u_{isv}(\mathbf{z})$ if $k = N_{sv}$ and zero otherwise. Since $u_{isv}(\mathbf{z})$, and thus also u_{isv}^k, might be greater than one the probability of success decreases with the number of defensive channels assigned to the engagement path. This is realistic as the defensive channels can be seen as independent, and the probability for a resource to survive two channels should be the probability of surviving them both. The general formula (1) now becomes

$$p_{stw}^{\text{kill}}(\mathbf{z}) = 1 - \prod_{j=1}^{V_t} \left[1 - p_{st} \prod_{i \in \bar{S} \setminus \{s\}} \prod_{k=1}^{K} \left(p_{isw_j}^k\right)^{u_{isw_j}^k}\right]^{m_t}.$$

The values of the variables u_{isv}^k are dependent on the entire attack plan $\mathbf{z}$. Once their values are known, it is, however, straightforward to evaluate the objective of the generic model.

4.3 An Illustrative Example

Consider a single defender i, as illustrated in Fig. 6. We name all paths (s,v) intersecting the area of defense in accordance with their rank, that is, the path with rank 1 is named path 1, and so on. Notice that one of the engagement paths never intersects the area of defense, and it is therefore never considered when the residual defensive channels are assigned. We assume that at most 3 channels might be used against a single engagement path (i.e., $F_i = 3$).

Assume first that at most 4 different engagement paths might be engaged (i.e., $G_i = 4$), and that there are 5 residual defensive channels (i.e., $D_i = 5$). Consider the case where all four paths passing by target i are active (i.e., $B_i = 4$), that is, at least one resource is following each path. Under the given assumptions, all paths should be engaged and first each path gets one defensive channel locked against it.

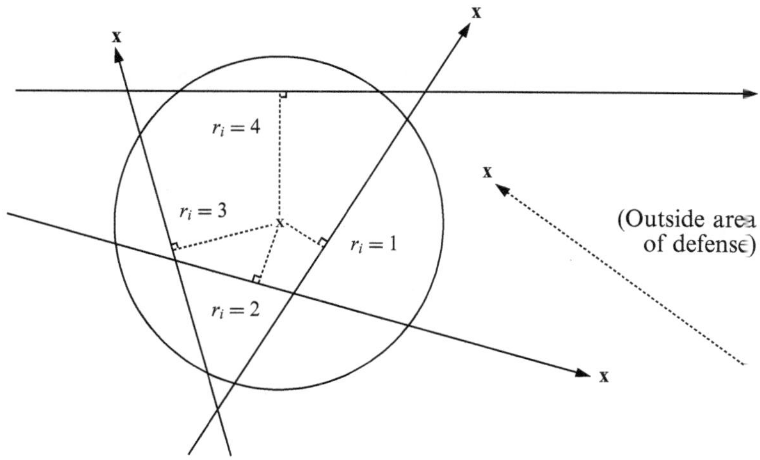

Fig. 6 A situation where multiple engagement paths intersect the area of defense for a target i

The remaining channel is assigned to the path closest to the target, which is path 1. The variables u_{isv} here take the values $u_{i1} = 2$, $u_{i2} = 1$, $u_{i3} = 1$ and $u_{i4} = 1$.

In the case that B_i or G_i decreases to 3, target i can only engage 3 engagement paths. For $B_i = 4$ and $G_i = 3$, the path most far away will no longer be engaged. The residual defensive channels are then distributed as follows: $u_{i1} = 2$, $u_{i2} = 2$, $u_{i3} = 1$ and $u_{i4} = 0$. If $B_i = 3$ and $G_i = 3$ (or 4), then only three engagement paths are active. Depending on which path that is not active, the other paths are assigned defensive channels like before, with respect to rank. Assume that, for example, path 2 is not active, in which case we get: $u_{i1} = 2$, $u_{i2} = 0$, $u_{i3} = 2$ and $u_{i4} = 1$.

Finally, if $B_i < 2$, all defensive channels cannot be assigned to an engagement path, since $F_i = 3$. With only one (or none) active path, at most $B_i \cdot F_i \leq 1 \cdot 3 = 3$ channels could be assigned. For example, if only path 3 is active, we obtain: $u_{i1} = 0$, $u_{i2} = 0$, $u_{i3} = 3$ and $u_{i4} = 0$.

5 Heuristic Solution Methods

A problem like this, with only a few constraints (one attack per target and shared resources) and a non-convex objective function, is well suited for meta-heuristics. Throughout this section, we base our work on the following assumptions:

1. The number of available resources is limited, that is, it is not possible to use the maximal number of resources against every target.
2. It is optimal to use all available resources.

The first assumption is reasonable, since otherwise the problem is reduced to choosing between tactics 3 and 5, either assigning all resources on the same path or splitting them on three different paths. (One would, however, still need to figure out the optimal combination of tactics and angle of attack for each target, and this would be a non-trivial problem.) The second assumption is very reasonable and simplifies the work of defining neighbourhoods and setting up heuristic schemes.

5.1 Local Search

Given a feasible solution to the generic model, $\mathbf{z}$, found by some heuristic scheme, one could try to improve it locally, that is, to perform a local search.

For this problem, where a solution $\mathbf{z}$ states which tactic t and angle w to be used for each target s, it is straightforward to test all feasible angles $w \in W_{st}$ for the assigned tactic t, one target at a time, and save the best improvement (if any). Then, if an improvement is made, one can repeat the same process again (since one target is now attacked from a different angle, and further improvements might be possible) until the process converges.

At the same time as one tests all angles, one can also switch between the tactics that use the same number of resources, hence conserving the overall usage of resources (assumed to be at its upper limit).

A limitation of this local search procedure is that the allocation of resources to targets is never changed. Even so, this procedure has proven to be an effective tool for finding good solutions, for almost any starting solution, as long as the allocation of resources to targets is close to the optimal one.

5.2 A Constructive Heuristic

An intuitive strategy is to iteratively augment a partial solution, adding one extra resource in each iteration. It seems plausible that the optimal solution using, say, 8 resources is close to the optimal solution for 7 resources.

Provided a feasible solution using $k \geq 0$ resources, denoted $\mathbf{z}_k$, we seek a solution $\mathbf{z}_{k+1}$. This is done by considering one target at a time, adding one resource if not $K = 3$ resources are already in use for this target, and then performing a local search. The best such augmentation, over all targets, is saved and returned as the new solution $\mathbf{z}_{k+1}$. The augmentation with one resource at a time is repeated until the available number of resources is reached. The cost of the heuristic will increase with respect to the number of targets, since it performs one local search per target.

Note that this constructive heuristic can be applied to any feasible starting solution. Further, if the initial solution is near-optimal for k resources, then it is likely that the augmented solution is also near-optimal, but now for $k+1$ resources.

As a bonus, this approach will generate Pareto-like solutions, stating the expected outcome of an attack for different numbers of resources, which also yields marginal values for additional resources with respect to the expected outcome.

This information is useful when choosing the number of resources to use for an attack. As will be seen in the forthcoming results, the gain in expected outcome of an additional resource decreases as a function of the number of resources already in use.

5.3 Simulated Annealing

A popular meta-heuristic, which is easy to implement, is simulated annealing. The basic idea, which makes it a meta-heuristic and not a local search method, is to accept solutions which are non-improving in order to escape local optima. This is done by chance, and the probability to accept a non-improving value is related to the change in objective value from the current solution to the new one.

Also, in order to assure finding a local optimum, the probability of accepting worse solutions decreases over time. This is done by a temperature parameter, which decreases as the iterations goes by. A simulated annealing approach is successfully used for a weapon-target allocation problem in [1].

In order to apply a simulated annealing approach, we need to define a neighborhood for a solution $\mathbf{z}$. Under the assumptions stated above, all we need is to work with feasible attack plans $\mathbf{z}$ that use all available resources. Hence we define five neighborhoods of an attack plan $\mathbf{z}$, denoted $N_k(\mathbf{z})$, in the following ways.

1. The angle of attack w is changed for one target s and tactic t in the attack plan, that is, $z_{stw} \rightarrow z_{st\bar{w}}$.
2. The tactic against one target is changed by switching between one angle and multiple angles, that is, $z_{stw} \rightarrow z_{s\bar{t}\bar{w}}$. If necessary, the reference angle w is adjusted. For example, instead of two resources attacking from the same angle, they attack from different angles. Notice that the number of resources involved in the attack is still the same though.
3. Pick two targets at random and switch their tactics and angle of attack. For example, variables $z_{s_1t_1w_1}$ and $z_{s_2t_2w_2}$ become $z_{s_1t_2w_2}$ and $z_{s_2t_1w_1}$ instead.
4. Pick two targets at random and exchange their angle of attack. For example, variables $z_{s_1t_1w_1}$ and $z_{s_2t_2w_2}$ become $z_{s_1t_1w_2}$ and $z_{s_2t_2w_1}$ instead.
5. Pick two targets at random, which do not use the same number of resources, and change to new tactics which increase/decrease the number of resources used, respectively. For example, one target is changed to be attacked by two resources instead of one, while another target is attacked by two resources instead of three.

The use of multiple neighborhoods provides diversity to the search, and by repeatedly changing between them all feasible solutions can be reached. Notice that neighborhood N_5 is crucial, since without it the number of resources allocated against each target would remain fixed to that of the initial solution throughout the search.

The implemented simulated annealing heuristic consists of outer and inner iterations. At the end of each outer iteration the temperature is decreased (from the initial temperature 0.9 and with the cooling factor 0.7). In each outer iteration, we cycle once through the different neighborhoods, according to the sequence $\{5,2,1,3,4,1,5,2,1\}$. For each of these, we perform 100 evaluations of neighbors. During the search, we keep track of the overall best found solution.

6 Numerical Experiments

The optimistic and pessimistic models presented in Sects. 3.2 and 3.3, respectively, are easily solved using a linear integer programming solver, in our case CPLEX. They provide upper and lower bounds on the true optimal value, and these are found in fractions of a second. In order to improve the lower bound, the pessimistic solution provided by the solver is simply evaluated using the true objective function. This step improves the bound significantly and is also instant. Moreover, if a local search, as described above, is performed from the pessimistic solution, an even better solution can be found. This is fairly inexpensive and improves the bound in most cases.

The constructive heuristic is initiated with the locally improved pessimistic solution obtained for $k = 2$ resources. It is then applied to find a solution with the available number of resources. The procedure should generate near-optimal solutions to the cost of at most one application of the local search procedure for each target and each new resource.

The simulated annealing method is applied as described in Sect. 5.3. This is a fairly time-consuming method, but is likely to produce the solutions of best quality.

6.1 Case 105

The test case, called Case 105, includes 2 targets with defense and 5 other targets, which are positioned as shown in Fig. 7. One unit step in the picture corresponds to 1 km. The targets with defense are positioned 10 km apart, and each of them has a defensive radius of 10 km. The distances between the targets are 300–500 m. When modelling the problem, a coarse angle discretization of 12 angles is used.

We define three different reward settings for the targets. In setting I, $r_s = 0$ for $s \in \bar{S}$ and $r_s = 1$ for $s \in S \setminus \bar{S}$, that is, there is no reward for the defenders and the same reward for every other target. Although this setting does not reward the targets with defense, it might still be optimal to attack the defenders in order to reduce their defensive capabilities and thus increase the overall reward of the attack. In reward setting II, $r_s = 1$ for $s \in \bar{S}$ and $r_s = 2$ for $s \in S \setminus \bar{S}$, so that the defenders are also considered valuable but only second to the other targets. In setting III, $r_s = 1$ for $s \in \bar{S}$ and $r_s = 5$ for $s \in S \setminus \bar{S}$, which differentiates the two types of targets more. Below, we present and analyze the result for the different reward settings.

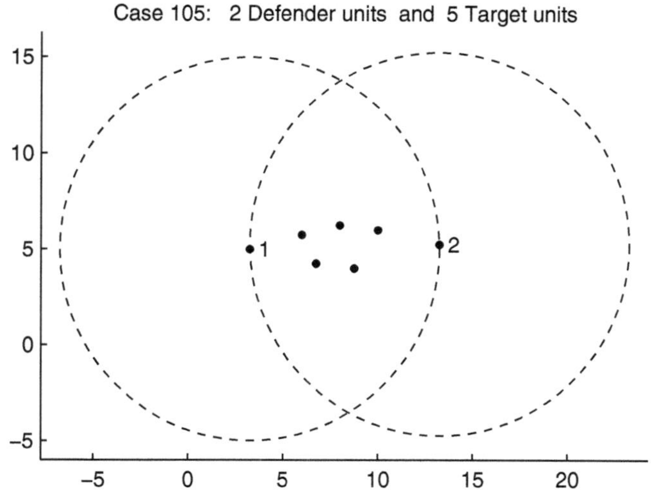

Fig. 7 Test case 105, with 2 defenders and 5 other targets

6.2 *Results for Case 105*

In Fig. 8 we see a graphical representation of the best found solution for Case 105 with reward setting III and 14 resources available. Both defenders, numbers 1 and 2, are attacked by tactic 5 which means 3 resources from different directions. Targets 5 and 6 are attacked using tactic 4, where 2 resources attack from opposite directions. Target 3 is attacked using tactic 2, that is, 2 resources from the same direction, indicated by the dashed line. Finally, targets 4 and 7 are attacked by single resources.

The solutions are not always intuitive at first glance. For example, one of the attack paths toward target 1 intersects the defensive area of target 2 for a long distance, and vice versa. Is it not better to attack with all 3 resources from the same angle and avoid the defense of the other defenders? The explanation is logical. Consider the resource attacking defender 1. By travelling inside the defensive area of defender 2, some of the defender's defensive capability will be allocated against this resource. As one of three resources taking part of the attack against target 1, the total expected probability of success will be quite high even though this specific resource faces great danger. In this way, the defensive capabilities available for target 1 to use against other resources are reduced, and the overall expected outcome will gain.

Figure 9 shows a graphical representation of the best found solution for the same case but with 17 resources available. The objective value is improved somewhat.

The use of reward setting I (i.e. reward 0 for defenders and reward 1 for other targets), render the result seen in Fig. 10. The x-axis represents the number of resources available and the y-axis the corresponding objective values.

The two outer dash-dotted lines represent the upper and lower bounds, respectively, found by CPLEX, where the pessimistic solutions have been evaluated

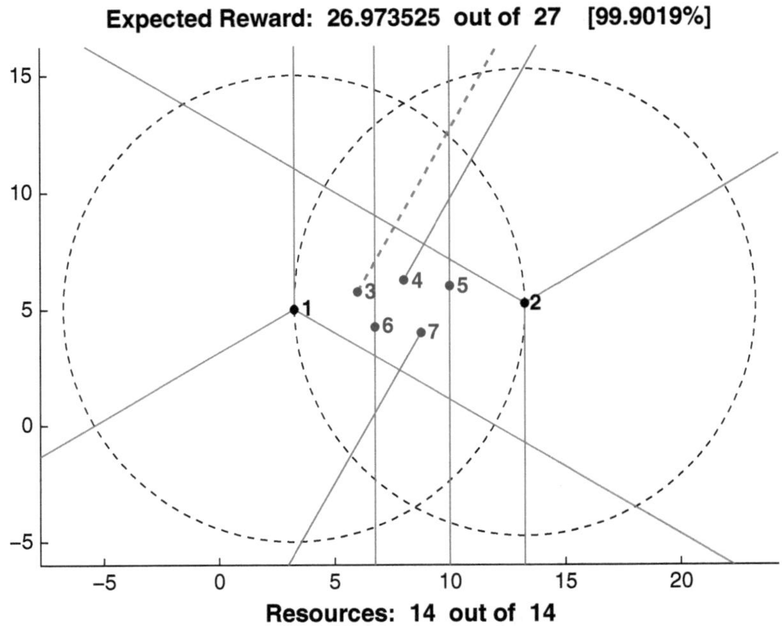

Fig. 8 Test case 105, with 2 defenders and 5 other targets. Best solution for 14 resources

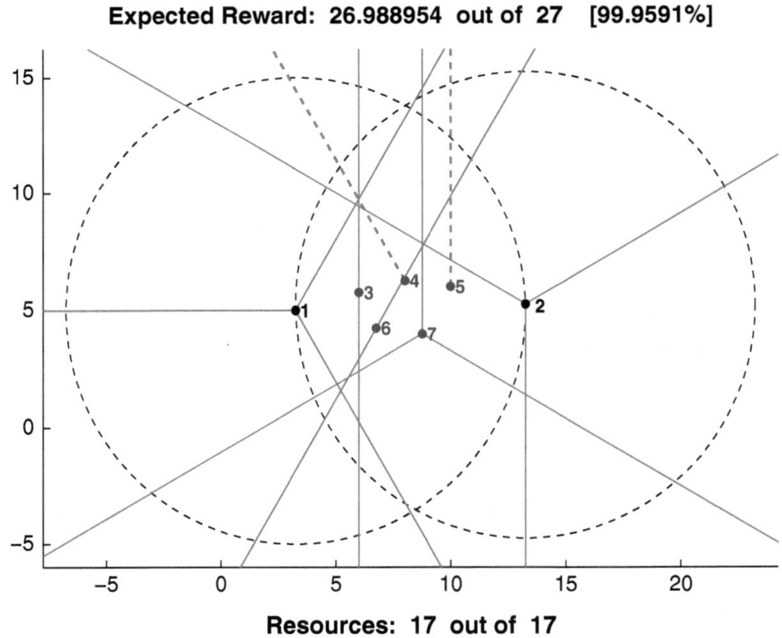

Fig. 9 Test case 105, with 2 defenders and 5 other targets. Best solution for 17 resources

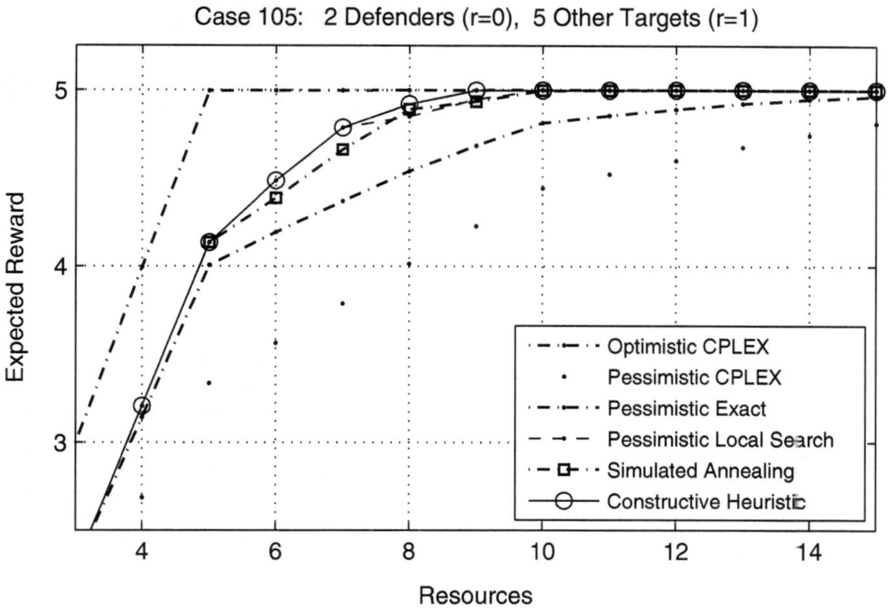

Fig. 10 Results for test case 105 with reward setting I

using the true objective function. The single dots represent the pessimistic values given by CPLEX. The dashed line with dots is the locally improved pessimistic solutions. We can see that the improvement is substantial for most numbers of resouces. The dash-dotted line with squares shows the best found solutions from the simulated annealing heuristic. The solid line with circles shows the result of the constructive heuristic. These solutions are in general the best ones found, but sometimes simulated annealing solutions are equally good.

For reward settings II and III, a similar behavior can be observed in Figs. 11 and 12, respectively. Obviously, the objective values differ due to the different reward settings, but the overall trend is the same.

We conclude this section with some remarks. The behavior is very similar for the different reward settings. The optimistic and pessimistic bounds are not tight for 5–10 resources, but a local search from the pessimistic solution improves the situation. For Case 105, with only 7 targets, using more than around 15 resources is not very interesting, and, as can be seen in the graphs, the optimistic and pessimistic bounds are then tight.

The simulated annealing algorithm performs very well and provides solutions comparable with the constructive heuristic approach, but it requires comparably long time even for a moderate number of resources. Mostly, the constructive heuristic finds the best found solution, and it is beaten by the simulated annealing method on only single occasions, but it requires even more time than the latter algorithm when considering many resources.

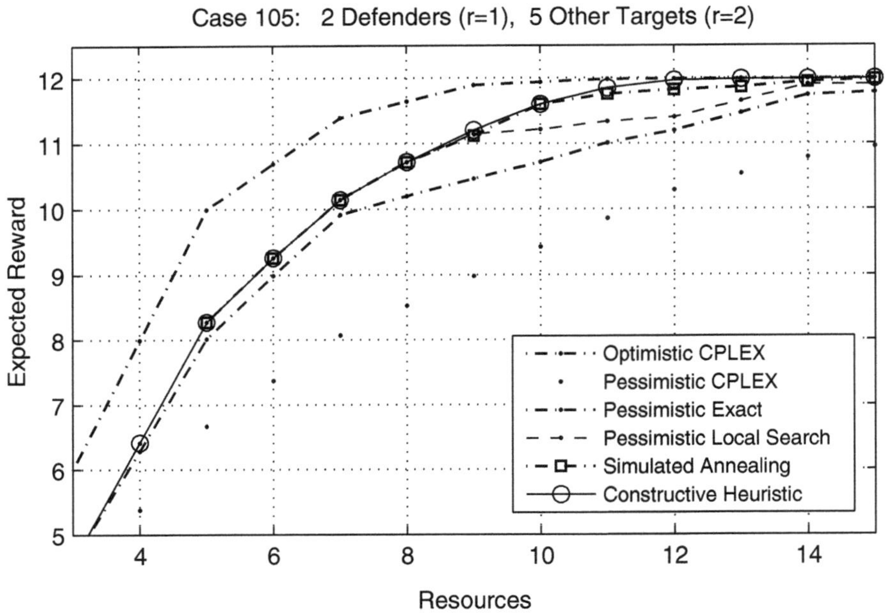

Fig. 11 Results for test case 105 with reward setting II

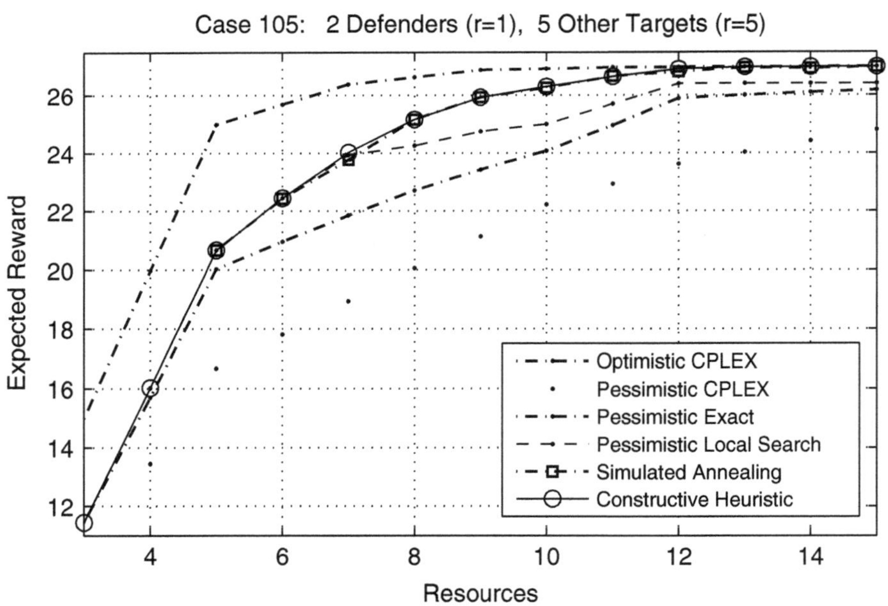

Fig. 12 Results for test case 105 with reward setting III

Table 1 Normalized mean objective values for each method

	Resources					
Method	5	10	15	20	25	30
Opt. CPLEX	1.0000	1.0000	1.0000	1.0000	1.0000	1.0000
Pess. exact	0.6702	0.7691	0.8399	0.8894	0.9209	0.9463
Pess. local	*0.6893*	0.8300	0.9023	0.9522	0.9737	0.9834
Constr. heur.	**0.6999**	**0.8599**	**0.9382**	**0.9852**	**0.9941**	**0.9988**
Sim. ann.	0.6845	*0.8545*	*0.9369*	*0.9840*	*0.9918*	*0.9940*

Best values are in boldface and second best values are *emphasized*

6.3 Results for Larger Instances

In addition to Case 105, a number of different cases have been studied. These ranges from 7 to 21 targets. Case 105 is a good representative for all of them, with respect to the behavior of the heuristic solution approaches. In Table 1, we give mean objective values for 12 different cases, with a varying number of resources. Here, the objective values are normalized with respect to the optimistic value found for each case.

For larger instances, with 10–20 targets, the quality of the optimistic and pessimistic bounds are not as good as for smaller instances. We suspect that the pessimistic bound is tight for up to 10 resources, and that the strength of the optimistic bound improves with an increasing number of resources. For instances where 10–20 resources are available, none of the bounds seems to be tight.

The constructive heuristic is the most stable of all solution methods, providing high quality solutions for all different scenarios and reward settings. The simulated annealing method is also very successful. Because of the long calculation times required for a single run of the simulated annealing method, it is only competitive with the constructive heuristic approach when seeking a single solution for a specific and quite large number of resources. Otherwise, the constructive heuristic provides both better calculation times and solution quality, with the important extra feature of providing a range of solutions, one for each number of resources. In all, the constructive heuristic is the clear winner.

7 Conclusions and Future Work

We have introduced and defined a mission planning problem. A generic mathematical model of the problem is presented, and the complex objective function is analyzed in detail. The generic model can be approximated in order to derive optimistic and pessimistic models. Such models are an important tool since they provide upper and lower bounds on the optimal value, hence limiting the uncertainty of the quality of solutions.

However, in order to solve problem instances of realistic sizes, it is necessary to use heuristic methods. We have proposed a constructive heuristic method and a simulated annealing heuristic to solve this difficult problem. The methods were tested on a set of problem instances, and the results are very promising. The constructive heuristic method has good solution times, while solution times are relatively long for the simulated annealing algorithm.

All methods are generic and can handle different scenarios for the defender's strategy. It is sufficient to provide a black-box function to call whenever the objective needs to be evaluated. Hence, if the assumptions in Sect. 4 are inadequate, or needs to be modified in any way, the given framework will still be applicable.

This paper has focused on the development of a planning system only considering target scene parameters such as target location and defense system description, and how the defense reacts upon attack. Resource performance is certainly included in the analysis but just in the sense of a static set-up of effect-on-target as a function of tactics, and the ability to survive in a surface-to-air defense system environment. This approach complies with future command and control doctrines which promote a separation of effect planning and resource allocation planning.

To extend the mission scope we can include planning aspects of the platform. Route planning can be conducted in a flexible way with its own objectives to conclude the overall mission success. Obvious aspects are minimizing radar cross section exposure during route phase, and minimize time to target, that is, to explore hiding possibilities or by clever surveillance tactics during the cruise phase. An obvious continuation from our work within this paper is to investigate the coupling between route and effect planning. If this is solved properly, a large step is taken to control and comprise vital aspects of ground attack planning.

Further, firing platforms must not be given in advance, instead maximizing the effect of the target area can be the driver to find the best platforms from a larger set. Based on this fact, future work could address at least two obvious scenarios. The first is when the target scene is known and there is a predefined number of platforms where route planning is included in the overall mission. A second scenario is to consider when several platforms are available. In this case we must allocate good firing units from a set of platforms but also decide firing position and route planning.

References

1. Ciobanu, C., Marin, G.: On heuristic optimization. Analele Stiintifice ale Universitatii Ovidius Constanta **9**, 17–30 (2001)
2. Kaminer, B.I., Ben-Asher, J.Z.: A methodology for estimating and optimizing effectiveness of non-independent layered defense. Syst. Eng. **13**, 119–129 (2010)
3. Kwon, O., Lee, K., Kang, D., Park, S.: A branch-and-price algorithm for a targeting problem. Naval Res. Logist. **54**, 732–741 (2007)
4. Miercort, F.A., Soland, R.M.: Optimal allocation of missiles against area and point defenses. Oper. Res. **19**, 605–617 (1971)
5. Quttineh, N.H.: Models and Methods for Costly Global Optimization and Military Decision Support Systems. Linköping Studies in Science and Technology, Dissertations, No. 1450, Linköping University (2012)
6. Yost, K.A., Washburn, A.R.: Optimizing assignment of air-to-ground assets and BDA sensors. Military Oper. Res. **5**, 77–91 (2000)

Competitive Multilevel Capacity Allocation

A. Karakitsiou

Abstract In this paper we consider a supply chain where the purchasing behavior of final users of the product influences the decisions that are made. We particularly examine the effects of customers' competition for the offered service level on the facility location decisions. We consider two types of decision makers, the producer who tries to provide at facilities the best level of service at minimum cost and the customers who make their choices in order to minimize their perceived costs. We consider first the case where customers are assigned for service to the facilities by the producer. In such a case the producer could be considered as monopolist who dominates and tries, through the facility location decisions, to ensure the best, at his opinion, service level at minimum cost. Then we suppose that the customers are involved in a Nash type game in their effort to ensure the best level of services for themselves, i.e. we assume that they are involved in an oligopsony. In order to take into consideration the effects of this competition to the facilities location decisions we formulated the problem as a bilevel programming model. Next, we suppose that there are two producers operating in the network, who constitute a duopoly. The producers compete with each other with respect to the service level they offer in order to attract customers. We propose a bilevel model with two leaders in order to take into account both the competition between producers and the competition among customers.

A. Karakitsiou (✉)
Technological Educational Institute of Serres, Terma Magnesias, 62100 Serres, Greece
e-mail: karakitsiou@teiser.gr

A. Migdalas et al. (eds.), *Optimization Theory, Decision Making, and Operations Research Applications*, Springer Proceedings in Mathematics & Statistics 31,
DOI 10.1007/978-1-4614-5134-1_5,

1 Introduction

The location of facilities, i.e., the determination of the best places for production facilities warehouse or intermediate distribution centers, is a key issue of the strategic management covering the core of the supply chain planning.

Facilities Location Problem and capacity allocation to them in order to be able to serve customer demand is one of the most traditional areas of optimization. The basic task of all variants of facility location problems is the following: a company wants to open up a number of facilities to serve their customers. Both the opening of a facility at a specific location and the service of a particular customer through a facility incurs some cost. The goal is to minimize the overall cost associated with a specific way of opening up facilities and serving customers [7, 8].

Therefore, when creating a new facility, factors including the selection of appropriate capacity and area of the facility deserves separate attention and should be taken carefully.

The location of facilities affects not only the distance that users will travel to them, but also, in connection with decisions about capacity, the time customers spend on-site prior to their service. The conditions under which customers make their choice of service facility are complicated, but it is generally reasonable to assume that every customer will choose the facilities that minimize their total transportation and waiting cost.

Mathematical models dealing with such situation (for example [12–14]) use direct choice to assign customers to facilities, that is the assignment is done by the system and each customer and his demand is directed to the closest facility. The congestion in facilities is controlled by incorporating constraints in order to ensure a desired level of waiting time or a specific number of customers. Such constraints tend to equalize the level of congestion at different facilities, whether it is measured by the number of users waiting or by waiting time. But empirical studies [9, 10] have shown that when customers are traveling, they select the facility that minimizes the travel time and waiting time. It is therefore likely that a user may choose not to travel to the nearest facility, but in another which although it is further away it is less congested.

In this work, we examine a supply chain network where the producer wants to determine the number of facilities and the total production capacity in order to ensure a certain level of satisfaction to his customers while taking into account the waiting time of customers in the system.

In the mathematical models presented initially a central coordinator, we assume that the producer, has the ability to direct customers to distribution centers that would be located. Particularly, we assume that we are dealing with a centralized supply chain management. Considering the supply network as a single market, the manufacturer can be regarded as a monopolist who dominates the market and tries, through the location and capacity choices to ensure a certain level of service to customers.

Next, the mathematical models are extended to include the case where customers are able by themselves to determine the distribution center from which to seek satisfaction of their demand. The customer's choice is affected by the total personal costs incurred during the transaction process with the producer. Besides the price, the total cost includes the transportation cost required to be paid by the customer to ensure the product and the costs generated by the delay observed during the serving process. Therefore, the selection of distribution center is made competitively aiming at minimizing their personal cost. It is proven that the choices of the customers are different from the assignments of the central coordinator when they are competing each other for the received service levels. That is, the customers form an oligopsony. Consequently, this competitive behavior should be taken into consideration by the producer during his decision-making process. Assuming that customers observe the location decisions of the producer and that he is fully informed about the events at each distribution center, we formulate the facility location problem as bilevel programming model.

We examine two types of decision makers who have different objectives and they are in different levels of hierarchy. The first level, the producer (the leader), offers at the distribution centers the best at his opinion service level, including location, at minimum cost. The second level, the customers (the followers) make their choices competitively (a Nash type game) aiming at minimizing their personal expenses These two levels of hierarchy are involved in a Stackelberg game. First the leader, determining the service level that minimise his cost. The followers react, being fully informed about the leader's decision. The leader knows it and takes it into account before he announces his strategy.

The results obtained by the numerical analysis of the proposed models demonstrate that the oligopsonistic behavior of the customers, regarding the service level, improves the quality of the provided service at the distribution centers by imposing adjustments to the market needs. It improves therefore market's flexibility. This finding opens a new research area in the study of classical oligopsonistic market structure, which in recent decades has received a lot of criticism [15].

In the last part of this work, we assume that the distribution centers are owned by two producers. Specifically, we assume that the producers form a duopoly which compete for customers through the provided service levels involved in a Nash game.

Unlike the existing models in the literature dealing with competition of suppliers in the supply chain (e.g., [18]), we formulate the problem of the competition between suppliers for service level they offer as a bilevel problem with two competing leaders. Due to the competitive nature of duopoly, the bilevel formulation of the problems is more complicated compared to those with one leader. We also demonstrate that due to the behavior and choices of competing producers the bilevel game is significantly different from those of monopoly. The bilevel oligopoly game formulation of the competitive location and capacity allocation is to our knowelge proposed for the first time in the bibliography.

The rest of the paper is organized as follows: Section 2 gives a short overview of the bilevel programming problem. We present in Sect. 3 the models dealing with the centralized supply chain management. Particularly, the model of Sect. 3.1 is concerned with the case where the the producer makes the related decisions by ignoring the competitive behavior of customers, while in the location resulting from the model of Sect. 3.2 this behavior is taken into account directly. The comparison of these two different models is illustrated in Sect. 3.3. In Sect. 4 we examine the case where more than one producers are competing for the offered service level.

2 Bilevel Programming

Bilevel programming problems describe a hierarchical system involving two decision levels with different, often conflicting, objectives [1, 17]. In the first level, the leader controls the decision variable $\mathbf{x}$. In the second level, the follower controls the decision variable $\mathbf{y}$. The corresponding loss functions $\phi_{\mathrm{L}}(\mathbf{x},\mathbf{y})$ and $\phi_{\mathrm{F}}(\mathbf{x},\mathbf{y})$ describe the interaction between these two decisions variables. These two levels of hierarchy are usually involved in a Stackelberg game. The basic idea of this game can be described as follows:

The leader chooses the strategy $\mathbf{x}$ which minimizes his loss function $\phi_{\mathrm{L}}(\mathbf{x},\mathbf{y})$ and the follower, fully informed about the leader's decision, reacts by choosing the strategy which minimizes his own loss function $\phi_{\mathrm{F}}(\mathbf{x},\mathbf{y})$. Thus, follower's choices depend on the leader's choices, i.e., $\mathbf{y}=\mathbf{y}(\mathbf{x})$. The leader, on the other hand, is aware of the follower' s choices and he takes this reaction into account before announcing his strategy. The bilevel problem could be written in the following general form:

$$[\mathbf{P}]\ \min_{\mathbf{x}\in\mathscr{X}} \quad \phi_{\mathrm{L}}(\mathbf{x},\mathbf{y}) \tag{1}$$

$$\text{s.t} \quad \varphi_{\mathrm{L}}(\mathbf{x},\mathbf{y}) \le 0, \tag{2}$$

where $\mathbf{y}(\mathbf{x})$ solves

$$\min_{\mathbf{y}\in\mathscr{Y}} \quad \phi_{\mathrm{F}}(\mathbf{x},\mathbf{y}) \tag{3}$$

$$\text{s.t} \quad \varphi_{\mathrm{F}}(\mathbf{x},\mathbf{y}) \le 0, \tag{4}$$

where $\mathbf{x}\in\mathscr{X}\subset R, \mathbf{y}\in\mathscr{Y}\subset R$ and $\mathscr{X},\mathscr{Y}$ are closed subsets and $\varphi_{\mathrm{L}}:\mathscr{X}\times\mathscr{Y}\to R^p$, $\varphi_{\mathrm{F}}:\mathscr{X}\times\mathscr{Y}\to R^q$.

The upper level of the problem (1)–(2) is the leader's problem, whereas the lower level (3)–(4) corresponds to the follower's problem The set $\mathscr{S}=\{(\mathbf{x},\mathbf{y}):\mathbf{x}\in\mathscr{X},\mathbf{y}\in\mathscr{Y},\phi_{\mathrm{L}}(\mathbf{x},\mathbf{y})\le 0,\ \phi_{\mathrm{F}}(\mathbf{x},\mathbf{y})\le 0\}$ is called constraint set. The set $\mathscr{Y}(\mathbf{x})=\{\mathbf{y}\in\mathscr{Y}:\phi_{\mathrm{F}}(\mathbf{x},\mathbf{y})\le 0\}$ is the feasible set of the follower for every given $\mathbf{x}\in\mathscr{X}$. The set of all orthological reactions of the follower is the $\mathscr{R}(\mathbf{x})=\{\mathbf{y}\in\mathscr{Y}:\mathbf{y}\in\arg\min_{\mathbf{z}\in\mathscr{Y}(\mathbf{x})}\phi_{\mathrm{F}}(\mathbf{z},\mathbf{y})\}$, while the set $\mathscr{F}=\{\mathbf{x},\mathbf{y}:(\mathbf{x},\mathbf{y})\in\mathscr{S},\mathbf{y}\in\mathscr{R}(\mathbf{x})\}$ is the set of the feasible solution of the **[BP]**. A feasible point $(\mathbf{x}^\star,\mathbf{y}^\star)\in\mathscr{F}$ is a Stackelberg equilibrium if $\phi_{\mathrm{L}}(\mathbf{x}^\star,\mathbf{y}^\star)\le\phi_{\mathrm{L}}(\mathbf{x},\mathbf{y})\ \forall\ (\mathbf{x},\mathbf{y})\in\mathscr{F}$. This equilibrium point describes the optimal leader's strategy.

Originally formulated as a mathematical model by Bracken and McGill [2, 3], the **[BP]** has been studied extensively in the last decades due to its numerous applications [1, 11, 16, 17].

A bilevel model can be linear, nonlinear, quadratic, etc., depending on the problem it formulates. Consequently, various methodological approaches and algorithms have been developed for its solution. Some recent application of the bilevel programming can be found at [11, 16].

3 Optimal Facilities Location and Capacity Assignment Under Customer Competition

In this section we assume that the producer addresses the supply chain as a single market and therefore he may be considered as a monopolist who dominates the market. Consequently, given the total demand of the chain, he can determine the optimal location of distribution centers and their total capacity by applying a combination of both the principles of the monopolistic market and the facilities location models (see, e.g., [15] and [8], respectively).

As soon as the distribution centers are located and capacity is assigned to them, the procurement of the customers is limited due to these decisions, since the sales of the products to the customers is actually a capacity assignment process to these distribution centers.

Therefore, the main problem of the supply chain management system under consideration is to find an appropriate decision-making mechanism based on which the location of the distribution centers will be, at individual level, advantageous for all members of the network.

3.1 Optimal System Location

We assume that the producer tries to provide to the n customers the best, at his opinion, service level at minimum cost. The evaluation of the offered service is based on the delay faced by the customers at each distribution center i.

If x_{ij} is the amount that the customer j buys from the distribution center i, then the performance function $d_i(x_i)$ measures the level of service offered by the distribution center i, where $x_i = \sum_{j=1}^{n} x_{ij}$.

Suppose that m is the set of potential sites for the location of the distribution centers. We assume that the establishment of a distribution center to the candidate site i implies a fixed location cost F_i. Furthermore, suppose further r_j is the demand of the customer j $(j = 1, \ldots, n)$ for the product, p_i is the unit price paid by customers at every distribution center and q_i the capacity of the distribution center i $(i = 1, \ldots, m)$.

The producer (the central coordinator) should choose the location of the distribution center such that the optimal benefit of the system is achieved. The aim is to find the location of the facilities and to assign the customers to them so as to minimize the total system cost. The mathematical model can be formulated as:

$$(\mathbf{SO}-\mathbf{FL})\ \min\ \sum_{i=1}^{m} d_i(x_i)x_i + \sum_{i=1}^{m} p_i x_i + \sum_{i=1}^{m}\sum_{j=1}^{n} t_{ij}x_{ij} + \sum_{i=1}^{m} F_i y_i \tag{5}$$

$$\text{s.t}\ \sum_{i=1}^{m} x_{ij} = r_j,\ \forall j \tag{6}$$

$$x_i \leq y_i q_i,\ \forall i \tag{7}$$

$$x_i - \sum_{j=1}^{n} x_{ij} = 0,\ \forall i \tag{8}$$

$$y_i \in \{0,1\}, \forall i \tag{9}$$

$$x_{ij} \geq 0,\ \forall i,\ \forall j \tag{10}$$

The objective function of problem (5) minimizes the total cost consisting of the cost of the delay, plus the transportation and purchasing costs plus the cost involved in setting up a distribution center. The constraints (6) ensure that the quantities purchased by the customer j at all distribution centers meet his overall demand. The constraints (7) impose that the total amount of the product available at each distribution center i does not exceed its capacity. In addition, it enables that the assignment of the customers' demand only in sited distribution. The relations (8) are the defining constraints of the model, ensuring the maintenance of flow in the network.

Another important decision the producer should make is the determination of the capacity of the distribution center. If the capacity is set at a level higher than the demand faced by a distribution center, then the producer will bear the cost of the capital committed to the production of this excess capacity. On the other hand, a low level of capacity leads to an increasing service time and consequently to lost sales since customers will be forced to seek service from other distribution center.

In the model discussed above, the capacity of a potential distribution center is given in advance. However, the case where the capacity is not given but it must be decided during the configuration of the system can be examined. In such a case the performance function will depend not only on the total amount of the product x_i that the distribution center i sells but also on the decision made by the producer concerning the level of the capacity q_i i.e $d(x_i, q_i)$. Essentially, this means that the capacity assignment to a distribution center implies the location of this center, since zero capacity implies a non-located distribution center.

Hence, the producer should solve the following problem:

$$(\mathbf{SO-CA})\ \min\ \sum_{i=1}^{m} d_i(x_i,q_i)x_i + \sum_{i=1}^{m} p_i x_i + \sum_{i=1}^{m}\sum_{j=1}^{n} t_{ij}x_{ij} \tag{11}$$

$$\text{s.t}\ \sum_{i=1}^{m} x_{ij} = r_j,\ \forall j \tag{12}$$

$$x_i \le q_i,\ \forall i \tag{13}$$

$$x_i - \sum_{j=1}^{n} x_{ij} = 0,\ \forall i \tag{14}$$

$$\sum_{i=1}^{m} \alpha_i q_i \le B \tag{15}$$

$$0 \le q_i \le U_i,\ \forall i \tag{16}$$

$$x_{ij} \ge 0,\ \forall i,\ \forall j \tag{17}$$

The relations (15) impose the total amount of money spent in capacity investment to m distribution center must not exceed the available budget B, while the constraints (16) ensure that the capacity of the distribution center will not exceed U unit.

It is possible to add to the objective function a cost function similar to that of (5) instead of the constraint (15), i.e., $\sum_{i=1}^{m} F_i(q_i)$ where $F_i(\cdot)$ are continuous function, to depict the economies of scale produced by the different capacity assignment levels.

3.2 Bilevel Problem Formulation Under Customer Competition

The producer must take into account the reactions of customers in every decision he takes and determine the final location of the distribution centers based on these reactions. In other words, the producer should understand that he cannot control the choices of the customers. Consequently he should use the Nash game, in which the customers are involved, as an oracle in order to be able to predict their reaction. That is, he should compute the total cost of the system based on the reactions of the customer to every decision he makes and select the most satisfactory.

The problem can be formulated as bilevel programming model:

$$(\mathbf{BSO-FL})\ \min_{[y_i]} \textstyle\sum_{i=1}^{m} F_i y_i + \sum_{i=1}^{m} d_i(\bar{x}_i)\bar{x}_i$$

$$+\textstyle\sum_{i=1}^{m} p_i \bar{x}_i + \sum_{i=1}^{m}\sum_{j=1}^{n} t_{ij}\bar{x}_{ij} \tag{18}$$

$$\text{s.t}\ \ y_i \in \{0,1\},\ \forall i \tag{19}$$

where $[\bar{x}_i]$ and $[\bar{x}_{ij}]$ solve

$$(\textbf{UO}-\textbf{TP})\ \min \quad \sum_{i=1}^{m}\int_0^{x_i} d_i(t)\mathrm{d}t$$

$$+\sum_{i=1}^{m} p_i x_i + \sum_{i=1}^{m}\sum_{j=1}^{n} t_{ij}x_{ij} \tag{20}$$

$$\text{s.t} \quad \sum_{i=1}^{m} x_{ij} = r_j,\ \forall j \tag{21}$$

$$x_i \leq q_i y_i,\ \forall i \tag{22}$$

$$x_i - \sum_{j=1}^{n} x_{ij} = 0,\ \forall i \tag{23}$$

$$x_{ij} \geq 0\ \forall i,j \tag{24}$$

According to this model, the leader (producer) decides the location of distribution centers solving the problem (18)–(19), but he does not control the variables x_i and x_{ij} since they describe the choices of his customers. The values of the variables $[\bar{x}_i]$ and $[\bar{x}_{ij}]$ are derived from the model (20)–(24) corresponding to an oracle. In other words, the leader uses (20)–(24) as an oracle to discover trends / reactions of the customers in each potential location and tries to minimize the total cost of the system based on these discoveries.

The bilevel capacity assignment problem can be formulated analogously:

$$(\textbf{SO}-\textbf{CA})\ \min_{[q_i]} \quad \sum_{i=1}^{m} d_i(\bar{x}_i,q_i)\bar{x}_i \sum_{i=1}^{m} p_i\bar{x}_i + \sum_{i=1}^{m}\sum_{j=1}^{n} t_{ij}\bar{x}_{ij} \tag{25}$$

$$\text{s.t} \quad \sum_{i=1}^{m} \alpha_i q_i \leq B \tag{26}$$

$$0 \leq q_i \leq U_i,\ \forall i \tag{27}$$

where $[\bar{x}_i]$ and $[\bar{x}_{ij}]$ solve

$$(\textbf{UO}-\textbf{TP})\ \min \quad \sum_{i=1}^{m}\int_0^{x_i} d_i(t)d(t) + \sum_{i=1}^{m} p_i x_i + \sum_{i=1}^{m}\sum_{j=1}^{n} t_{ij}x_{ij} \tag{28}$$

$$\text{s.t} \quad \sum_{i=1}^{m} x_{ij} = r_j,\ \forall j \tag{29}$$

$$x_i \leq q_i,\ \forall i \tag{30}$$

$$x_i - \sum_{j=1}^{n} x_{ij} = 0,\ \forall i \tag{31}$$

$$x_{ij} \geq 0,\ \forall i,j \tag{32}$$

Table 1 Parameters of the example

	t_{ij}					p_i	q_i	F_i
	C_1	C_2	C_3	C_4	C_5			
F_1	15	16	15	13	17	14	2,350	950
F_2	19	25	30	26	17	16	1,350	1,600
F_3	27	15	21	29	16	15	1,000	1,700
	r_j							
	C_1	C_2	C_3	C_4	C_5			
	400	500	450	600	350			

According to this model, the producer determines the capacity of the distribution centers by solving the problem (25)–(27). The customers fully informed about the decisions of the producer, choose the distribution center which ensures them the optimal level of service by solving the problem (28)–(32). The leader knows that the selection of the customers is based on this criterion, and due to this reason he uses the problem (20)–(24) as a tool to predict the trends / reactions of customers in each potential location.

3.3 *Numerical Comparison of the Models*

The aim of this section is to clarify the differences in the decision-making process that are proposed by the models (5)–(10) and (18)–(24). Furthermore, the purpose of this section is to determine the negative results of the incorrect choice of location. For this reason we will employ a randomly generated numerical example.

We consider the case where the producer has 3 potential sites for the establishment of distribution centers that should satisfy the demand of 5 customers. These sites differ with each other not only in the available capacity but also in the fixed location cost.

The performance function is given by the equation:

$$d_i(x_i) = \frac{1}{q_i - x_i}, \tag{33}$$

and the total unit cost is calculated using the formula:

$$\tilde{c}_{ij}(x_i) = d_i(x_i) + x_i \frac{\partial d_i(x_i)}{\partial x_i} + p_i + t_{ij}. \tag{34}$$

Table 1 presents all the necessary parameters.

The Problems **(SO-FL)** and **(BSO-FL)** were modeled using the mathematical programming language AMPL and solved, after implementing a branch-and-bound scheme, by the MINOS 5.5 solver. Figure 1 depicts the flow of customers to distribution centers arising after the solution of the problem.

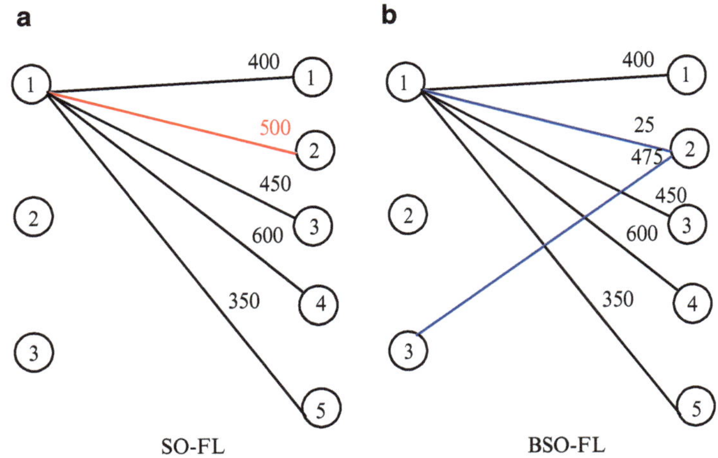

Fig. 1 Optimal system location vs. location based on customer competition

As it is shown in Fig. 1a, when the producer is interested in minimizing the "average" cost faced by the customers plus the location cost (i.e., when the **(SO-FL)** is solved), a single distribution center is located and all customers fullfil their demands there.

The basic question is then: Does this assignment satisfy all the customers? If we look at the unit cost of the solution in Table 2 we observe that the assignment, for example, of the C_2 to the distribution of center F_1 results to a unit cost of 30.60, which is not the minimum that could be faced by this particular customer, since the unit cost at the distribution center F_3 is lower (30 vs. 30.6).

Figure 1b shows the optimal location when the producer solves the problems (20)–(24) i.e., when the producer tries to identify the reaction/trends of the customers to his location options. The optimal location of this scenario indicates that a second distribution center must be opened and customer C_2 will choose to satisfy the maximum amount of his demand from this distribution center.

It is obvious that in terms of location cost, the solution proposed by the second model is more expensive. So, naturally arises the question "why should the producer take into account the behavior of the customers instead of accepting the solution of the first model?"

By comparing the two figures we can conclude that the customer 2 has "escape" trends in the sense that if a new distribution center will be opened (either by the producer himself or even worse by a competitor selling the same product) then the the larger part of the customer's demand will be lost. In the next section, we will examine in more detail the location decision under competition among producers.

Table 2 Solution results of the problems **(SO-FL)** and **(BSO-FL)**

	SO-FL		**BSO-FL**	
	Total cost: 68,630		Total Cost: 69,051.25	
	Unit cost	Quantity	Unit cost	Quantity
C_{11}	28.60	400	28.06	400
C_{21}	35.00	0	34.02	0
C_{31}	42.00	0	43.06	0
C_{12}	30.60	500	30.06	25
C_{22}	41.00	0	40.02	0
C_{32}	30.00	0	30.06	475
C_{13}	29.60	450	29.057	450
C_{23}	46.00	0	45.02	0
C_{33}	36.00	0	37.06	0
C_{14}	27.60	600	27.06	600
C_{24}	42.00	0	42.02	0
C_{34}	44.00	0	45.06	0
C_{15}	31.60	350	31.05	350
C_{25}	33.00	0	32.02	0
C_{35}	31.00	0	32.06	0

4 Duopoly

In a supply chain network where there are more than one producers, none of them has the power (monopolistic power) to direct customers to distribution centers. Thus, as a result, the offered service level and the customer satisfaction are the basic differentiation and discrimination components among economic units of the same sector.

In this section we examine the impact of the producers' competition for the customers attraction, first to the location decisions and second of setting capacity assignment to distribution centers. We examine a duopolistic supply chain network. We assume that producers compete by taking part in Nash game. They try to attack the customers' demand by providing the optimal service level, that is the one which minimizes the costs arising from the customers reaction to their decision.

Definition 1. A Nash equilibrium for this duopolistic game corresponds to a set of location and capacity choices (strategies), which ensure that none of the players are better of by unilaterally changing his strategy.

We assume further that the customers participate in a second Nash game in order to ensure the optimal service level for themselves.

We formulate the problem as a bilevel model, where the two producers determine the optimal location and capacity of the distribution centers, by taking into account the choices and the requirements set by the customers for the offered service level.

The competition among the members of the supply chain has been studied extensively in the literature [4–6, 18, 19]. The vast majority of this scientific work has be carried out in the framework of the classical economic theory of duopoly. The competition takes place either by fixing the price level or by determining production levels that maximize the profit of the producer. In addition, in all these models the decision-making process takes place in a single level.

The competition models that will be developed in the next section refer to the competition between two producers but can easily be extended in the case where more than two producers participate.

4.1 *Competitive Facility Location when Customers Participate in Their Competitive Game*

Let's assume that the potential location of distribution centers $i = 1,\dots,m$ are dispersed between the two producers who in turn are involved in a competition for customer attraction through the provided service level. Let M_1 and M_2 ($m = |M_1| + |M_2|$) be the nodes of the two producers, respectively. Then, under the assumption that both producers "announce their strategies simultaneously," we obtain a Nash game with two players who are dealing (for $K = 1,2$) with the following problems:

The facility location problem of the producer 1:

$$(\mathbf{CFL_1})\ \min \quad \sum_{i\in M_1} F_i y_i + \sum_{i\in M_1} d_i(\bar{x}_i)\bar{x}_i + \sum_{i\in M_1} p_i\bar{x}_i + \sum_{i\in M_1}\sum_{j=1}^{n} t_{ij}\bar{x}_{ij} \tag{35}$$

$$\text{s.t} \quad y_i \in \{0,1\}, \forall i \in M_1 \tag{36}$$

The facility location problem of the producer 2:

$$(\mathbf{CFL_2})\ \min \quad \sum_{i\in M_2} F_i y_i + \sum_{i\in M_2} d_i(\bar{x}_i)\bar{x}_i + \sum_{i\in M_2} p_i\bar{x}_i + \sum_{i\in M_2}\sum_{j=1}^{n} t_{ij}\bar{x}_{ij} \tag{37}$$

$$\text{s.t} \quad y_i \in \{0,1\}, \forall i \in M_2 \tag{38}$$

where $[\bar{x}_i]$ and $[\bar{x}_{ij}]$ solve (20)–(24)

Let $Y = \{y_i | y_i \in \{0,1\}, \forall i \in M_k\}$ be the feasible sets of the players for $k = 1,2$, $\mathbf{y}_k = [y_i]_{i\in M_k}$ and $\mathbf{y} = \begin{bmatrix} \mathbf{y}_1 \\ \mathbf{y}_2 \end{bmatrix}$. We have already mentioned the existence of optimal solutions $\bar{x}_i$ and $\bar{x}_{ij}$ for given capacity $[\bar{q}_i]$. Thus, there is a function from $\mathbb{R}^m$ to $\mathbb{R}^m$, such that for a given $\bar{\mathbf{y}}$ it returns the unique equilibrium point $[\bar{x}_i]$ from (20)–(24)

and a corresponding mapping from $\mathbb{R}^m$ to $\mathbb{R}^{m \cdot n}$ such that for a given $\bar{\mathbf{y}}$ it returns an optimal transportation plan $[\bar{x}_{ij}]$, which correspond to the equilibrium point $[\bar{x}_i]$, thus it holds that $\bar{x}_i = x_i(\bar{\mathbf{y}})$ and $\bar{x}_{ij} = x_{ij}(\bar{\mathbf{y}})$, respectively.

Hence problems $(\mathbf{CFL_k})$ could be formulated as a single-level problems:

$$(\mathbf{SCFL_k}) \quad \min_{\mathbf{y}_k \in Y_k} \quad \sum_{i \in M_k} d_i(x_i\mathbf{y}), y_i) x_i(\mathbf{y}) + \sum_{i \in M_k} p_i x_i(\mathbf{y}) \tag{39}$$

$$+ \sum_{i \in M_k} \sum_{j=1}^{n} t_{ij} x_{ij}(\mathbf{y}) \tag{40}$$

Each problem, $(\mathbf{SCFL_k})$ corresponds to player k who is involved into the Nash game.

Similarly, we can formulate the competitive capacity assignment of these two producers.

The problem of the first producer:

$$(\mathbf{P_1}) \min_{[q_i]} \sum_{i \in M_1} d_i(\bar{x}_i, q_i)\bar{x}_i \sum_{i \in M_1} p_i \bar{x}_i + \sum_{i \in M_1} \sum_{j=1}^{n} t_{ij} \bar{x}_{ij} \tag{41}$$

$$\text{s.t} \sum_{i \in M_1} \alpha_i q_i \le B, \tag{42}$$

$$0 \le q_i \le U_i, \ \forall i \in M_1 \tag{43}$$

The problem of the second producer:

$$(\mathrm{P}_2) \min_{[q_i]} \sum_{i \in M_2} d_i(\bar{x}_i, q_i)\bar{x}_i \sum_{i \in M_2} p_i \bar{x}_i + \sum_{i \in M_2} \sum_{j=1}^{n} t_{ij} \bar{x}_{ij} \tag{44}$$

$$\text{s.t} \sum_{i \in M_2} \alpha_i q_i \le B, \tag{45}$$

$$0 \le q_i \le U_i, \ \forall i \in M_2 \tag{46}$$

where $[\bar{x}_i]$ and $[\bar{x}_{ij}]$ solve (28)–(32)

Let $Q_k = \{q_i \in \mathbb{R} | \sum_{i \in M_k} a_i q_i \le B, \ 0 \le q_i \le U_i, \ \forall i \in M_k\}$ for $= 1, 2$, the feasible sets of the player, $\mathbf{q}_k = [q_i]_{i \in M_k}$ and $\mathbf{q} = \begin{bmatrix} \mathbf{q}_1 \\ \mathbf{q}_2 \end{bmatrix}$. We have already mentioned the existence of optimal solution $\bar{x}_i$ and $\bar{x}_{ij}$ for given capacity $[\bar{q}_i]$. Thus, there is a function from $\mathbb{R}^m$ to $\mathbb{R}^m$, such that for given $\bar{\mathbf{q}}$ it returns the unique equilibrium point $[\bar{x}_i]$ from (28)–(32). There is also a respective mapping for $\mathbb{R}^m$ to $\mathbb{R}^{m \cdot n}$ such that for a given $\bar{\mathbf{q}}$ it returns an optimal transportation plan $[\bar{x}_{ij}]$ which corresponds to $[\bar{x}_i]$.

We can then write that $\bar{x}_i = x_i(\bar{\mathbf{q}})$ and $\bar{x}_{ij} = x_{ij}(\bar{\mathbf{q}})$. Therefore the problem $(\mathbf{P_k})$ could be stated as single-level problems. For $k = 1,2$ we will have:

$$(\mathbf{SLP_k}) \ \min_{\mathbf{q}_k \in Q_k} \sum_{i \in M_k} d_i(x_i(\mathbf{q}), q_i) x_i(\mathbf{q}) + \sum_{i \in M_k} p_i x_i(\mathbf{q}) + \sum_{i \in M_k} \sum_{j=1}^{n} t_{ij} x_i(\mathbf{q}) \qquad (47)$$

where the problem $(\mathbf{SLP_k})$ is faced by the player k of the Nash game.

4.2 The Impact of the Duopoly in the Service Level

In this part we extend the analysis of Sect. 3.3 in order to make inferences on the impact of the competition of producers with respect to the service level offered. We assume that both producers have the opportunity to locate distribution centers in the same area, where they face exactly the same fixed location cost.

The corresponding problems were specified using the parameters listed in Table 1 and were modeled using the mathematical programming language AMPL and solved by the MINOS 5.5 solver.

In order to be able to find the Nash equilibrium points of the game, it is useful to transfer it into its normal bimatrix form.

It should be mentioned that a bimatrix game in strategic or normal form formulates a non-repeatable situation where rational players choose their strategies independently and simultaneously, having full information about the game details. Specifically, each player knows (a) the number of the players, (b) the pure strategies available to each player, and (c) all the possible outcomes of the game. This knowledge is common i.e., each player knows that all other players are rational and all players know that all players know this and so on. Since players decide simultaneously, none of them knows the choice of others when deciding. In other words, when a player chooses his strategy he does not know in advance and with certainty the choices of his competitors but he can assume that his opponents, being rational, are reasong along the same lines.

In our case the available strategies for the players are the choices for the location or not of a distribution center in the candidate region i, $i = 1,2,\ldots,8$. Assuming that the regions are listed in numerical order, Table 3 presents all the strategies for each player where 1 means that the player k opens the corresponding distribution center while 0 that the corresponding center do not open and so player does not pay the fixed costs.

Table 3 The players strategies

S_{k1}	S_{k2}	S_{k3}	S_{k4}	S_{k5}	S_{k6}	S_{k7}	S_{k8}
(0,0,0)	(1,0,0)	(0,1,0)	(0,0,1)	(1,1,0)	(1,0,1)	(0,1,1)	(1,1,1)

Table 4 Loss function of producer 1

	S_{21}	S_{22}	S_{23}	S_{24}	S_{25}	S_{26}	S_{27}	S_{28}
S_{11}	–	**0**	–	–	0	0	0	0
S_{12}	**68630**	33828.75	68630	53104.29	33828.75	33828.75	53104.29	33828.75
S_{13}	–	1600	47722.5	**52071.46**	1600	1600	1600	1600
S_{14}	–	16327.14	**36806.27**	–	16327.14	11509.86	23405.71	11509.86
S_{15}	70230	35428.75	70230	54704.29	35428.75	35428.75	54704.29	35428.75
S_{16}	69081.43	35528.75	69081.43	59392.99	35528.75	35528.75	59392.99	35528.75
S_{17}	88877.73	17577.14	62351.44	63041.43	18041.29	12823.92	44773.36	12823.92
S_{18}	70681.43	37128.75	70681.43	60962.93	37128.75	37128.75	60962.93	37128.75

Table 5 Loss function of producer 2

	S_{21}	S_{22}	S_{23}	S_{24}	S_{25}	S_{26}	S_{27}	S_{28}
S_{11}	–	**68630**	–	–	70230	69556.4	88227.7	71156.43
S_{12}	**0**	34428.75	1600	15977.14	36028.75	35528.75	17577.14	44853.75
S_{13}	–	68630	47772.5	**36806.27**	70230	70330	70980	71930
S_{14}	–	52754.29	**52071.46**	–	54354.29	59375.99	63041.43	61122.28
S_{15}	0	34428.75	1600	15977.14	36028.75	36128.75	17577.14	37728.75
S_{16}	0	34428.75	1600	11356.92	36028.75	36128.75	12926.86	42482.61
S_{17}	0	53104.29	26464.53	23405.71	54704.29	59495.93	43223.36	61095.93
S_{18}	0	34428.75	1600	11356.92	36028.75	36128.75	12956.92	37728.75

Tables 4 and 5 contain the solution results of the $\mathbf{CFL}_k$ (for $k = 1,2$) for all pairs of strategies, that is, they are the loss matrices of players 1 and 2, respectively. Thus, they present the cost paid by each producer for every possible outcome of the game. Tables 4 and 5 suggest that there are 4 Nash equilibria:

1. The pair of strategies (S_{11}, S_{22}) according to which only player 2 opens a distribution center in region 1.
2. The pair of strategies (S_{12}, S_{21}) according to which only player 1 opens a distribution center in region 1.
3. The pair of strategies (S_{13}, S_{24}) which suggests that producer 1 opens a distribution center in region 2 and producer 2 in the region 3.
4. The pair of strategies (S_{14}, S_{23}) which suggests that producer 1 opens a distribution center in region 3 and producer 2 in the region 2.

The Nash equilibria appear symmetrical, which is a natural result of players' symmetry not only in strategies but also in costs functions.

Figure 2 shows all possible outcomes of the game in a Cartesian space. These points are represented by the EV and EP. Points EP correspond to the four Nash equilibrium points. It is evident from the Fig. 2 that no equilibrium point is dominated by another. The *indeterminacy* phenomenon, often inherent to Nash games, does not allow us to conclude directly about the actual outcome of the game. In fact *an disequilibrium* is possible, if, for example, the first player persists in the strategy S_{12}, while the second player persist in the equilibrium S_{24}. The pair of strategies (S_{12}, S_{24}) does not correspond to an equilibrium.

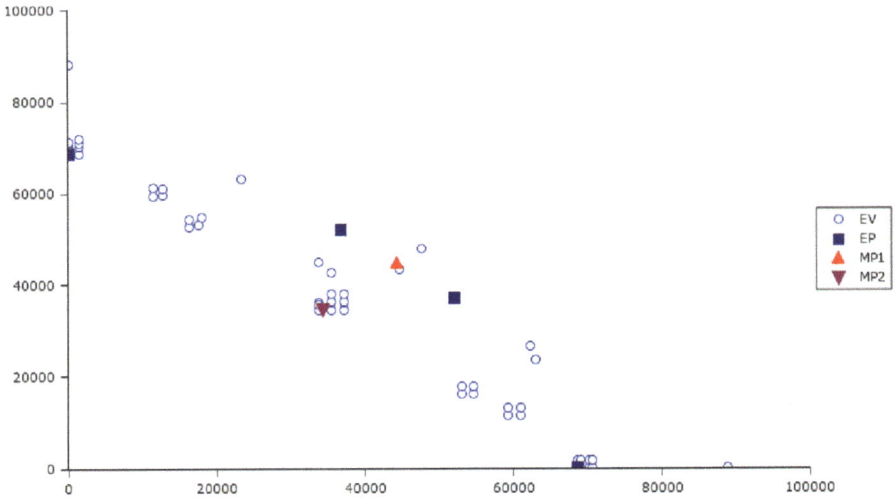

Fig. 2 Set of outcomes. Points EV and EP not MP1 and MP2

Nevertheless, we can assume that in a real situation, players will focus on some of the equilibrium point and they will ignore others. The equilibrium points in which the players will focus their attention are referred to as *focal equilibrium points.*

In our case it is easy to report that points (S_{11}, S_{22}) and (S_{12}, S_{21}), where only one player should install a single distribution center may not be focal equilibrium point. One reason is the mere fact that a player examines *where* and *how* distribution centers will open makes the decision of his opponent to open *somewhere some* center to be almost taken. In other words, by putting himself in his opponent's position he will reject the possibility to allow the opening of the distribution center only by his competitor. Indeed, by making a complete analysis of the game he will realize that having a single distribution center does not satisfy all customers. Additionally, he understands that there will be tendency to escape by some of them, as we have already seen in Sect. 3.3 and is demonstrated in Fig. 1 and consequently his competitor will take advantage of this escape. Therefore the focal equilibrium point of the game are points (S_{13}, S_{24}) and (S_{14}, S_{23}).

According to this analysis, the first two pairs of strategies although they are equilibrium points they will never be followed by the player. Players participate in the game, having already decided to enter in the network. Thus the choice of either strategy S_{11} of the producer 1 or the S_{21} of the producer 2 is not compatible with such a decision. Consequently, under real circumstances, only the equilibrium points (S_{13}, S_{24}) and (S_{14}, S_{23}) are possible outcomes of the game.

Observe that producers, taking into account the competition of customers with respect to the service, will choose to operate their distribution centers in different locations trying to attract non satisfied customers of their competitor. It should be noted that, the equilibrium points guarantee the fulfillment of customer demand at the minimum cost while satisfying their preferences about the service level they receive. Table 6 confirms this finding.

Table 6 The effect of the equilibrium point (S_{13}, S_{24}) and (S_{14}, S_{23}) to the game of the customers

	Equilibrium point (S_{13}, S_{24})					Equilibrium point (S_{14}, S_{23})			
	Producer 1		Producer 2			Producer 1		Producer 2	
	Unit cost	Quantity	Unit cost	Quantity		Unit cost	Quantity	Unit cost	Quantity
C_{11}	35.76	0	35.76	0	C_{11}	35.76	0	35.76	0
C_{21}	35.76	400	35.76	0	C_{21}	35.76	0	35.76	400
C_{31}	35.76	0	35.76	0	C_{31}	35.76	0	35.76	0
C_{12}	32.76	0	32.76	0	C_{12}	32.76	0	32.76	0
C_{22}	32.76	0	32.76	0	C_{22}	32.76	0	32.76	0
C_{32}	32.76	0	32.76	500	C_{32}	32.76	500	32.76	0
C_{13}	38.76	0	38.76	0	C_{13}	38.76	0	38.76	0
C_{23}	38.76	0	38.76	0	C_{23}	38.76	0	38.76	0
C_{33}	38.76	0	38.76	450	C_{33}	38.76	450	38.76	0
C_{14}	42.76	0	42.76	0	C_{14}	42.76	0	42.76	0
C_{24}	42.76	600	42.76	0	C_{24}	42.76	0	42.76	600
C_{34}	42.76	0	42.76	0	C_{34}	42.76	0	42.76	0
C_{15}	33.76	0	33.76	0	C_{15}	33.76	0	33.76	0
C_{25}	33.76	311	33.76	0	C_{25}	33.76	0	33.76	311
C_{35}	33.76	0	33.76	39	C_{35}	33.76	39	33.76	0

The comparison of Tables 2, 4 and 5 certifies that the competitive location of the distribution centers is, in terms of total cost, more beneficial for each producer. No matter which of these two equilibria will be chosen the cost that each producer is going to deal with is smaller in comparison with the total cost of the system optimum and the cost of the monopolistic solution which takes into consideration the competition of the customers.

However, the indeterminacy among the focal equilibrium points cannot easily be eliminated. These two focal equilibrium are symmetric in terms of total cost (36,806, 52071.46) and (52071.46, 36,806), respectively. As illustrated in Fig. 3 both structures of the network distribute symmetrically the customers' demand. Table 6 demonstrates that this distribution is a robust equilibrium for the customers' game, since none of them wants to deviate from it. Therefore, the existence of those loyal customer for any of the focal outcome should satisfy the competing producers.

It should be mentioned that we do not consider equilibrium in mixed strategies although we could examine their existence using the free and open source software Gambit [20]. The reason we ignore mixed strategies is that they may propose expected cost for the competing producers which generally do not correspond to an equilibrium of the customers. For the same reason we can ignore the Nash arbitration solution. A simple analysis which takes into consideration points MP1 and MP2 in Fig. 2 can convince us. First, the point MP1 is pareto dominated by MP2, therefore it cannot be a Nash arbitration point. On the other hand, point MP2 which represents the pair of total cost (33,315, 34,315) does not correspond to costs (35) and (37) which are calculated for an equilibrium of customers. Such a decision of the competitors would lead to a situation where customers would tend to escape. The nearest point to (34,315, 34,315), which has been estimated based on

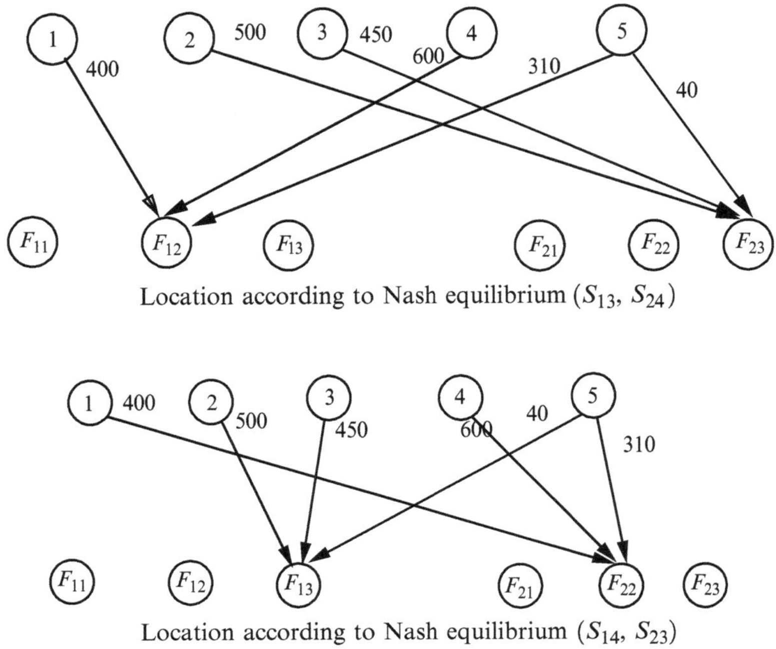

Fig. 3 Competitive facility location

an equilibrium of customers, is the point (33,828, 34428.75) corresponding the pair (S_{12}, S_{22}) open a center, the first one. However, in a competitive environment, the competitors should not have opportunities for such agreements.

5 Conclusion

The aim of this work was to formulate problems of facility location and capacity assignment resulting from the competitive behavior of economic unit operating in a supply chain network. Within this context we examined two different types of models. In the beginning we considered the case where the producer controls the overall supply network and tries to choose the facility location and capacity allocation plan that minimizes the total system cost. Next, we expand the formulation in order to take into account the purchasing behavior of customers. In this case the problem is formulated as bilevel programming problem.

In addition we proposed a bilevel problem with two leaders. This model describes the competition between the two producers in order to attract customers through the quality of the service they provide. The results of the analysis of a random example indicate that the competitive location decisions proposed by the model are the most effective since it minimizes the cost of the system as perceived by producers while they handle of customer behavior.

In conclusion, we could say that the competition of the producers in terms of the service level they provide competition increases the benefit of the purchasers since it ensures:

1. High service level
2. Low cost with declining trends.
3. Improvement of the producer flexibility and their adaptation to the market needs.

References

1. Bard, F.J.: Practical Bilevel Optimization, Algorithms and Applications. Kluwer, Dordrecht (1998)
2. Bracken, J., McGill, J.M.: Mathematical programs with optimization problems in the constraints. Oper. Res. **21**, 37–44 (1973)
3. Bracken, J., McGill, J.M.: A method for solving mathematical problems with nonlinear programs in the constrains. Oper. Res. **22**, 1097–1101 (1973)
4. Cachon, G.P.: Supply chain with contracts. In: Graves, S., de Kok, T. (eds.) Handbooks in Operations Research and Management Science: Supply Chain Management, pp. 229–340. North-Holland, Amsterdam (2003)
5. Cachon, G.P., Lariviere, M.: Capacity choice and allocation: strategic behavior and supply chain performance. Manag. Sci. **45**, 1091–1108 (1999)
6. Chinchuluun, A., Karakitsiou, A., Mavrommati, A.: Game theory models and their application in inventory management and supply chain. In: Migdalas, A., Pardalos, P.M., Pitsoulis, L. (eds.) Pareto Optimality, Game Theory and Equilibria, Springer, NY, pp. 833–865 (2008)
7. Daskin, M.S.: Network and Discrete Location: Models Algorithms and Applications. Wiley, New York (1995)
8. Drezner, Z., Hamacher H.W.: Facility Location: Applications and Theory. Springer, Berlin (2003)
9. Grossman, T.A., Jr., Brandeau M.L.: Optimal pricing for service facilities with self-optimizing customers. Eur. J. Oper. Res. **141**, 39–57 (2002)
10. Heinhold, M.: An operational research approach to allocation of clients to a certain class of service institutions. J. Oper. Res. Soc. **29**, 273–276 (1978)
11. Karakitsiou, A., Prokopye, O.A.: Special issue in multilevel optimization: algorithms and applications. J. Global Optim. **38**(4), 507–666 (2007)
12. Marianov, V., Rios, M.: A probabilistic quality of service constraint for a location model of switches in ATM communications networks. Ann. Oper. Res. **96**, 237–243 (2000)
13. Marianov, V., Serra D.: Probabilistic maximal covering location-allocation for congested systems. J. Regional Sci. **38**, 401–424 (1998)
14. Marianov, V., Serra, D.: Hierarchical location-allocation models for congested systems. Eur. J. Oper. Res. **135**, 196–209 (2001)
15. Mas-Colell, A., Whinston, M.D., Green, J.R.: Microeconomic Theory. Oxford University Press, New York (1995)
16. Migdalas A., Pardalos, P.M.: Special issue in global optimization and hierarchical decision making. J. Global Optim. **8**(3), pp. 209–215 (1996)
17. Migdalas, A., Pardalos, P.M., Värbrand, P.: Multilevel Optimization - Algorithms and Applications. Kluwer Academic (1997)
18. Nagurney, A., Dong J., Zhang, D.: A supply chain network equilibrium model. Transport. Res. E **38**, 281–303 (2002)
19. Van Mieghem, J., Dada, M.: Price versus production postponement: capacity and competition. Manag. Sci. **45**, 1631–1649 (1999)
20. http://econweb.tamu.edu/gambit, LAST VISITED 25/6/2010

A Hybrid Particle Swarm Optimization Algorithm for the Permutation Flowshop Scheduling Problem

Yannis Marinakis and Magdalene Marinaki

Abstract This paper introduces a new hybrid algorithmic nature inspired approach based on Particle Swarm Optimization, for successfully solving one of the most computationally complex problems, the Permutation Flowshop Scheduling Problem. The Permutation Flowshop Scheduling Problem (PFSP) belongs to the class of combinatorial optimization problems characterized as NP-hard and, thus, heuristic and metaheuristic techniques have been used in order to find high quality solutions in reasonable computational time. The proposed algorithm for the solution of the PFSP, the Hybrid Particle Swarm Optimization (HybPSO), combines a Particle Swarm Optimization (PSO) Algorithm, the Variable Neighborhood Search (VNS) Strategy and a Path Relinking (PR) Strategy. In order to test the effectiveness and the efficiency of the proposed method we use a set of benchmark instances of different sizes.

Key words Permutation flowshop scheduling problem • Particle swarm optimization • Variable neighborhood search • Path relinking

1 Introduction

Particle swarm optimization (PSO) is a population-based swarm intelligence algorithm that was originally proposed by Kennedy and Eberhart [14]. PSO simulates the social behavior of social organisms by using the physical movements of the individuals in the swarm. Its mechanism enhances and adapts to the global and local exploration. Most applications of PSO have concentrated on the optimization in continuous space but in the last years the PSO algorithm is used also in discrete

Y. Marinakis (✉) • M. Marinaki
Technical University of Crete, Department of Production Engineering and Management
Industrial Systems Control Laboratory, 73100 Chania, Greece
e-mail: marinakis@ergasya.tuc.gr; magda@dssl.tuc.gr

A. Migdalas et al. (eds.), *Optimization Theory, Decision Making, and Operations Research Applications*, Springer Proceedings in Mathematics & Statistics 31,
DOI 10.1007/978-1-4614-5134-1_6,

optimization problems. Recent complete surveys for the PSO can be found in [1, 2, 24]. The PSO is a very popular optimization method and its wide use, mainly during the last years, is due to the number of advantages that this method has, compared to other optimization methods. Some of the key advantages are that this method does not need the calculation of derivatives that the knowledge of good solutions is retained by all particles and that particles in the swarm share information between them. PSO is less sensitive to the nature of the objective function, can be used for stochastic objective functions and can easily escape from local minima. Concerning its implementation, PSO can easily be programmed, has few parameters to regulate and the assessment of the optimum is independent of the initial solution.

In this paper, we would like to develop a competitive Nature Inspired method based on PSO for the solution of the Permutation Flowshop Scheduling Problem (PFSP) and to test its efficiency. Thus, in this paper, we demonstrate how a nature inspired intelligent technique, the PSO [14] and two metaheuristic techniques, the Variable Neighborhood Search (VNS) [11] and the Path Relinking (PR) [8] can be incorporated in a hybrid scheme in order to give very good results for the PFSP. The VNS is used to improve the solution of each particle and, as it is desired, to keep the computational time as low as possible while the PR strategy is used in order to improve the solution of the best particle in each iteration.

The rest of the paper is organized as follows: In the next section a description of the PFSP is presented. In the third section the proposed algorithm, the Hybrid Particle Swarm Optimization (HybPSO) is presented and analyzed in detail. Computational results are presented and analyzed in the fourth section while in the last section conclusions and future research are given.

2 The Permutation Flowshop Scheduling Problem

The flowshop scheduling problem proposed by Johnson [13] is an important scheduling problem [23] and has been extensively studied. In a flowshop scheduling problem there is a set of n jobs, tasks or items to be processed in a set of m machines or processors in the same order, i.e. first in machine 1, then on machine 2 and so on until machine m. At any time, each job can be processed on at most one machine and each machine can process at most one job. Also, once a job is processed on a machine, it cannot be terminated before completion. The objective is to find a sequence for the processing of the jobs in the machines so that a given criterion is optimized. No preemption is allowed, i.e. the processing of a job i on a machine j cannot be interrupted. All jobs are independent and are available for processing at time 0. The set-up times of the jobs on machines are negligible and therefore can be ignored. The machines are continuously available. In the literature, the most common criterion is the minimization of the makespan ($C_{\max}$), i.e the minimization of the maximum completion time.

The makespan problem for flow shops has been the most studied by far in the literature. This is partly because:

- Makespan is a simple and useful criterion for heavily loaded shops when long-term utilization should be maximized
- Makespan is the only objective function simple enough to have available some analytic results for multi-machine problems and to make some branch-and-bound methods practical for medium-sized problems

The minimization of the makespan objective is to a certain extent equivalent to the maximization of the utilization of the machines. The models, however, tend to be of such complexity that makespan results are already relatively hard to obtain. Even harder to analyze are the flow time and the due-date-related objectives.

In the permutation flowshop scheduling problem (PFSP) [22, 28], solutions are represented by the permutation of n jobs, i.e., $\pi = \{\pi_1, \pi_2, \ldots, \pi_n\}$. Each job is composed of m operations, and every operation is performed by a different machine. Thus, given the processing time p_{jk} for the job j on the machine k (these times are fixed, known in advance and non-negative), the PFSP is to find the best permutation of jobs $\pi^* = \pi_1^*, \pi_2^*, \ldots, \pi_n^*$ to be processed on each machine subject to the makespan criterion. Let $C(\pi_j, m)$ denote the completion time of the job π_j on the machine m. Then, given the job permutation π, the completion time for the n-job, m-machine problem is calculated as follows:

$$C(\pi_1, 1) = p_{\pi_1,1} \tag{1}$$

$$C(\pi_j, 1) = C(\pi_{j-1}, 1) + p_{\pi_j,1}, j = 2, \ldots, n \tag{2}$$

$$C(\pi_1, k) = C(\pi_1, k-1) + p_{\pi_1,k}, k = 2, \ldots, m \tag{3}$$

$$C(\pi_j, k) = \max\{C(\pi_{j-1}, k), C(\pi_j, k-1) + p_{\pi_j,k}\}, \quad j = 2, \ldots, n, k = 2, \ldots, m \tag{4}$$

So, the makespan of a permutation π can be formally defined as the completion time of the last job π_n on the last machine m, i.e.:

$$C_{\max}(\pi) = C(\pi_n, m). \tag{5}$$

Therefore, the PFSP with the makespan criterion is to find the optimal permutation π^* in the set of all permutations Π such that:

$$C_{\max}(\pi^*) \leq C(\pi_n, m) \text{ for each permutation } \pi \text{ belonging to } \Pi. \tag{6}$$

The computational complexity of the PFSP has been proved to be NP-hard by [7, 27]. Due to this fact, the solution procedure for the PFSP is often either heuristic or metaheuristic. A number of heuristic and metaheuristic algorithms have been developed in the past for this problem. Some recent works are presented in the following. In [29] an iterated greedy algorithm is used for the permutation flowshop scheduling problem. In [35] cooperative metaheuristic methods are proposed for the permutation flowshop scheduling problem. A hybrid metaheuristic that comprises three components, an initial population generation method based on a greedy randomized constructive heuristic, a genetic algorithm for solution evolution, and a variable neighborhood search to improve the population is used in [39] for the permutation flowshop scheduling problems. A tabu search technique with a specific neighborhood definition which employs a block of jobs notion is used for the permutation flowshop scheduling problem in [21]. In [9] a tabu search algorithm is presented for the permutation flowshop problem. In [33] an algorithm that hybridizes the genetic algorithm and a local search scheme that combines two local search methods, the Insertion Search and the Insertion Search with Cut-and-Repair, is used for solving the permutation flowshop scheduling problem. An algorithm that hybridizes the genetic algorithm and the tabu search is used for solving the permutation flowshop scheduling problem in [34]. In [3] a self-guided genetic algorithm is used for the permutation flowshop scheduling problem. In [30] genetic algorithms are used for solving the permutation flow shop scheduling problem. In [22] a discrete differential evolution algorithm is used for the permutation flowshop scheduling problem. In [16] a discrete version of particle swarm optimization is used for the flowshop scheduling problem. A hybrid alternate two phases PSO algorithm which combines the PSO with genetic operators and annealing strategy is proposed in [38] to solve the flowshop scheduling problem. In [18] a particle swarm optimization-based memetic algorithm is proposed for the permutation flowshop scheduling problem. An algorithm that combines the particle swarm optimization algorithm with genetic operators is proposed in [37] for the flowshop scheduling problem. A particle swarm optimization algorithm applied for permutation flowshop scheduling is used in [15]. In [6] an ant-colony algorithm has been developed in order to solve the flowshop scheduling problem. Two ant-colony optimization algorithms are proposed and analyzed for solving the permutation flowshop scheduling problem in [25, 26]. A hybrid discrete artificial bee colony algorithm is presented in [19] for the solution of the permutation flowshop scheduling problem. More analytical reviews of approaches applied for the solution of the flowshop scheduling problem are given in [10, 28].

3 Hybrid Particle Swarm Optimization Algorithm

3.1 General Description

In this paper, a hybrid PSO (HybPSO) algorithm is used for the solution of the PFSP. In PSO algorithm, initially a set of particles is created randomly where each particle corresponds to a possible solution. Each particle has a position in the space of solutions and moves with a given velocity. One of the key issues in designing a successful PSO for the Permutation Flowshop Scheduling Problem is to find a suitable mapping between PFSP solutions and particles in PSO. Each particle is recorded via the permutation π of jobs. For example, if we have a particle (solution) with ten jobs, a possible permutation representation is the following:

2 3 8 5 4 10 9 6 1 7

As the calculation of the velocity of each particle is performed by (7) (see below), the above-mentioned representation should be transformed appropriately. We transform each element of the solution into a floating point in the interval (0,1], calculate the velocities and the positions of all particles and, then, convert back the particles' positions into the integer domain using relative position indexing [17].

The position of each particle is represented by a d-dimensional vector in problem space $x_i = (x_{i1}, x_{i2}, \ldots, x_{id})$, $i = 1,2,\ldots,N$ (N is the population size and n is the number of the vector's dimension), and its performance is evaluated on the predefined fitness function. The velocity v_{ij} represents the changes that will be made to move the particle from one position to another. Where the particle will move depends on the dynamic interaction of its own experience and the experience of the whole swarm. There are three possible directions that a particle can follow: to follow its own path, to move towards the best position it had during the iterations ($pbest_{ij}$) or to move to the best particle's position ($gbest_j$). The velocity and position equations are updated as follows (constriction PSO) [4]:

$$v_{ij}(t+1) = \chi(v_{ij}(t) + c_1 rand_1(pbest_{ij} - x_{ij}(t)) + c_2 rand_2(gbest_j - x_{ij}(t))) \quad (7)$$

and

$$x_{ij}(t+1) = x_{ij}(t) + v_{ij}(t+1) \quad (8)$$

where

$$\chi = \frac{2}{|2 - c - \sqrt{c^2 - 4c}|} \text{ and } c = c_1 + c_2, c > 4 \quad (9)$$

t is the iterations' counter, c_1 and c_2 are the acceleration coefficients, $rand_1$ and $rand_2$ are two random variables in the interval (0, 1). A local search strategy based on the Variable Neighborhood Search (VNS) algorithm [11] is applied in each particle in the swarm in order to improve the solutions produced from the particle swarm optimization algorithm. Finally, a path relinking strategy [8] with starting solution the best particle and target solution one of the other particles of the swarm is applied. During the path relinking procedure, if a better solution than the current best solution is found, then the current best solution is replaced by this solution (see Sect. 3.3). In each iteration of the algorithm the optimal solution of the whole swarm and the optimal solution of each particle are kept. The algorithm stops when a maximum number of iterations have been reached. A pseudocode of the proposed algorithm is presented in the following.

Initialization
Select the number of Swarms
Select the number of Particles for each swarm
Generate the initial population
Evaluate the fitness function of each particle
Apply Variable Neighborhood Search in each particle
Keep Optimum particle of the whole swarm
Keep Optimum solution of each particle
Main Phase
Do until the maximum number of generations has not been reached:
    Calculate the velocity of each particle
    Evaluate the new fitness function of each particle
    Apply Variable Neighborhood Search in each particle
    Update the best solution of each particle
    Update the best particle
    Apply a Path Relinking strategy in the best particle
    if a new best solution is found **then**
        Update the best particle
    endif
Enddo
Return the best particle (the best solution).

3.2 Variable Neighborhood Search

A variable neighborhood search (VNS) [11] algorithm is applied in order to optimize the particles. The basic idea of the method is the successive search in a number of neighborhoods of a solution. With the term neighborhood it is meant different number of local search algorithms. The search is applied either with random or with a more systematical manner in order to escape the solution from a local minimum. This method takes advantage of the fact that different local search algorithms will lead to different local minimums. In this paper, the VNS algorithm is used with the following way. Initially, the number of local search algorithms is selected. The local search strategies for the Permutation Flowshop Scheduling Problem are the 2-opt, 3-opt, 1–0 insert, 1–1 interchange and threshold accepted [5] neighborhoods.

As we do not want to increase the complexity of the algorithm, it is decided to apply in each particle one local search combination of algorithms per iteration. For this reason, a VNS operator C_{VNS} is selected that controls which local search algorithm is applied. The C_{VNS} value is compared with the output of a random number generator, $rand_i(0,1)$. If the random number is less or equal to the C_{VNS}, then the first local search algorithm is used. Then, if the random number is less or equal to the $2 * C_{\mathrm{VNS}}$, then the second local search algorithm is used, and so on. As we would like to have not only simple local search algorithms but also their combinations we select ten local search algorithms, the five previously mentioned methods and five combinations (2-opt and 3-opt, 2-opt and 1–1 interchange, 2-opt and 1–0 insert, 3-opt and 1–0 insert and, finally, 2-opt, 1–1 interchange, 3-opt and

1–0 insert). Finally if the local search algorithm is stuck in a local optimum a dynamic iterated local search [20] is applied periodically. Thus, the C_{VNS} operator is set equal to 0.1.

3.3 Path Relinking

This approach generates new solutions by exploring trajectories that connect high-quality solutions—by starting from one of these solutions, called the *starting solution* and generating a path in the neighborhood space that leads towards the other solution, called the *target solution* [8]. The roles of starting and target solutions can be interchangeable. In one case, the worst among the two solutions plays the role of the starting solution and the other plays the role of the target solution. In another case, the roles are changing. There is the possibility the two paths to simultaneously be explored. In the proposed algorithm, after the completion of an iteration, a path relinking algorithm is applied for exploring trajectories between the best particle and a number of other particles of the swarm. In this algorithm the best particle plays the role of the starting solution and in each iteration the other random particles play the role of target solutions. We are using random particles for the target solutions in order to give to the best particle more exploration abilities by combining not only the best particle with its neighbor particles but also with equal probabilities with all the particles in the swarm. If a better solution than the current best solution is found, then the current best solution is replaced by this solution.

4 Results and Discussion

The algorithm was implemented in Fortran 90 and was compiled using the Lahey f95 compiler on a Intel Core 2 DUO CPU T9550 at 2.66 GHz, running Suse Linux 9.1. The algorithm was tested on the 90 benchmark instances of Taillard [31]. In these instances there are different sets having 20, 50, and 100 jobs and 5, 10, or 20 machines. There are 10 problems inside every size set. In total there are 9 sets and these are: 20×5 (i.e., 20 jobs and 5 machines), 20×10, 20×20, 50×5, 50×10, 50×20, 100×5, 100×10 and 100×20. The parameters of the proposed algorithm are selected after thorough testing. A number of different alternative values were tested and the ones selected are those that gave the best computational results concerning both the quality of the solution and the computational time needed to achieve this solution. The selected parameters are: number of particles equal to 50, number of generations equal to 1,000, $c_1 = c_2 = 2.05$ and the number of local search iterations is equal to 20. The efficiency of the HybPSO algorithm is measured by the quality of the produced solutions. The quality is given in terms of the relative deviation from the best known solution, that is $\omega = \frac{(c_{\mathrm{HybPSO}} - c_{\mathrm{BKS}})}{c_{\mathrm{BKS}}}\%$, where c_{HybPSO} denotes the cost of the solution found by HybPSO and c_{BKS} is the cost of the best known solution. In Table 1 for everyone of the 9 sets we have averaged the quality of the 10 corresponding instances. There are a number of heuristic and metaheuristic algorithms that have been applied for the finding of the makespan in a PFSP. Table 1 presents the average quality of the solutions of the proposed algorithm (HybPSO) and the average quality of other 12 algorithms from the literature. The first one is the most important heuristic algorithm, the NEHT which is the classic NEH algorithm together with the improvement that was presented by Taillard [31]. A Simple Genetic Algorithm (SGA) [3], a Mining Genetic Algorithm [3], an Artificial Chromosome with Genetic Algorithms [3], a Self-Guided Genetic Algorithm (SGGA) [3], three versions of Particle Swarm Optimization algorithms (PSO1) [32], (CPSO) [12] and (PSO2) [16], a Discrete Differential Evolution (DDE) algorithm [22], a hybridization of Genetic Algorithm with Variable Neighborhood Search (GA-VNS) [39] and an Ant Colony Optimization algorithm (ACS) [36] are given.

From Table 1, a number of important conclusions about HybPSO are derived. First of all, the algorithm finds the optimum in the first set in all instances. This happens only in one more algorithm, in the GA-VNS. Both algorithms have as local search phase a Variable Neighborhood Search algorithm which lead us to the conclusion that the combination of a population-based algorithm (like PSO in the proposed algorithm and a genetic algorithm in GA-VNS) with a very strong local search technique like VNS increases both the exploration and exploitation abilities of the algorithm. The algorithm performs better in 10 out of the 12 other algorithms used for the comparisons. As it was expected, the algorithm performs better than the only heuristic algorithm, the NEHT. The algorithms that can deal with only discrete values in the representation of the solutions have in general advantage from the algorithms that use transformations from continuous to discrete values and vice versa. Thus, a genetic algorithm should have an advantage when compared with a Particle Swarm Optimization algorithm. However, in this NP-hard problem studied in this paper, the proposed algorithm performs better than the 5 out of 6 versions of the genetic algorithms that are used in the comparisons (only GA-VNS performs better). When the proposed algorithm is compared with the other PSO algorithms and the DDE algorithm, it performs better. This fact shows that the proposed implementation of PSO is very efficient for the solution of this kind of problems. Finally, the comparison with the Ant Colony Optimization algorithm shows that these two algorithms perform equally well to the selected instances as in five sets the proposed algorithm performs better while for the other four sets the ACS algorithm gives better results.

5 Conclusions

In this paper, a new algorithm based on the Particle Swarm Optimization for the solution of the Permutation Flowshop Scheduling Problem is presented. This algorithm is a hybridization of the Particle Swarm Optimization algorithm with

Table 1 Comparisons of the results of HybPSO in Taillard benchmark instances for the PFSP

Problems	HybPSO	NEHT	SGA	MGGA	ACGA	SGGA	PSO1	DDE	CPSO	GMA	PSO2	GA-VNS	ACS
20 × 5	0	3.35	1.02	0.81	1.08	1.1	1.75	0.46	1.05	1.14	1.25	0	1.19
20 × 10	0.15	5.02	1.73	1.4	1.62	1.9	3.25	0.93	2.42	2.3	2.17	0	1.7
20 × 20	0.31	3.73	1.48	1.06	1.34	1.6	2.82	0.79	1.99	2.01	2.09	0	1.6
50 × 5	0.2	0.84	0.61	0.44	0.57	0.52	1.14	0.17	0.9	0.47	0.47	0	0.43
50 × 10	2.2	5.12	2.81	2.56	2.79	2.74	5.29	2.26	4.85	3.21	3.6	0.77	0.89
50 × 20	3.81	6.26	3.98	3.82	3.75	3.94	7.21	3.11	6.4	4.97	4.84	0.96	2.71
100 × 5	0.19	0.46	0.47	0.41	0.44	0.38	0.63	0.08	0.74	0.42	0.35	0	0.22
100 × 10	1.33	2.13	1.67	1.5	1.71	1.6	3.27	0.94	2.94	1.96	1.78	0.08	1.22
100 × 20	4.36	5.23	3.8	3.15	3.47	3.51	8.25	3.24	7.11	4.68	5.13	1.31	2.22

the Variable Neighborhood Search algorithm and with the Path Relinking Strategy. As a number of different variants of the Particle Swarm Optimization algorithm have been published, mainly using a different equation for the calculation of the velocities, we used the constriction Particle Swarm Optimization. Another issue that we have to deal with was the fact that the PSO algorithm is suitable for continuous optimization problems. Thus, it was a challenge to find an effective transformation of the solutions of PSO in discrete values without losing information from this procedure. The algorithm was tested in 90 benchmark instances that are usually used in the literature and gave very good results. This fact demonstrates the efficiency of the algorithm when it is used for the solution of an NP-hard problem, like PFSP. In the future, this algorithm will be used for the solution of other NP-hard combinatorial optimization problems.

References

1. Banks, A., Vincent, J., Anyakoha, C.: A review of particle swarm optimization. Part I: background and development. Nat. Comput. **6**(4), 467–484 (2007)
2. Banks, A., Vincent, J., Anyakoha, C.: A review of particle swarm optimization. Part II: hybridisation, combinatorial, multicriteria and constrained optimization, and indicative applications. Nat. Comput. **7**, 109–124 (2008)
3. Chen, S.H., Chang, P.C., Cheng, T.C.E., Zhang, Q.: A Self-guided genetic algorithm for permutation flowshop scheduling problems. Comp. Oper. Res. **39**, 1450–1457 (2012)
4. Clerc, M., Kennedy, J.: The particle swarm: explosion, stability and convergence in a multidimensional complex space. IEEE Trans. Evol. Comput. **6**, 58–73 (2002)
5. Dueck, G., Scheurer, T.: Threshold accepting: A general purpose optimization algorithm appearing superior to simulated annealing. J. Comput. Phys. **90**, 161–175 (1990)
6. Gajpal, Y., Rajendran, C.: An ant-colony optimization algorithm for minimizing the completion-time variance of jobs in flowshops. Int. J. Prod. Econ. **101**(2), 259–272 (2006)
7. Garey, M.R., Johnson, D.S., Sethi, R.: The complexity of flowshop and jobshop scheduling. Math. Oper. Res. **1**, 117–129 (1976)
8. Glover, F., M. Laguna, M., Marti, R.: Scatter search and path relinking: advances and applications. In: Glover, F., Kochenberger, G.A. (eds.) Handbook of Metaheuristics, pp. 1–36. Kluwer, Boston (2003)
9. Grabowski, J., Wodecki, M.: A very fast tabu search algorithm for the permutation flow shop problem with makespan criterion. Comp. Oper. Res. **31**, 1891–1909 (2004)
10. Gupta, J.N.D., Stafford, E.F., Jr.: Flowshop scheduling research after five decades. Eur. J. Oper. Res. **169**, 699–711 (2006)
11. Hansen, P., Mladenovic, N.: Variable neighborhood search: principles and applications. Eur. J. Oper. Res. **130**, 449–467 (2001)
12. Jarboui, B., Ibrahim, S., Siarry, P., Rebai, A.: A combinatorial particle swarm optimisation for solving permutation flow shop problems. Comp. Ind. Eng. **54**, 526–538 (2008)
13. Johnson, S.: Optimal two-and-three stage production schedules with setup times included. Naval Res. Logist. Q. **1**, 61–68 (1954)
14. Kennedy, J., Eberhart, R.: Particle swarm optimization. In: Proceedings of 1995 IEEE International Conference on Neural Networks, vol. 4, Perth, WA and the paper ISBN: 0-7803-2768-3 pp. 1942–1948 (1995)
15. Lian, Z., Gu, X., Jiao, B.: A similar particle swarm optimization algorithm for permutation flowshop scheduling to minimize makespan. Appl. Math. Comput. **175**(1), 773–785 (2006)

16. Liao, C.J., Tseng, C.T., Luarn P.: A discrete version of particle swarm optimization for flowshop scheduling problems. Comp. Oper. Res. **34**, 3099–3111 (2007)
17. Lichtblau, D.: Discrete optimization using Mathematica. In: Callaos, N., Ebisuzaki, T., Starr, B., Abe, J.M., Lichtblau, D. (eds.) Proceedings of World Multi-Conference on Systemics, Cybernetics and Informatics (SCI 2002). International Institute of Informatics and Systemics, vol. 16, pp. 169–174 Orlando, Florida, USA (2002)
18. Liu, B., Wang, L., Jin, Y.H.: An effective PSO-based memetic algorithm for flow shop scheduling. IEEE Trans. Syst. Man Cybern. B Cybern. **37**(1), 18–27 (2007)
19. Liu, Y.F., Liu, S.Y.: A hybrid discrete artificial bee colony algorithm for permutation flowshop scheduling problem. Appl. Soft Comput. (2011) doi:10.1016/j.asoc.2011.10.024
20. Lourenco, H.R., Martin, O., Stützle, T.: Iterated local search. In: Handbook of Metaheuristics, vol. 57. Operations Research and Management Science, pp. 321–353. Kluwer, Boston (2002)
21. Nowicki, E., Smutnicki, C.: A fast tabu search algorithm for the permutation flow-shop problem. Eur. J. Oper. Res. **91**, 160–175 (1996)
22. Pan, Q.K., Tasgetiren, M.F., Liang, Y.C.: A discrete differential evolution algorithm for the permutation flowshop scheduling problem. Comp. Ind. Eng. **55**, 795–816 (2008)
23. Pinedo, M.: Scheduling. Theory, Algorithms, and Systems. Prentice Hall, Englewood Cliffs (1995)
24. Poli, R., Kennedy, J., Blackwell, T.: Particle swarm optimization. An overview. Swarm Intell. **1**, 33–57 (2007)
25. Rajendran, C., Ziegler, H.: Ant-colony algorithms for permutation flowshop scheduling to minimize makespan/total flowtime of jobs. Eur. J. Oper. Res. **155**(2), 426–438 (2004)
26. Rajendran, C., Ziegler, H.: Two ant-colony algorithms for minimizing total flowtime in permutation flowshops. Comp. Ind. Eng. **48**(4), 789–797 (2005)
27. Rinnooy Kan, A.H.G.: Machine Scheduling Problems: Classification, Complexity, and Computations. Nijhoff, The Hague (1976)
28. Ruiz, R., Maroto, C.: A comprehensive review and evaluation of permutation flowshop heuristics. Eur. J. Oper. Res. **165**, 479–494 (2005)
29. Ruiz, R., Stützle, T.: A simple and effective iterated greedy algorithm for the permutation flowshop scheduling problem. Eur. J. Oper. Res. **177**, 2033–2049 (2007)
30. Ruiz, R., Maroto, C., Alcaraz, J.: Two new robust genetic algorithms for the flowshop scheduling problem. Omega **34**, 461–476 (2006)
31. Taillard, E.: Benchmarks for basic scheduling problems. Eur. J. Oper. Res. **64**, 278–285 (1993)
32. Tasgetiren, M., Liang, Y., Sevkli, M., Gencyilmaz, G.: A particle swarm optimization algorithm for makespan and total flow time minimization in the permutation flowshop sequencing problem. Eur. J. Oper. Res. **177**, 1930–1947 (2007)
33. Tseng, L.Y., Lin, Y.T.: A hybrid genetic local search algorithm for the permutation flowshop scheduling problem. Eur. J. Oper. Res. **198**, 84–92 (2009)
34. Tseng, L.Y., Lin, Y.T.: A genetic local search algorithm for minimizing total flowtime in the permutation flowshop scheduling problem. Int. J. Prod. Econ. **127**, 121–128 (2010)
35. Vallada, E., Ruiz, R.: Cooperative metaheuristics for the permutation flowshop scheduling problem. Eur. J. Oper. Res. **193**, 365–376 (2009)
36. Ying, K.C., Liao, C.J.: An ant colony system for permutation flow-shop sequencing. Comp. Oper. Res. **31**, 791–801 (2004)
37. Zhang, C., Sun, J., Zhu, X., Yang, Q.: An improved particle swarm optimization algorithm for flowshop scheduling problem. Inform. Process. Lett. **108**, 204–209 (2008)
38. Zhang, C., Ning, J., Ouyang, D.: A hybrid alternate two phases particle swarm optimization algorithm for flow shop scheduling problem. Comp. Ind. Eng. **58**, 1–11 (2010)
39. Zobolas, G.I., Tarantilis, C.D., Ioannou, G.: Minimizing makespan in permutation flow shop scheduling problems using a hybrid metaheuristic algorithm. Comp. Oper. Res. **36**, 1249–1267 (2009)

Optimization Over Stochastic Integer Efficient Set

Djamal Chaabane and Fatma Mebrek

Abstract In this paper we study the problem of optimizing a linear function over an integer efficient solution set of a Multiple objective Stochastic Integer Linear Programming problem (MOSILP). Once the problem is converted into a deterministic one by adapting the 2-levels recourse approach, a new pivoting technique is applied to generate an optimal efficient solution without having to enumerate all of them. This method combines two techniques, the L-Shaped method and the combined method developed in [Kall, Stochastic Linear Programming (1976)]. A detailed didactic example is given to illustrate different steps of our algorithm.

Key words Multi-objective programming • Stochastic programming • 2-levels recourse model • Efficient solutions

1 Introduction

In real life problems the fundamental assumption for multi-objective linear programming that the problem entries—except for the variables x—known as fixed data does not often happen. This is probably the case that some parameters are taken as estimates from some statistical samples or we know that they are random variables from the model design. These situations can be modeled by linear programming problems whose objectives and constraints depend on uncertain parameters. The

D. Chaabane (✉)
USTHB University, B.P. 32, Bab-Ezzouar, El-Alia 16111, Algiers, Algeria
e-mail: chaabane_dj@yahoo.fr

F. Mebrek
ENSTP-KOUBA, Algiers, Algeria
e-mail: mebrek_f@yahoo.fr

A. Migdalas et al. (eds.), *Optimization Theory, Decision Making, and Operations Research Applications*, Springer Proceedings in Mathematics & Statistics 31,
DOI 10.1007/978-1-4614-5134-1_7,

presence of these two aspects, namely, multiple objective and stochastic leads to a multiple objective stochastic linear programming (MOSLP). Few methods are proposed in the literature and most of them are using the techniques of goal programming (interactive) and the stochastic programming. In this context, we cite some methods: PROTRADE [5], PROMISE [13] in the continuous case and the discrete, STRANGE-MOMIX Method [11, 12] and search and cut technique [1, 2].

In many situations, the decision maker faces a large number of different efficient solutions and the selection of his/her preferred solutions becomes a very hard task, a way of assessing some preferred solution is by optimizing a function (utility function written as a function of decision variables), particularly linear, over the efficient set, an appropriate approach that has been used by many authors (see [3, 4]) in continues and discrete case, respectively.

We focus on solving MOSLP problem in presence of integer decision variables. We therefore propose a technique that combines L-Shaped method [7] and a method that consists of optimizing an arbitrary linear function over the whole set of discrete deterministic efficient solutions of Multiple Objective Integer Linear Programming (MOILP) problem without enumerate explicitly all non-dominated solutions. We follow the notations used by Peter Kall and Janos Mayer [8]. Given a Multiple Objective Integer Stochastic Linear Programming problem (MOISLP):

$$(P_{SC})\begin{cases} \text{“min''} Z_k = C_k x; & k = 1..K \\ s.t. \quad Ax = b \\ \quad Tx = h \\ \quad x \in \mathbb{N} \end{cases} \tag{1}$$

where x is the decision vector variable of dimension $(n \times 1)$. C_k, T, h are random matrices of dimensions $(1 \times n), (m_1 \times n)$ and $(m_1 \times 1)$, respectively, with a known joint probability distribution which is not influenced by the choice of the decision x and defined on a probability space (Ω, Ξ, P). We assume that A, b are fixed known integer data (deterministic) of dimensions $(m \times n)$ and $(m \times 1)$, respectively. Our main purpose is to solve the following problem:

$$(P_E(SC))\begin{cases} \min \phi(x) = dx \\ s.t. \, x \in E(P_{SC}) \end{cases} \tag{2}$$

where d is a random line vector of dimension n and $E(P_{SC})$ is the solution set of (P_{SC}) problem, without having to enumerate all the elements of $E(P_{SC})$.

In the next section we introduce the associate deterministic problem, we show some basic results concerning the L-shaped decomposition method. In Sect. 3, we give some important results that deal with optimization over efficient set. Section 4 presents the proposed method illustrated by a didactic example, and finally we conclude by suggesting some other issues.

2 Passage to Deterministic Equivalent Problem of MOISLP

A mono-criteria stochastic linear programming problem does not have meaning since some of the parameters are not well known at the moment the decision maker has to take decision about the values of the variables, the sense of the optimization cannot be preserved. In addition when many criteria are to be taken all together at the same time the task becomes extremely difficult. We suppose some probability space (Ω, Ξ, P) that defines the probabilistic aspect of random parameters in our problem. The decision on x has to be taken before the realization of the random variables is known. But consequently, after the observation of the random variables realization, it may turn out that $Tx \neq h$, i.e. x lies out of the admissibility region. In this case, it may be necessary to compensate for the deficiency, i.e. for $h - Tx$, after its observation. This can be done by the introduction of recourse defining the constraints $Wy = h - Tx, y \geq 0$. We suppose a fixed recourse matrix W and the recourse costs is taken as linear function $q'y$. Obviously we want to achieve this compensation with minimal costs. Hence the recourse problem is defined by:

$$(P_1) \begin{cases} Q(x;T,h) = \min q'y \\ s.t. \qquad Wy = h - Tx \\ \qquad\qquad y \geq 0 \end{cases} \tag{3}$$

The expectation value of the kth criterion is $\widetilde{Z}_k = E\left(Z_k + Q(x,\xi)\right);\ k = 1,\ldots,K$ and the deterministic multiple objective integer linear programming problem with the main problem can be stated as follows:

$$(P) \begin{cases} \text{“min}'' \widetilde{Z}_k = \mathbb{E}\left(Z_k + Q(x;T,h)\right) \quad k = 1,..,K \\ s.t. \qquad Ax = b \\ \qquad\qquad x \in \mathbb{N} \end{cases} \tag{4}$$

$$(P_E) \begin{cases} \min \widetilde{\phi} = \mathbb{E}\left(d\right)x \\ s.t. \qquad x \in E\left(P_{SC}\right) \end{cases} \tag{5}$$

The relaxed problem can be stated as follows:

$$(P_R) \begin{cases} \min \widetilde{\phi} = \mathbb{E}\left(\phi(x) + Q(x;T,h)\right) \\ s.t. \qquad x \in D = \{x \in \mathbb{R}^n | Ax = b, x \in \mathbb{N}\} \end{cases} \tag{6}$$

We suppose that the penalties $q^r = q(\xi^r)$ of the constraint violations are given. Here a recourse function $Q(x,\xi^r)$ is added to each criterion $Z_k^{(r)}$ for scenario r and the corresponding penalty is given by:

$$Q(x,\xi^r) = \min_z \left\{(q^r)^t z | W(\xi^r)z = h(\xi^r) - T(\xi^r)x; z \geq 0\right\}. \tag{7}$$

3 Theoretical Results

In this section we present two sets of most important results. The first set concerns the feasibility and optimality tests as was introduced in L-shaped technique (see [7,9]) and the second set contains some basic definitions and results in multiple objective integer linear programming theory and optimization over efficient solutions set [3,10].

3.1 *Feasibility Test*

We use the dual problem of (7) stated below to test whether a given solution x^0 will yield feasible second-stage problems for all possible realizations of ξ.

$$(F_{test})\begin{cases}\max\ \pi^t\,(h(\xi^r)-T(\xi^r)x)\\ s.t.\ \ \pi^t W\le (q^r)^t\\ \qquad \pi\in\mathbb{R}\end{cases} \tag{8}$$

According to Farkas's lemma $\left\{z|Wz=h(\xi^r)-T(\xi^r)x^0,\ z\ge 0\right\}\neq\varnothing$ if and only if:

$$u^tW\le 0 \text{ implies } u^t\left[h(\xi^r)-T(\xi^r)x^0\right]\le 0$$

Therefore, $Q(x^0,\xi^r)$ is infeasible if and only if $P=\{\pi|\pi^tW\le(q^r)^t\}$ has an extreme ray u such that $u^t\left[h(\xi^r)-T(\xi^r)x^0\right]>0$.

To check out for feasibility of the second stage-problems, we solve the following problem:

$$(F_{\text{Dual}})\begin{cases}\max\ u^t\left(h(\xi^r)-T(\xi^r)x^0\right)\\ s.t.\ \ u^tW\le 0\\ \qquad \|u\|_1\le 1\\ \qquad u\in\mathbb{R}\end{cases} \tag{9}$$

The constraints $\|u\|_1\le 1$ bounds the value of u.

In case where $u_r^t\left[h(\xi^r)-T(\xi^r)x^0\right]>0$ for some ξ^r; $r\in\{1,\ldots,R\}$ and u_r is an optimal solution of problem (9); We add feasibility cut:

$$u_r^t\left[h(\xi^r)-T(\xi^r)x^0\right]\le 0 \tag{10}$$

3.2 *Optimality Test*

Resolution of the problem (P_R') permit to test the optimality of a given solution x^0 with penalty θ

$$(P'_R)\begin{cases}\min \widetilde{\phi} = E(\phi)+\theta; \\ s.t. \qquad x \in D = S \cap \mathbb{Z} \\ \qquad\qquad \theta \geq Q(x) \\ \qquad\qquad \theta \in \mathbb{R}^+\end{cases} \tag{11}$$

where $S = \{x \in \mathbb{R}^n | Ax = b, u_r^t T(\xi^r)x \geq u_r^t h(\xi^r), r = 1,\ldots,R, x \geq 0\}$ and

$$\theta \geq Q(x) \tag{12}$$

is the optimality cut.

$\widetilde{S} = \{x \in \mathbb{R} | \widetilde{A}x = \widetilde{b}, x \geq 0\}$ is a non-empty, compact polyhedron in $\mathbb{R}^n$ and $Q(x) = E(Q(x,\xi)) = \sum_{r=1}^{R} p^r Q(x,\xi^r) = \sum_{r=1}^{R} p^r (q^r)^t z^r$.

We solve problem (P'_R) and we obtain a feasible solution. Efficiency and non-dominance are defined as follows:

Definition 1. A point $\overline{x} \in S$ is an *efficient* solution for problem (P) if and only if there is no $x \in S$ such that $\widetilde{z}_i(x) \leq \widetilde{z}_i(\overline{x})$ for all $i \in \mathrm{Im} = \{1,2,\ldots,K\}$ and $\widetilde{z}_i(x) < \widetilde{z}_i(\overline{x})$ for at least one $i \in \mathrm{Im}$.

Otherwise, $\overline{x}$ is not efficient and the corresponding vector $(\widetilde{z}_1(\overline{x}), \widetilde{z}_2(\overline{x}), \ldots, \widetilde{z}_p(\overline{x}))$ is said to be *dominated*. The set of efficient solutions is denoted by $E(P)$.

The next theorem provides another characterization of an efficient solution that is integrated as a test-procedure in our method.

Theorem 1. *Let x^0 be an arbitrary element of the region D; $x^0 \in E(P)$ if and only if the optimal value of the objective function $\Psi(\psi,x)$ is null in the following integer linear programming problem:*

$$(P(x^0))\begin{cases}\max \quad \Psi(\psi_1,\ldots,\psi_K,x_1,\ldots,x_n) = \sum_{i=1}^{K} \psi_i \\ s.t. \begin{cases}\sum_{j=1}^{n} \widetilde{c}_j^i x_j + \psi_i = \sum_{j=1}^{n} \widetilde{c}_j^i x_j^0 \quad \forall i \in \{1,\ldots,K\} \\ x = (x_1,\ldots,x_n) \in D \\ \psi_i : \text{ are real nonnegative integer variables} \\ \text{for all } i \in \{1,\ldots,K\} \\ \widetilde{c}_j^i : \text{ is the } j^{th} \text{ component of row vector} \\ \qquad \widetilde{c}^i \text{ in problem } (P)\end{cases}\end{cases} \tag{13}$$

The proof of the theorem is omitted; it can be found in [1]. Its utilization guarantees that the feasible solution is either efficient or otherwise provides an efficient solution for problem (P). Once an efficient solution is generated and added to the current list where the expected objective $E(\phi(x))$ is evaluated, new constraints over the feasible set D of the relaxed problem (P_R) are imposed, discarding from further consideration not only efficient solutions generated previously, but also any other feasible solutions with dominated objectives vectors. The algorithm terminates when the current feasible space becomes empty.

Assuming that all coefficients of matrix C are integers. Afterwards, at iteration k, using Sylva and Crema's idea, see [10], the feasible set D is reduced gradually by eliminating all dominated solutions by $C\hat{x}^k$. The resolution of the following problem enables us to perform this elimination.

$$(P_R^\ell) \equiv \min\left\{dx, x \in D - \bigcup_{s=1}^{\ell} D_s\right\}$$

where $D_s = \{x, x \in \mathbb{Z}_+^n, Cx \leq Cx^s\}$ and $\{Cx^s\}_{s=1}^{\ell}$ are non-dominated criteria solutions of (P) obtained at iterations $1, 2, \ldots, \ell$, respectively.

$$D - \bigcup_{s=1}^{\ell} D_s = \left\{\begin{array}{l} c^i x \leq (c^i x^s - 1)y_i^s + M_i(1 - y_i^s), \\ i = 1, \ldots, K; s = 1, \ldots, \ell; \\ y_i^s \in \{0,1\}; i = 1, \ldots, K; s = 1, 2, \ldots, \ell \\ \sum_{i=1}^{K} y_i^s \geq 1;\ s = 1, \ldots, \ell \\ x \in D \end{array}\right\} \tag{14}$$

M_i is an upper bound for any feasible value of the ith objective function. The associate variables y_i^s, $i = 1, \ldots, K$ of $\hat{x}^s$ and additional constraints are added to impose an improvement on at least objective function. Note that when $y_i^s = 0$, the constraint is not restrictive and when $y_i^s = 1$, a strict improvement is forced in the ith objective function evaluated at $\hat{x}^s$.

4 The Method

In this section we give two descriptions of our technique. One is purely technique and the second describes informally different integrated steps that yield to a solution of problem $(P_E(SC))$ stated in (2). Initially, we solve the relaxed problem (P_R) associated with problem (P_{SC}), its feasible set is defined by deterministic constraints of problem $(MOILP)$ without any feasibility or optimality cut. The obtained optimal solution x^* passes through three tests, feasibility, optimality, and efficiency, respectively. The efficient solution $\overline{x}^\ell$ issued from the efficiency test is considered as a first efficient solution; we initialize $X_{\text{opt}} := \overline{x}^\ell$ and $\phi_{\text{opt}} := d\overline{x}^\ell$.

We solve the problem $(P_R^\ell) \equiv \min\{\mathbb{E}(d)x; x \in D - \bigcup_{s=1}^{\ell} D_s\}$. The obtained optimal solution, x^ℓ, produces a minimum value of the criterion $\widetilde{\phi}$ in the reduced domain. The process continue in this manner until the current feasible space becomes empty or $\widetilde{\phi}_l \geq \widetilde{\phi}_{\text{opt}}$. The technical presentation of the proposed method is outlined in the following algorithm:

5 The Algorithm

Algorithm 1: Optimizing a Linear Function over Integer Efficient Set

Input Data
$\downarrow K, m, m_1, n, n_1$: The dimensions of the problem;
$\downarrow S, Pr$: Number of scenarios and their probability vector;
$\downarrow A_{(m\times n)}, b_{(m\times 1)}$: Deterministic constraints parameters;
$\downarrow T_{(m_1\times n)}(sc), h_{(m_1\times 1)}(sc)\, \forall sc \in \{1,\cdots,S\}$: Stochastic constraints parameters;
$\downarrow d_{(1\times n)}(sc),\, \forall sc \in \{1,\cdots,S\}$: Stochastic main criterion vector;
$\downarrow Cr_{(K\times n)}(sc),\, \forall sc \in \{1,\cdots,S\}$: Stochastic criteria matrix;
$\downarrow W_{(m_1\times n_1)}$: a fixed recourse matrix; $q'_{(1\times n_1)}$: the penalties of the constraint violations;
Output
$\uparrow X_{opt}$: optimal solution of the problem (P_E);
$\uparrow \mathbb{E}\left(\tilde{\phi}_{opt}\right)$: optimal value of criterion $\tilde{\phi}$;
Initialization $\mathbb{E}\left(\tilde{\phi}_{opt}\right) \leftarrow +\infty$, $\ell \leftarrow 1$, $End \leftarrow false$, $\theta \leftarrow -\infty$ and $D^1 \leftarrow D$;
Solve the deterministic relaxed problem $\left(P_R^\ell(D^\ell)\right)$;
$\uparrow x_0, \uparrow \mathbb{E}(z_0) =$ Pb_relaxed $(\downarrow d, \downarrow A, \downarrow b)$;
if *the problem does not have a feasible solution* **then**
  problem (P) is not feasible: **Terminate**;
else
  while *End=false* **do**
    Let $x\ell$ be an optimal solution of $\left(P_R^\ell(D^\ell)\right)$
    Feasibility and Optimality Test
    for $sc \leftarrow 1$ **to** S **do**
      $Feasibility \leftarrow false$;
      while $Feasibility \neq true$ **do**
        Solve problem (F_{dual}) (9);
        $\beta \leftarrow u^t\left[h(\xi^r) - T(\xi^r)x^\ell\right]$;
        if $\beta > 0$ **then**
          $D^\ell = D^\ell \cup \{$ feasibility cut (10) $\}$
          Solve $\left(P_R^\ell(D^\ell)\right)$
          Let x^ℓ be an optimal solution
        else
          $Feasibility \leftarrow true$;
        end
      end
    end
    $Q \leftarrow 0$
    for $sc \leftarrow 1$ **to** S **do**
      Solve problem (F_{test}) *(8);*
      $Q \leftarrow Q + Pr \times Q\left(x^\ell; sc\right)$;
    end
    while $\theta < Q$ **do**
      $D^\ell_{opt} = D^\ell \cup \{$Optimality constraint (12)$\}$;
      Solve $\left(P_R^\ell(D^\ell)\right)$, let x^ℓ be an optimal solution with a penalty value θ;
    end
    EFFICIENCY TEST
    Solve $\left(P\left(x^\ell\right)\right)$; Ψ is the optimal solution criteria;
    if $\Psi \neq 0$ **then**
      x^ℓ is not efficient;
      $\overline{x}^\ell$ an optimal solution of $\left(P\left(x^\ell\right)\right)$ is efficient;
      Feasibility and Optimality Test for the solution $\overline{x}^\ell$;
      $X_{opt} \leftarrow \overline{x}^\ell$, $\phi_{opt} \leftarrow \tilde{\phi}(\overline{x}^\ell)$;
      $\ell \leftarrow \ell + 1$;
    else
      x^ℓ is an efficient solution;
      $X_{opt} = x^\ell$, $\tilde{\phi}_{opt} = \tilde{\phi}(x^\ell)$;
      $\ell \leftarrow \ell + 1$;
    end
    $D^\ell \leftarrow D^\ell - \bigcup_{s=1}^{\ell-1} D_s$ *and Solve* $\left(P_R^\ell(D^\ell)\right)$;
    if $D^\ell = \varnothing$ ***Or*** $\tilde{\phi}(x^j) \geq \tilde{\phi}_{opt}$ **then**
      X_{opt} is an optimal efficient solution with value $\tilde{\phi}_{opt}$;
      Terminate; $End \leftarrow true$;
    else
      $End \leftarrow false$;
    end
  end
end

6 Didactic Example

Consider the following multiple objective integer linear programming stochastic problem:

Scenario 1

$$(P_{sc1})\begin{cases} \text{"max"}\, Z_1 = 9x_1 + 3x_2 \\ \text{"max"}\, Z_2 = 3x_1 + 5x_2 \\ D\begin{cases} -2x_1 + 5x_2 \le 23 \\ 4x_1 + x_2 \le 31 \\ x_1 - x_2 \le 4 \\ -x_1 - 3x_2 \le -8 \\ -3x_1 - x_2 \le -8 \\ 6x_1 - 5x_2 \le 21 \\ 10x_1 + 3x_2 \ge 30 \\ x_1\,,\ x_2 \in \mathbb{Z}_+^* \end{cases} \end{cases}$$

Scenario 2

$$(P_{sc2})\begin{cases} \text{"max"}\, Z_1 = -3x_1 + 3x_2 \\ \text{"max"}\, Z_2 = -6x_1 - 4x_2 \\ D\begin{cases} -2x_1 + 5x_2 \le 23 \\ 4x_1 + x_2 \le 31 \\ x_1 - x_2 \le 4 \\ -x_1 - 3x_2 \le -8 \\ -3x_1 - x_2 \le -8 \\ 4x_1 + x_2 \ge 12 \\ 5x_1 - 1x_2 \le 20 \\ x_1\,,\ x_2 \in \mathbb{Z}_+^* \end{cases} \end{cases}$$

$$(PE_{sc1})\begin{cases} \max \phi_1(x) = -5x_1 + 4x_2 \\ s.t. \\ D\begin{cases} -2x_1 + 5x_2 \le 23 \\ 4x_1 + x_2 \le 31 \\ x_1 - x_2 \le 4 \\ -x_1 - 3x_2 \le -8 \\ -3x_1 - x_2 \le -8 \\ 6x_1 - 5x_2 \le 21 \\ 10x_1 + 3x_2 \ge 30 \\ x_1\,,\ x_2 \in \mathbb{Z}_+^* \end{cases} \end{cases}$$

$$(PE_{sc2})\begin{cases} \max \phi_2(x) = 4x_1 - 8x_2 \\ s.t. \\ D\begin{cases} -2x_1 + 5x_2 \le 23 \\ 4x_1 + x_2 \le 31 \\ x_1 - x_2 \le 4 \\ -x_1 - 3x_2 \le -8 \\ -3x_1 - x_2 \le -8 \\ 4x_1 + x_2 \le 12 \\ 5x_1 - x_2 \ge 20 \\ x_1\,,\ x_2 \in \mathbb{Z}_+^* \end{cases} \end{cases}$$

$$q_1' = (1\,0\,6\,2); \quad P_1 = \frac{1}{3} \qquad\qquad q_2' = (5\,3\,2\,1); \quad P_2 = \frac{2}{3}$$

The recourse matrix and the deterministic constraints matrices are given for both scenarios by:

$$W = \begin{pmatrix} -2 & -1 & 2 & 1 \\ 3 & 2 & -5 & -6 \end{pmatrix} \qquad A = \begin{pmatrix} -2 & 5 \\ 4 & 1 \\ 1 & -1 \\ -1 & -3 \\ -3 & -1 \end{pmatrix} \qquad b = \begin{pmatrix} 23 \\ 31 \\ 4 \\ -8 \\ -8 \end{pmatrix}$$

The stochastic constraints matrices are given for both scenarios by:

$$T^1 = \begin{pmatrix} 6 & -5 \\ 10 & 3 \end{pmatrix} \qquad h^1 = \begin{pmatrix} 21 \\ 30 \end{pmatrix}; \quad T^2 = \begin{pmatrix} 4 & 1 \\ 5 & -1 \end{pmatrix} \qquad h^2 = \begin{pmatrix} 12 \\ 20 \end{pmatrix}$$

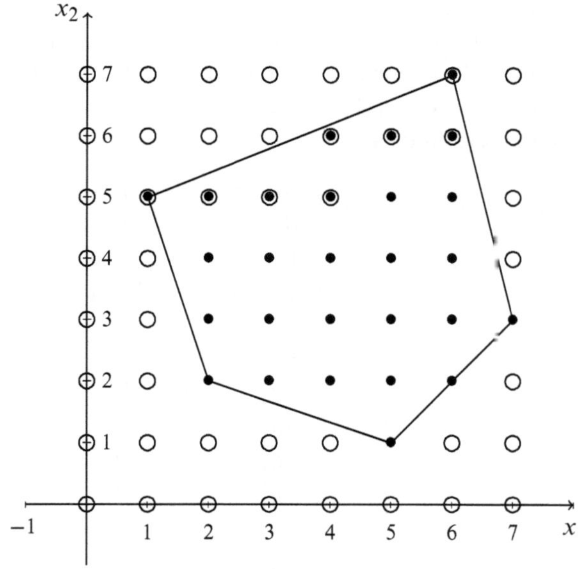

Fig. 1 Decision space

The deterministic multiple objective integer linear programming problem

$$(P)\begin{cases} \text{“max}'' E(z_1) = \quad x_1 + 3x_2 \\ \text{“max}'' E(z_2) = -3x_1 - \quad x_2 \\ s.t. \qquad\qquad\qquad x \in \quad D \end{cases} \tag{15}$$

where

$$D = \{x \in \mathbb{R}^n | -2x_1 + 5x_2 \leq 23, 4x_1 + x_2 \leq 31, x_1 - x_2 \leq 4, -x_1 - 3x_2 \leq -8,$$
$$-3x_1 - x_2 \leq -8, x_1, x_2 \in \mathbb{N}\}$$

The main deterministic relaxed problem is defined by:

$$(P_R)\begin{cases} \max E(\phi) = x_1 - 4x_2 \\ s.t. \qquad\qquad x \in \quad D \end{cases} \tag{16}$$

Step 0: $l = 0$, $\phi_{\text{opt}} = -\infty$, $H^0 = D$
Step 1: The relaxed problem (P_R) is solved.

$$(P_R^0)\begin{cases} \min E(\phi) = x_1 - 4x_2 \\ s.t. \qquad\qquad x \in \ H^0 \end{cases}$$

Initial iteration

- Lower bounds of the objective functions are $M_1 = 0$, $M_2 = -25$;

First iteration

We solve the relaxed problem $(P_R^0) \equiv \max\{\widetilde{\phi} | x \in D\}$

An optimal solution is $x^1 = (5\ 1)'$ (Fig. 1). Let $z^1 = Cx^1 = (8\ -16)'$ its image in the outcome space criteria. $\widetilde{\phi}_{\sup} = dx^1 = 1$.

Step 2: feasibility test

$$h^1 - T^1 x^0 = \begin{pmatrix} 21 \\ 30 \end{pmatrix} - \begin{pmatrix} 6 & -5 \\ 10 & 3 \end{pmatrix} \begin{pmatrix} 5 \\ 1 \end{pmatrix} = \begin{pmatrix} -4 \\ -23 \end{pmatrix}$$

$$h^2 - T^2 x^0 = \begin{pmatrix} 12 \\ 20 \end{pmatrix} - \begin{pmatrix} 4 & 1 \\ 5 & -1 \end{pmatrix} \begin{pmatrix} 5 \\ 1 \end{pmatrix} = \begin{pmatrix} -9 \\ -4 \end{pmatrix}$$

$$(Pu_1) \begin{cases} \max\ -4u_1^1 - 23u_2^1 \\ s.t.\ \ -2u_1^1 + 3u_2^1 \le 0 \\ \quad -1u_1^1 + 2u_2^1 \le 0 \\ \quad 2u_1^1 - 5u_2^1 \le 0 \\ \quad u_1^1 - 6u_2^1 \le 0 \\ \quad u_1^1 + u_2^1 \le 1 \\ \quad u_1^1,\ u_2^1 \in \mathbb{R} \end{cases} \text{, maximum is at:} \begin{pmatrix} u_1^1 \\ u_2^1 \end{pmatrix} = \begin{pmatrix} 0 \\ 0 \end{pmatrix}$$

$$(Pu_2) \begin{cases} \max\ -9u_1^2 - 4u_2^2 \\ s.t.\ \ -2u_1^2 + 3u_2^2 \le 0 \\ \quad -u_1^2 + 2u_2^2 \le 0 \\ \quad 2u_1^2 - 5u_2^2 \le 0 \\ \quad u_1^2 - 6u_2^2 \le 0 \\ \quad u_1^2 + u_2^2 \le 1 \\ \quad u_1^2,\ u_2^2 \in \mathbb{R} \end{cases} \text{, maximum is at:} \begin{pmatrix} u_1^2 \\ u_2^2 \end{pmatrix} = \begin{pmatrix} 0 \\ 0 \end{pmatrix}$$

$u_1 = u_2 = 0$. The solution is feasible for both first and second scenario.

Optimality test of $x^0 = (5\ 1)^t$

$$(P(\pi_1)) \begin{cases} \max\ -4\pi_1^1 - 23\pi_2^1 \\ s.t.\ \ -2\pi_1^1 + 3\pi_2^1 \le 1 \\ \quad -\pi_1^1 + 2\pi_2^1 \le 0 \\ \quad 2\pi_1^1 - 5\pi_2^1 \le 6 \\ \quad \pi_1^1 - 6\pi_2^1 \le 2 \\ \quad \pi_1^1,\ \pi_2^1 \in \mathbb{R} \end{cases} \text{, maximum is at:} \begin{pmatrix} \pi_1^1 \\ \pi_2^1 \end{pmatrix} = \begin{pmatrix} -1 \\ -0.5 \end{pmatrix}$$

$$(P(\pi_2)) \begin{cases} \max\ -9\pi_1^2 - 4\pi_2^2 \\ s.t.\ \ -2\pi_1^2 + 3\pi_2^2 \le 5 \\ \quad -\pi_1^2 + 2\pi_2^2 \le 3 \\ \quad 2\pi_1^2 - 5\pi_2^2 \le 2 \\ \quad \pi_1^2 - 6\pi_2^2 \le 1 \\ \quad \pi_1^2,\ \pi_2^2 \in \mathbb{R} \end{cases} \text{, maximum is at:} \begin{pmatrix} \pi_1^2 \\ \pi_2^2 \end{pmatrix} = \begin{pmatrix} -3.67 \\ -0.78 \end{pmatrix}$$

Penalty value at $(5\ 1)^t$
$Q(x,sc1)=\pi_1\left[h^1-T^1x^0\right]=\frac{31}{2}$,
$Q(x,sc2)=\pi_2\left[h^2-T^2x^0\right]=36.11$,
$Q(x)=P_1Q(x,sc1)+P_2Q(x,sc2)=\frac{1}{3}(\frac{31}{2})+\frac{2}{3}(36.11)=29.24$
We test the efficiency of x^0 by solving the problem

$$(P(x^0))\begin{cases}\max\Psi = & \psi_1+\ \psi_2\\ s.t. & x\in H^{(0)}\\ & x_1+\ 3x_2-\psi_1= \quad 8\\ & -3x_1-\ x_2-\psi_2= \ -16\\ & x_1,\ x_2\in\mathbb{N},\psi_1,\ \psi_2\in\mathbb{R}^+\end{cases}$$

We obtain $\max\Psi=16\neq 0$; indicating that x^0 is not efficient but $\overline{x}^0=(1\ 5)^t$ produced by this program is an efficient, corresponding to $\widetilde{\phi}(\overline{x}^0)=-19$.

The equivalent efficient solutions program

$$(P(\overline{x}^0))\begin{cases}\min\widetilde{\phi}= & x_1-\ 4x_2\\ s.t. & x\in H^{(0)}\\ & x_1+\ 3x_2=\ 16\\ & -3x_1-\ x_2=-8\\ & x_1,\ x_2\in\ \mathbb{N}\end{cases}$$

An optimal solution is $\overline{x}^0$. We test its feasibility as above,

$$\begin{pmatrix}u_1^1\\u_2^1\end{pmatrix}=\begin{pmatrix}\frac{5}{7}\\\frac{2}{7}\end{pmatrix};\quad\begin{pmatrix}u_1^2\\u_2^2\end{pmatrix}=\begin{pmatrix}\frac{2}{3}\\\frac{1}{3}\end{pmatrix}$$
$$u_1^t\left[h^1-T^1\overline{x}^0\right]=\left(\tfrac{5}{7}\ \tfrac{2}{7}\right)\begin{pmatrix}28\\-15\end{pmatrix}=\tfrac{110}{7}$$

x^0 is not feasible for first scenario. We construct the feasible co▪straint, $u_1^t\left[h^1-T^1x^0\right]>0$, then we work out the feasibility cut as

$$\begin{pmatrix}\frac{5}{7}&\frac{2}{7}\end{pmatrix}\begin{pmatrix}6&-5\\10&3\end{pmatrix}\begin{pmatrix}x_1\\x_2\end{pmatrix}\geq\begin{pmatrix}\frac{5}{7}&\frac{2}{7}\end{pmatrix}\begin{pmatrix}21\\30\end{pmatrix}=\frac{165}{7}\Longleftrightarrow 50x_1-19x_2\geq 165;$$

The cut is added to the first problem(P_R^0)

$$(P_R^{01})\begin{cases}\max\widetilde{\phi}=x_1-\ 4x_2\\ s.t.\qquad x\in H^{(01)}\end{cases}$$

where $H^{(01)}=H^0\cap\{x\in\mathbb{N}^n|\,50x_1-19x_2\geq 165\}$

We get a new integer solution $x^{01} = (5\ 1)^t$; we test its feasibility

$$h^1 - T^1 x^{01} = \begin{pmatrix} 21 \\ 30 \end{pmatrix} - \begin{pmatrix} 6 & -5 \\ 10 & 3 \end{pmatrix} \begin{pmatrix} 5 \\ 1 \end{pmatrix} = \begin{pmatrix} -4 \\ -23 \end{pmatrix}$$

$$h^2 - T^2 x^{01} = \begin{pmatrix} 12 \\ 20 \end{pmatrix} - \begin{pmatrix} 4 & 1 \\ 5 & -1 \end{pmatrix} \begin{pmatrix} 5 \\ 1 \end{pmatrix} = \begin{pmatrix} -13 \\ -4 \end{pmatrix}$$

$u_1 = u_2 = 0$. The solution is feasible for both first and second scenario.
The penalty of $x^{01} = (5\ 1)$ is found by solving (P_{π_1}) and (P_{π_2}) gives:

$$\text{maximum is at: } \begin{pmatrix} \pi_1^1 \\ \pi_2^1 \end{pmatrix} = \begin{pmatrix} -1 \\ -0.5 \end{pmatrix}$$

$$\text{maximum is at: } \begin{pmatrix} \pi_1^2 \\ \pi_2^2 \end{pmatrix} = \begin{pmatrix} -3.67 \\ -0.78 \end{pmatrix}$$

$$Q(x, sc1) = \pi_1 \left[h^1 - T^1 x^1\right] = 15.5,$$

$$Q(x, sc2) = \pi_2 \left[h^2 - T^2 x^1\right] = 36.11$$

$$Q(x) = P_1 Q(x, sc1) + P_2 Q(x, sc2) = \frac{1}{3}(15.5) + \frac{2}{3}(36.11) = 29.24$$

$\theta = Q(x) = 29.24$. Then $x^{01} = (5\ 1)^t$ is an optimal feasible solution.
$\widetilde{\phi}_{\text{opt}} = -\infty, \widetilde{\phi}(x^{01}) = 1 > \widetilde{\phi}_{\text{opt}}, X_{\text{opt}} = (5\ 1)^t, \widetilde{\phi}_{\text{opt}} = 1$
$\ell := \ell + 1 = 1$ and we solve problem (P_R^1)

$$(P_R^1) \begin{cases} \max \widetilde{\phi} = x_1 - 4x_2 \\ s.t. \quad x \in H^{(1)} \end{cases} \tag{17}$$

where

$$H^{(1)} = H^{(01)} \cap \begin{cases} x_1 + 3x_2 - 9y_1^1 \geq 0\ ,\ -3x_1 - 1x_2 - 10y_2^1 \geq -25 \\ y_1^1 \leq 1,\ y_1^2 \leq 1 \\ y_1^1 + y_2^1 \geq 1 \\ y_1^1,\ y_2^1 \in \mathbb{N} \end{cases}$$

An optimal solution is $x^1 = (6\ 2)^t$, $\widetilde{\phi}(x^1) = -2$, $\widetilde{\phi}(x^1) < \widetilde{\phi}_{\text{opt}}$.

$\widetilde{\phi}(x^1) = -2 < \widetilde{\phi}_{\text{opt}}$. The algorithm terminates with $x_{\text{opt}} = (5\ 1)^t$ and $\widetilde{\phi}_{\text{opt}} = 1$.

7 Conclusion

We have presented an exact method that optimizes a linear function over an integer efficient solutions set in stochastic environment. We achieve this objective by combining two techniques : one uses L-shaped method and the second explores progressively the admissible region going through only efficient solutions that ameliorates the main linear criteria; then the domain is being reduced consequently until it becomes empty. The problem with deterministic parameters is known to be very hard, resolving it under uncertainty becomes harder. As far as we know, the problem has not been yet studied in the literature, we suggest development of new benchmarks in order to make a rational comparative study. Concerning the complexity, as we are obliged to transform the stochastic problem into deterministic one, the problem remains very hard as was stated in deterministic case by N.C. Guyen [6].

References

1. Abbas, M., Bellahcene, F.: Cutting plane method for multiple objective stochastic integer linear programming. Eur. J. Oper. Res. **168**, 967–984 (2006)
2. Abbas, M., Chaabane, D.: Optimizing a linear function over an integer efficient set. Eur. J. Oper. Res. **174**, 1140–1161 (2006)
3. Chaabane, D., Pirlot, M.: A method for optimizing over the integer efficient set. J. Ind. Manag. Optim. **6**(4), 811–823 (2010)
4. Ecker, J.G., Song, H.G.: Optimizing a linear function over an efficient set. J. Optim. Theor. Appl. **83**(3), 541–563 (1994)
5. Goicoechea, A., Dukstein, L., Bulfin, RL.: Multiobjective stochastic programming the PROTRADE-method. Operation Research Society of America, San Francisco (1967)
6. Guyen, N.C.: An Algorithm for Optimizing a Linear Function Over the Integer Efficient Set. Konrad-Zuse-Zentrum fur Informationstechnik Berlin (1992)
7. Kall, P.: Stochastic Linear Programming. Springer, Berlin (1976)
8. Kall, P., Mayer, J.: Stochastic Liner Programming Models, Theory, and Computation. Kluwer, Boston (2005)
9. Kall, P., Wallace, S.W.: Stochastic programming. Wiley interscience series in systems and optimization, John Wiley & Sons, New York (1994)
10. Sylva, J., Crema, A.: A method for finding well-dispersed subsets of non-dominated vectors for multiple objective mixed integer linear programs. Eur. J. Oper. Res. **180**, 1011–1027 (2007)
11. Teghem, J.: Strange-Momix An interactive method for mixed integer linear programming. In: Slowinski, R., Teghem, J. (eds.) Stochastic Versus Fuzzy Apprroaches to Multiobjective Mathematical Programming Under Uncertainty, pp. 101–115. Kluwer, Dordrecht (1990)
12. Teghem, J. Kunsch, P.L.: A survey of techniques for finding efficient solutions to multi-objective integer linear programming. Asia-Pac. J. Oper. Res. **3**, 95–108 (1986)
13. Urli, B., Nadeau, R.: Multiobjective stochastic linear programming with incomplete information. A general methodology. In: Slowinski, R., Teghem, J. (eds.) Stochastic versus fuzzy approaches to multiobjective mathematical programming under uncertainty, Kluwer Academic Publishers, Dordrecht pp. 131–161 (1990)

Open-Pit Mining with Uncertainty: A Conditional Value-at-Risk Approach

Henry Amankwah, Torbjörn Larsson, and Björn Textorius

Abstract The selection of a mine design is based on estimating net present values of all possible, technically feasible mine plans so as to select the one with the maximum value. It is a hard task to know with certainty the quantity and quality of ore in the ground. This geological uncertainty and also the future market behavior of metal prices and foreign exchange rates, which are always uncertain, make mining a high risk business.

Value-at-Risk (VaR) is a measure that is used in financial decisions to minimize the loss caused by inadequate monitoring of risk. This measure does, however, have certain drawbacks such as lack of consistency, nonconvexity, and nondifferentiability. Rockafellar and Uryasev [J. Risk 2, 21–41 (2000)] introduce the Conditional Value-at-Risk (CVaR) measure as an alternative to the VaR measure. The CVaR measure gives rise to a convex optimization problem.

An optimization model that maximizes expected return while minimizing risk is important for the mining sector as this will help make better decisions on the blocks of ore to mine at a particular point in time. We present a CVaR approach to the uncertainty involved in open-pit mining. We formulate investment and design models for the open-pit mine and also give a nested pit scheduling model based on CVaR. Several numerical results based on our models are presented by using scenarios from simulated geological and market uncertainties.

H. Amankwah (✉)
Department of Mathematics and Statistics, University of Cape Coast, Ghana
e-mail: khamankwah@yahoo.com

T. Larsson • B. Textorius
Department of Mathematics, Linköping University, Sweden
e-mail: torbjorn.larsson@liu.se; bjorn.textorius@liu.se

A. Migdalas et al. (eds.), *Optimization Theory, Decision Making, and Operations Research Applications*, Springer Proceedings in Mathematics & Statistics 31,
DOI 10.1007/978-1-4614-5134-1_8, © Springer Science+Business Media New York 2013

1 Introduction

Open-pit mining is a surface mining operation whereby ore, or waste, is excavated from the surface of the land. In the process of digging the surface of the land, a deeper and deeper pit is formed until the mining operation ends. The entire mining volume is usually partitioned into regular three-dimensional blocks. By using information from drill holes the mining industry is able to estimate the value of each block of the orebody in the ground. Before the mining operation begins, it is desirable to determine the ultimate pit contour that maximizes the net present value. We also have to know the order in which to mine the blocks of the orebody over the lifetime of the mine. Thus, the pit design and mine scheduling are important tasks to the mining industry. Another aspect of importance in the mining evaluation process is the consideration of uncertainty and risk. Further, the aspect of uncertainty and risk is becoming increasingly important, simply because the best high-grade and low-cost orebodies in the world have already been mined [16], so that the orebodies to be mined in the future require more cautious evaluations and planning with respect to the trade-off between expected return and risk.

Mine projects are complex businesses that demand a constant assessment of risk. This is because the value of a mine project is typically influenced by many underlying economic and physical uncertainties, such as metal prices, metal grades, costs, schedules, quantities and environmental issues, among others, which are not known with certainty [13]. In the prefeasibility stage of a mining project, the geology and ore distribution in the mineral deposit are estimated from the information derived from the exploration drilling samples. Since the information obtained from the samples is not representative of the entire (3-D) ore deposit, the geology of the ore deposit represents one of the most critical sources of technical and financial uncertainty in a mine operation. One consequence of this lack of information is the misclassification of resources, where economic ore can be dispatched to the waste dump and non-economic ore can be sent to processing.

The selection of a mine design is based on estimating net present values of all possible, technically feasible mine plans so as to select the one with the maximum value. However, mine planners cannot know with certainty the quantity and quality of ore in the ground. This, Abdel Sabour et al. [1] termed the *geological uncertainty*. It is recognized among practitioners that mining is a high risk business and the geological uncertainty is a major source of risk. There are other sources of uncertainties. The future market behavior of metal prices and foreign exchange rates are impossible to be known with certainty, and therefore, they are also sources of risks affecting mine project profitability. Abdel Sabour et al. use the term *market uncertainty* for these sources of risks and classify them as the second major source of risk. The existence of uncertainties can lead to a high probability that the actual cash flows throughout the lifetime of the mining project will be different from those expected. An optimization model that maximizes expected return while minimizing risk is therefore important for the mining sector as this will help make better decisions on the blocks of ore to mine at a particular point in time.

One of the standard procedures to deal with uncertainty and risk in the mining industry is to perform the evaluation process at different scenarios for the project key variables [1]. A popular approach to deal with risk is to apply conventional Monte Carlo simulation, in which case a distribution for the mine value is obtained rather than a single expected value. From this distribution, the risk associated with a long-term production scheduling can be explored by defining the range for the expected value at a certain degree of confidence [1].

Dimitrakopoulos et al. [4] are the first to introduce the integration of geological uncertainty into open-pit mine planning [1]. They are of the view that geological uncertainty as an element in key parameters of open-pit mining projects can be quantified by conditional simulation combined with open-pit optimization studies. They further claim that having an accurate assessment of uncertainty arising from grade variability in the ore reserve allows risk in a mining project to be quantified and considered in decision-making processes. In their opinion, further integration of uncertainty in the optimization process is needed to enhance the interaction and efficacy of open-pit optimization and risk assessment.

It has therefore become necessary for the mining industry to explore decision support for minimizing risk while maximizing expected returns. In a paper presented at a symposium in Chile in 2009, Lai et al. [8] pointed out that *Value-at-Risk* (VaR) has been used by banks and financial institutions to minimize financial loss caused by inadequate monitoring of risk. This, they say, is a method which is particularly effective in assessing risk of investments with large loss potential. In simple terms, VaR is the maximum likely loss incurred over a specified period of time at a given confidence level. The chosen confidence level depends on the purpose of the exercise and the risk tolerance level of management. The focus of their research is on the development of a VaR method for the open-pit mine slope stability risk assessment. To ensure that the impacts of major slope failures are fully accounted for, failure costs are included in the mining schedule to estimate their effect on the rate of return for a given project. They demonstrate how VaR could be used to assess the economic risk and return associated with different pit slope designs.

We will in this paper introduce the *Conditional Value-at-Risk* (CVaR) measure [14] as a tool for taking both geological and market uncertainty into account in the planning of open-pit mining. In order to introduce the reader to the CVaR concept, we will below outline how it is motivated and used in an easily described application, namely portfolio optimization under uncertainty, as described by Lim et al. [11].

Measure of risk plays a critical role in the optimization of portfolios under the presence of uncertainties. Among various risk criteria, the CVaR is a popular measurement of risk representing the percentile of the loss distribution with a specified confidence level [11]. Let $\alpha \in (0,1)$ denote the confidence level and $f(x,y)$ a loss function associated with a portfolio x (a vector indicating the fraction of instrument of some available budget in each of n financial instruments) and an instrument price (or return) vector $y \in R^n$. [It should be noted that $f(x,y) < 0$ means a positive return.] Then, the VaR function $\zeta(x,\alpha)$ is given by the smallest number satisfying $\Phi(x,\zeta(x,\alpha)) = \alpha$, where $\Phi(x,\zeta)$ is the probability that the loss $f(x,y)$

does not exceed a threshold value ζ, that is, $\Phi(x,\zeta) = \mathtt{Pr}[f(x,y) \le \zeta]$. So, for any portfolio $x \in R^n$ and any given confidence level α, VaR is interpreted as the value of ζ such that the probability of the loss not exceeding ζ is α.

The VaR measure has drawbacks, among them is lack of consistency. This is because it is not subadditive, which means that the risk of a portfolio can be higher than the sum of the risks of its individual components. Furthermore, in practice, the VaR measure is nonconvex and nondifferentiable, and hence, it is difficult to find a global minimum [11]. Criticisms of the VaR approach resulted in new proposals for ways to measure risk in portfolios.

Rockafellar and Uryasev [14] introduce the CVaR measure as an alternative to VaR, since the CVaR gives rise to a convex problem. The CVaR measure is further developed in [15]. It is considered to be more consistent than the VaR, and it is defined as the mean loss by which the VaR is exceeded [3]. In other words, the CVaR is the conditional expected loss of a portfolio at a confidence level, given that the loss to be accounted for exceeds or equals the VaR. By definition, the VaR at a given confidence level is never higher than the corresponding CVaR. Andersson et al. and Mansini et al. [2, 12], also support the use of CVaR over VaR. Other names for CVaR are *mean excess loss*, *mean shortfall*, or *tail VaR*.

For continuous distributions, CVaR is the conditional expected loss given that the loss exceeds VaR. That is, CVaR is given by

$$\Phi_\alpha(x) = (1-\alpha)^{-1} \int_{f(x,y)>\zeta(x,\alpha)} f(x,y)p(y)\mathrm{d}y, \tag{1}$$

where $p(y)$ is a probability density function of y. To avoid complications caused by an implicitly defined function $\zeta(x,\alpha)$, Rockafellar and Uryasev [14] give an alternative function,

$$F_\alpha(x,\zeta) = \zeta + (1-\alpha)^{-1} \int_{f(x,y)>\zeta} [f(x,y)-\zeta]p(y)\mathrm{d}y, \tag{2}$$

for which they show that minimizing $F_\alpha(x,\zeta)$ with respect to (x,ζ) yields the minimum CVaR and its solution.

When applied to portfolio optimization, x_i is the portion of the total investment that is made in a certain security. If the probability distribution of y is not available we can exploit price scenarios, which can be obtained from past price data. Assume that every price data is equally likely (for example, from random sampling from a joint price distribution). For a given price data y^j, $j = 1,\ldots,J$, we can approximate $F_\alpha(x,\zeta)$ by

$$\widetilde{F}_\alpha(x,\zeta) = \zeta + [(1-\alpha)J]^{-1} \sum_{j=1}^{J} \max\left\{f^j(x)-\zeta,\, 0\right\}, \tag{3}$$

with $f^j(x) \equiv f(x,y^j)$. The function $\widetilde{F}_\alpha(x,\zeta)$ is convex when each $f^j(x)$ is convex, and nondifferentiable at the points where $f^j(x) - \zeta = 0$ hold. A portfolio that approximately minimizes CVaR is found by minimizing $\widetilde{F}_\alpha(x,\zeta)$ over the set of feasible compositions of the portfolio.

The purpose of our study is to formulate optimization models that can be used to maximize expected profit while minimizing risk in the open-pit mining industry. Section 2 is mainly devoted to our problem formulations. In this section, we formulate investment and design models for the open-pit mine and also give a nested pit contour model based on CVaR. Several numerical examples based on our models are presented in Sect. 3, where we use scenarios from simulated geological and market uncertainties. We give a conclusion in the last section.

2 Problem Formulation

The following notations will be used.

V_o	$=$	Set of blocks containing ore.
V_w	$=$	Set of waste blocks.
V	$=$	Set of all blocks that can be mined (i.e., $V = V_o \cup V_w$).
V_s	$=$	Top layer of blocks ($V_s \subset V$).
A	$=$	Set of pairs (i,j) of blocks such that block j is a neighboring block to i that must be removed before block i can be mined.
M_i	$=$	Stochastic variable describing metal content in block $i \in V_o$.
μ_i	$=$	Expected metal content (expectation of M_i).
P	$=$	Stochastic variable describing metal price.
π	$=$	Expected metal price (expectation of P).
K	$=$	Investment cost plus required return on investment.
c_i	$=$	Cost of mining and processing block i.
Q	$=$	Upper bound on CVaR.
G	$=$	Lower bound on expected variable profit (i.e., expected profit excluding K).
α	$=$	Confidence level ($0 < \alpha < 1$).
J	$=$	Number of scenarios.
v	$=$	$[(1-\alpha)J]^{-1}$.
y_i^j	$=$	Metal content in block $i \in V_o$ for scenario $j \in J$.
p^j	$=$	Metal price for scenario $j \in J$.

The decision variables of the optimization models are

$$x_o = \begin{cases} 1 & \text{if any mining is made,} \\ 0 & \text{otherwise,} \end{cases}$$

and

$$x_i = \begin{cases} 1 & \text{if block } i \text{ is mined,} \\ 0 & \text{otherwise,} \end{cases} \qquad i \in V.$$

In the ideal situation, the metal content and the price of the metal are known in advance, but we study the case when they are both considered to be uncertain. If P is considered to be known, we have only geological uncertainty, while if M_i is considered to be known, we have only price uncertainty. The cost for mining and processing each block, c_i, is assumed to be known in advance. (The models to be presented can, however, if needed, also include uncertainty in mining costs.)

The optimization model for maximizing expected profit (see, e.g., [5, 6]) is given by

$$\begin{aligned} \text{maximize} \quad & \sum_{i \in V_o} \pi \mu_i x_i - \sum_{i \in V} c_i x_i \\ \text{subject to} \quad & \\ & x_i \leq x_j, \qquad (i,j) \in A \\ & x_i \in \{0,1\}, \qquad i \in V. \end{aligned} \tag{4}$$

The first term in the objective function is the expected revenue derived from the ore blocks and the second term is the mining cost for all the blocks that are mined. The precedence constraints capture both the pit slope and the immediate block precedence restrictions.

2.1 A CVaR Open-Pit Investment Model

In this section, we propose two CVaR open-pit investment models. These models will enable us to decide whether or not to carry out the mining process. The starting point is the model below, which can be used to determine the maximum expected variable profit that can be made.

$$\begin{aligned} \text{maximize} \quad & \sum_{i \in V_o} (\pi \mu_i - c_i) x_i - \sum_{i \in V_w} c_i x_i - K x_o \\ \text{subject to} \quad & \\ & x_i \leq x_j, \qquad (i,j) \in A \\ & x_i \leq x_o, \qquad i \in V_s \\ & x_i \in \{0,1\}, \qquad i \in V \\ & x_o \in \{0,1\} \end{aligned} \tag{5}$$

Note that the role of the variable x_o, and also of the entire model, is only to indicate whether any mining at all is made, or not, given that the investment cost plus required return on investment is K. We will below construct corresponding models, for the case when the metal contents and prices are described by scenarios.

Introducing the linear loss function

$$f^j(x) = Kx_o - \sum_{i \in V_o} \left(p^j y_i^j - c_i \right) x_i + \sum_{i \in V_w} c_i x_i, \tag{6}$$

which is obtained for a certain mining decision, given that scenario j occurs, an open-pit investment model that approximately minimizes CVaR is given by the following nondifferentiable optimization problem, cf. (3).

$$\begin{aligned}
&\text{minimize} \quad \zeta + v \sum_{j=1}^{J} \max \left\{ Kx_o - \sum_{i \in V_o} \left(p^j y_i^j - c_i \right) x_i + \sum_{i \in V_w} c_i x_i - \zeta,\ 0 \right\} \\
&\text{subject to} \\
&\quad x_i \le x_j, \qquad (i,j) \in A \\
&\quad x_i \le x_o, \qquad i \in V_s \\
&\quad x_i \in \{0,1\}, \qquad i \in V \\
&\quad x_o \in \{0,1\}
\end{aligned} \tag{7}$$

Here, ζ is the VaR. (Strictly speaking, the objective value obtained here is an approximation of the true value of the CVaR, because a finite number of scenarios is used. For simplicity, we will, however, anyway refer to the value obtained as CVaR.) By introducing an auxiliary variable vector $z \in R^J_+$, Problem (7) can be reformulated as the linear problem

$$\begin{aligned}
&\text{minimize} \quad \zeta + v \sum_{j=1}^{J} z_j \\
&\text{subject to} \\
&\quad z_j \ge Kx_o - \sum_{i \in V_o} \left(p^j y_i^j - c_i \right) x_i + \sum_{i \in V_w} c_i x_i - \zeta, \qquad j = 1, \ldots, J \\
&\quad x_i \le x_j, \qquad (i,j) \in A \\
&\quad x_i \le x_o, \qquad i \in V_s \\
&\quad x_i \in \{0,1\}, \qquad i \in V \\
&\quad x_o \in \{0,1\} \\
&\quad z_j \ge 0, \qquad j = 1, \ldots, J.
\end{aligned} \tag{8}$$

Typically, there is a conflict between profit and risk. Carneiro et al. [3] argue that in many cases it is better to maximize returns with risk constraints. This supports the proposal of Krokhmal et al. [7] that, instead of minimizing CVaR, it is better to maximize expected returns and specify a maximum level of risk. In this regard, the mathematical formulation that represents the uncertainties using J scenarios in the case of open-pit mining is given as

$$
\begin{aligned}
\text{maximize} \quad & \sum_{i \in V_o} (\pi \mu_i - c_i) x_i - \sum_{i \in V_w} c_i x_i - K x_o \\
\text{subject to} \quad & \\
& \zeta + v \sum_{j=1}^{J} z_j \le Q x_o \\
& z_j \ge K x_o - \sum_{i \in V_o} \left(p^j y_i^j - c_i \right) x_i + \sum_{i \in V_w} c_i x_i - \zeta, \qquad j = 1, \ldots, J \\
& x_i \le x_j, \qquad (i,j) \in A \\
& x_i \le x_o, \qquad i \in V_s \\
& x_i \in \{0,1\}, \qquad i \in V \\
& x_o \in \{0,1\} \\
& z_j \ge 0, \qquad j = 1, \ldots, J.
\end{aligned} \tag{9}
$$

At an optimal solution (x^*, ζ^*, z^*) to Problem (9), CVaR is at most Qx_o^*. Note that the higher the value of z_j^*, the higher impact of scenario j on the CVaR.

2.2 A CVaR Open-Pit Design Model

We now assume that the overall decision to mine has been taken, and turn to the issue of designing an optimal pit in the presence of uncertainty. To this extent, we introduce the alternative loss function

$$
f^j(x) = - \sum_{i \in V_o} p^j \left(y_i^j - \mu_i \right) x_i, \tag{10}
$$

which has the interpretation of loss incurred by scenario j relative to the expected revenue. This gives rise to the CVaR open-pit design model

$$
\begin{aligned}
\text{maximize} \quad & \sum_{i \in V_o} (\pi \mu_i - c_i) x_i - \sum_{i \in V_w} c_i x_i \\
\text{subject to} \quad & \\
& \zeta + v \sum_{j=1}^{J} z_j \le Q \\
& z_j \ge - \sum_{i \in V_o} p^j \left(y_i^j - \mu_i \right) x_i - \zeta, \qquad j = 1, \ldots, J \\
& x_i \le x_j, \qquad (i,j) \in A \\
& x_i \in \{0,1\}, \qquad i \in V \\
& z_j \ge 0, \qquad j = 1, \ldots, J,
\end{aligned} \tag{11}
$$

which amounts to maximizing variable profit while imposing a limit on risk. To determine the number of scenarios that will be enough to include in Problem (11) in order to find a pit design that is likely to be the one that is indeed optimal, the following formulation will be used.

$$
\begin{aligned}
&\text{minimize} \quad \zeta + v \sum_{j=1}^{J} z_j \\
&\text{subject to} \\
&\quad \sum_{i \in V_o} (\pi \mu_i - c_i) x_i - \sum_{i \in V_w} c_i x_i \geq G \\
&\quad z_j \geq - \sum_{i \in V_o} p^j \left(y_i^j - \mu_i \right) x_i - \zeta, \qquad j = 1, \ldots, J \\
&\quad x_i \leq x_j, \qquad (i,j) \in A \\
&\quad x_i \in \{0,1\}, \qquad i \in V \\
&\quad z_j \geq 0, \qquad j = 1, \ldots, J
\end{aligned}
\tag{12}
$$

We can alternatively use a weighted objective, where we maximize the expected variable profit and at the same time minimize the CVaR, that is,

$$
\begin{aligned}
&\text{maximize} \quad \lambda \left(\sum_{i \in V_o} (\pi \mu_i - c_i) x_i - \sum_{i \in V_w} c_i x_i \right) - (1-\lambda) \left(\zeta + v \sum_{j=1}^{J} z_j \right) \\
&\text{subject to} \\
&\quad z_j \geq - \sum_{i \in V_o} p^j \left(y_i^j - \mu_i \right) x_i - \zeta, \qquad j = 1, \ldots, J \\
&\quad x_i \leq x_j, \qquad (i,j) \in A \\
&\quad x_i \in \{0,1\}, \qquad i \in V \\
&\quad z_j \geq 0, \qquad j = 1, \ldots, J,
\end{aligned}
\tag{13}
$$

where $0 < \lambda < 1$. By varying the value of this parameter, we can then study the trade off between the two objectives: maximizing the expected variable profit and minimizing the CVaR.

Suppose we solve Problem (13), for some value of λ, and get an optimal solution $x(\lambda)$. One can then calculate the corresponding expected variable profit and CVaR, which we refer to as $G(\lambda)$ and $Q(\lambda)$, respectively. By observing that the objective function in (13) can alternatively be stated as

$$
\text{maximize} \quad \left(\sum_{i \in V_o} (\pi \mu_i - c_i) x_i - \sum_{i \in V_w} c_i x_i \right) - \frac{1-\lambda}{\lambda} \left(\zeta + v \sum_{j=1}^{J} z_j \right) \tag{14}
$$

or

$$\text{minimize} \quad \zeta + v\sum_{j=1}^{J} z_j - \frac{\lambda}{1-\lambda}\left(\sum_{i\in V_o} (\pi\mu_i - c_i)x_i - \sum_{i\in V_w} c_i x_i\right), \tag{15}$$

and interpreting $-(1-\lambda)/\lambda$ and $-\lambda/(1-\lambda)$ as Lagrangian multipliers for the CVaR constraint in (11) and the expected profit constraint in (12), respectively, it follows from Everett's theorem (e.g., [9, p. 402]) that $x(\lambda)$ solves Problem (11) with $Q = Q(\lambda)$, and that $x(\lambda)$ solves Problem (12) with $G = G(\lambda)$.

2.3 *Nested Pit Contours Based on the CVaR Concept*

We finally turn to the problem of creating nested pit contours, in a way similar to that proposed by Lerchs and Grossmann [10]. The importance of these nested contours lies in that they provide an approximate mining schedule.

In order to produce nested pit contours we create artificial costs by adding a parameter $\gamma \in R_+$ to the costs c_i, $i \in V$, in the objective of Problem (13). The resulting formulation is given below, where ε is a positive number and sufficiently small. An increase in γ will increase the total cost and by so doing less blocks will be mined. The parameter γ can be viewed as a penalty, as in Lerchs and Grossmann [10]. The reason for having the second term in the objective function is to make the value of the CVaR well defined. For a given confidence level, α, we can then determine the blocks to be mined for a range of values of γ and obtain the expected profit and the CVaR for this range.

$$\begin{aligned}
&\text{maximize} \quad \left(\sum_{i\in V_o} [\pi\mu_i - (c_i+\gamma)]x_i - \sum_{i\in V_w} c_i x_i\right) - \varepsilon\left(\zeta + v\sum_{j=1}^{J} z_j\right)\\
&\text{subject to}\\
&\qquad z_j \geq -\sum_{i\in V_o} p^j\left(y_i^j - \mu_i\right)x_i - \zeta, \qquad j = 1,\ldots,J \qquad\qquad (16)\\
&\qquad x_i \leq x_j, \qquad (i,j) \in A\\
&\qquad x_i \in \{0,1\}, \qquad i \in V\\
&\qquad z_j \geq 0, \qquad j = 1,\ldots,J
\end{aligned}$$

3 Numerical Study

This section is devoted to numerical study of the models. The numerical results were obtained using the AMPL modelling language and the CPLEX solver, with all the scenarios being generated from MATLAB. The computational times for solving the models range between a second and 30 s, with the observation that the time

0	0	0	0	12	16	16	4	0	0	0	0	0	0	0	0	0	0
71	72	73	74	75	76	77	78	79	80	81	82	83	84	85	86	87	88
	0	0	0	4	16	16	12	0	0	0	0	0	0	0	0	0	
	55	56	57	58	59	60	61	62	63	64	65	66	67	68	69	70	
		0	0	0	12	16	16	4	0	0	0	0	0	0	0		
		41	42	43	44	45	46	47	48	49	50	51	52	53	54		
			0	0	4	16	16	12	0	0	0	0	0	0			
			29	30	31	32	33	34	35	36	37	38	39	40			
				0	0	12	16	16	4	0	0	0	0				
				19	20	21	22	23	24	25	26	27	28				
					0	4	16	16	12	0	0	0					
					11	12	13	14	15	16	17	18					
						0	12	16	16	4	0						
						5	6	7	8	9	10						
							4	16	16	12							
							1	2	3	4							

Fig. 1 A 2-D pit model

0	0	0	0	12	16	16	4	0	0	0	0	0	0	0	0	0	0
71	72	73	74	75	76	77	78	79	80	81	82	83	84	85	86	87	88
	0	0	0	4	16	16	12	0	0	0	0	0	0	0	0	0	
	55	56	57	58	59	60	61	62	63	64	65	66	67	68	69	70	
		0	0	0	12	16	16	4	0	0	0	0	0	0	0		
		41	42	43	44	45	46	47	48	49	50	51	52	53	54		
			0	0	4	16	16	12	0	0	0	0	0	0			
			29	30	31	32	33	34	35	36	37	38	39	40			
				0	0	12	16	16	4	0	0	0	0				
				19	20	21	22	23	24	25	26	27	28				
					0	4	16	16	12	0	0	0					
					11	12	13	14	15	16	17	18					
						0	12	16	16	4	0						
						5	6	7	8	9	10						
							4	16	16	12							
							1	2	3	4							

Fig. 2 Optimal pit for the 2-D pit model in Fig. 1

decreases with an increase in the value of the confidence level, α, and increases with an increase in the number of scenarios, J.

3.1 Test Problem

We have carried out the numerical study using the 2-D pit with 88 blocks shown in Fig. 1. Each cell of the pit represents a block with a given expected revenue, the value on top, while the number at the bottom is the block index. This pit model is the same as that of Lerchs and Grossmann [10], with the interpretation that the cost for mining and processing each block is 4 (i.e., $c_i = 4$, for all $i \in V$). The optimal contour for this pit is shown in Fig. 2, with a maximum profit of 108.

The numerical study is based on simulations of geological and price uncertainties, obtained from randomly generated scenarios. For creating a scenario, j, of geological uncertainty, the metal content of each ore block $i \in V_o$, y_i^j, is drawn from a uniform distribution over the range of $\pm 10\%$ of the expected metal content, μ_i. (Since the ore block revenues in Fig. 1 correspond to metal contents times a metal price, the geological scenarios reduce to making independent perturbations of the ore block revenues with at most $\pm 10\%$.) The waste blocks, V_w, were excluded because each of these blocks has zero expected revenue and, moreover, such blocks are not processed since they are not expected to contain any valuable ore. For the case of price uncertainty, a natural assumption is to use the log-normal distribution to generate scenarios for the metal price P. Therefore, each price scenario, p^j, is drawn from a log-normal distribution, with expectation one and variance 0.002. (Each price scenario clearly reduces to multiplying each of the block revenues in Fig. 1 with a common log-normally distributed random number.)

We will in the numerical study consider geological and price uncertainty separately, in order to investigate if they have different characteristics and because they together span any kind of combined uncertainty.

3.2 Number of Scenarios Needed

Problem (8), with $x_o = 1$, was solved with 9,000 scenarios of the simulated geological uncertainty, for $K = 102$ and for $\alpha = 0.98$, 0.95, and 0.90. The result of CVaR against the number of scenarios is given in Fig. 3. For all numbers of scenarios, we shall here mine all the 36 blocks shown in Fig. 2. It is realized that we need at least 2,000 scenarios to guarantee a stable result for this instance of data. The same problem was solved using 9,000 scenarios of the simulated price uncertainty, for $K = 100$. A similar graph shown in Fig. 4 gives an indication that we need at least 1,000 scenarios to ensure stability for such data. It was observed in this case that for any number of scenarios, when $\alpha = 0.90$ we shall mine all the 36 blocks, while we shall mine only 25 blocks when $\alpha = 0.98$. However, the situation was different when $\alpha = 0.95$. For this case, all the 36 blocks, as above, are mined when the number of scenarios $J < 500$, while 25 blocks are mined when $J \geq 500$. Figure 5 depicts the pit with 25 blocks being mined. The total expected profit for this pit (including the investment cost plus required return on investment, K) is 4.

To find the number of scenarios that will be enough to give a reliable result in Problem (11), we solved Problem (12) with 9,000 scenarios each, of the simulated geological uncertainty and the simulated price uncertainty, for $K = 100$ and for $\alpha = 0.98$, 0.95, and 0.90. About 1,000 scenarios will be enough in both cases, as can be seen in Figs. 6 and 7. For this choice of K, only 25 blocks are mined in both cases for all numbers of scenarios, except that for some values of $J < 100$, all the 36 blocks are mined when the simulated geological uncertainty was used.

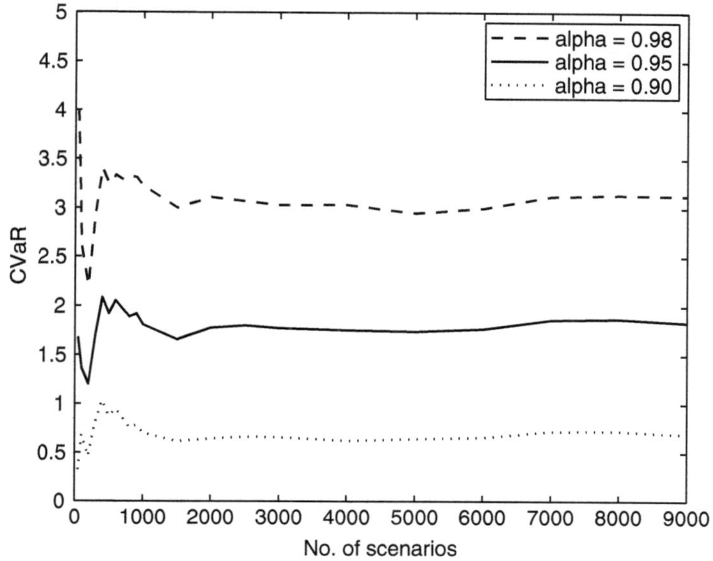

Fig. 3 CVaR against the number of scenarios of the simulated geological uncertainty, for $K = 102$

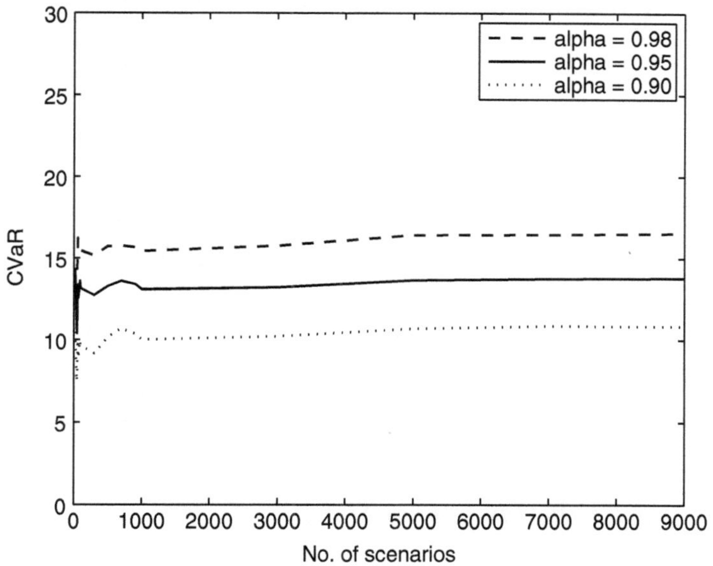

Fig. 4 CVaR against the number of scenarios of the simulated price uncertainty, for $K = 100$

0 71	0 72	0 73	0 74	12 75	16 76	16 77	4 78	0 79	0 80	0 81	0 82	0 83	0 84	0 85	0 86	0 87	0 88
	0 55	0 56	0 57	4 58	16 59	16 60	12 61	0 62	0 63	0 64	0 65	0 66	0 67	0 68	0 69	0 70	
		0 41	0 42	0 43	12 44	16 45	16 46	4 47	0 48	0 49	0 50	0 51	0 52	0 53	0 54		
			0 29	0 30	4 31	16 32	16 33	12 34	0 35	0 36	0 37	0 38	0 39	0 40			
				0 19	0 20	12 21	16 22	16 23	4 24	0 25	0 26	0 27	0 28				
					0 11	4 12	16 13	16 14	12 15	0 16	0 17	0 18					
						0 5	12 6	16 7	16 8	4 9	0 10						
							4 1	16 2	16 3	12 4							

Fig. 5 A pit of 25 blocks being mined

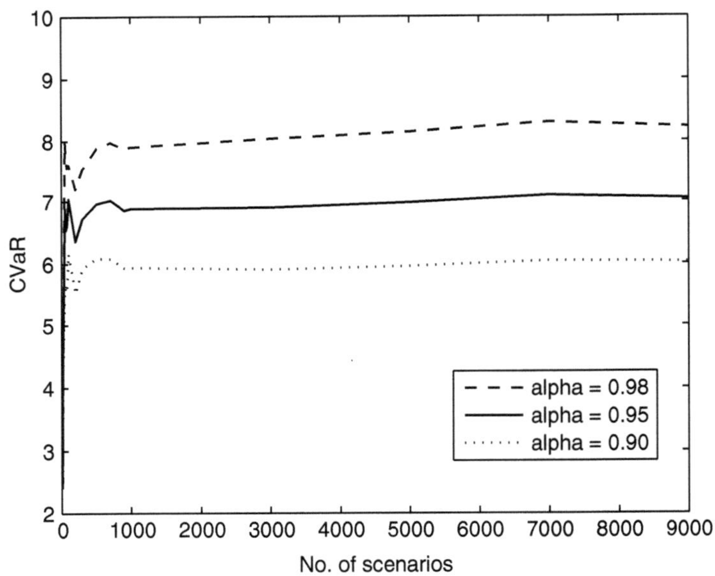

Fig. 6 CVaR against the number of scenarios of the simulated geological uncertainty, for $K = 100$

3.3 *Results for the Investment Model*

By using 3,000 scenarios of the simulated geological uncertainty in Problem (8), the breakpoints of K were found for different values of α. A breakpoint here refers to the transition between the cases $x_o^* = 0$ and $x_o^* = 1$, that is, between no mining and mining. The result is given in Fig. 8. When the value of K is below or exactly on the curve we shall mine all the 36 blocks, while no block is mined when the K value is above the curve. The result shows that a higher value of α implies that one will be avoiding risk, as expected.

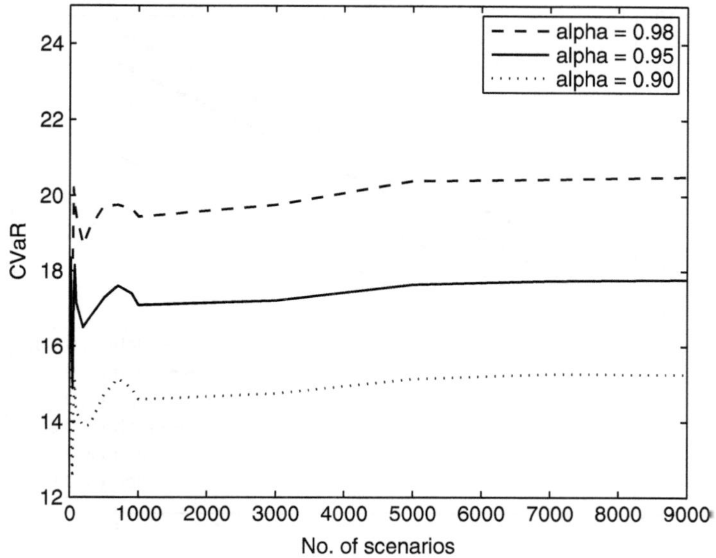

Fig. 7 CVaR against the number of scenarios of the simulated price uncertainty, for $K = 100$

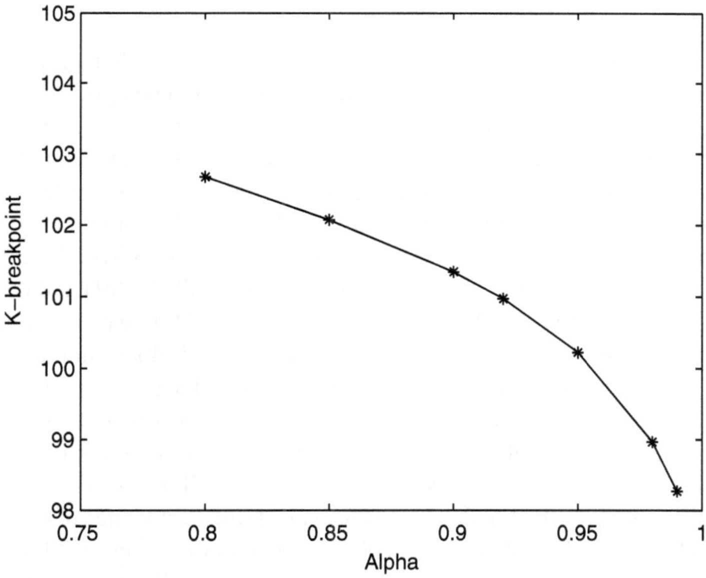

Fig. 8 Breakpoints of K against α

The same number of scenarios of the simulated geological uncertainty was then used in Problem (9) to find the breakpoints of K, while varying Q for a fixed α. Figure 9 is the outcome for $\alpha = 0.99$, 0.98, 0.95, 0.90, 0.85, and 0.80. The linear relations are a consequence of the linearity of Problem (9). For each value of α,

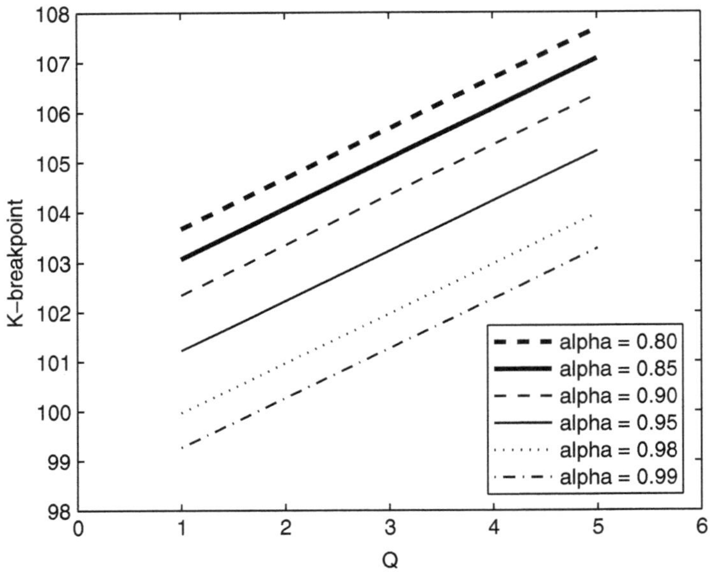

Fig. 9 Breakpoints of K against Q for 3,000 scenarios of geological uncertainty

a breakpoint indicates the value of K at which we shall either mine all the 36 blocks or mine nothing. Using 2,000 scenarios of the simulated price uncertainty in Problem (9) and varying Q in a similar manner, for $\alpha = 0.99$, 0.95, 0.90, and 0.80, gives the result shown in Fig. 10. It should be noted that there are actually six lines in this figure. The situation here is rather interesting, in terms of the breakpoints and the number of blocks mined, for a given value of α. When $\alpha = 0.80$ and 0.90 there is a single breakpoint for a given Q and we shall mine either all the 36 blocks or mine nothing. When $\alpha = 0.99$, we have two cases called 0.99a and 0.99b, with two different respective breakpoints of K. In the case 0.99a we shall mine either 25 blocks or nothing, while we shall mine either 36 or 25 blocks in the case 0.99b. When $\alpha = 0.95$ the situation was found to be similar to that of $\alpha = 0.99$, except that the two lines are very close in the figure. This shows how sensitive the situation is for this value of α. Here also, we shall mine either 25 blocks or nothing in the case 0.95a, while in the case 0.95b either 36 or 25 blocks shall be mined.

In Fig. 11, we give the breakpoints of K against different values of α, for a fixed value of Q. The result is obtained by solving Problem (9) using 3,000 scenarios of the simulated geological uncertainty. In all cases, we shall mine either all the 36 blocks or mine nothing. When a value of K is below or on a curve then 36 blocks are mined while no block is mined when the value of K is above a curve, for a given Q. Using 2,000 scenarios of the simulated price uncertainty, Problem (9) is then solved with different values of α and a fixed value of Q. The result, as shown in Fig. 12, gives different curves for each of the values of Q. Further, each of these curves ends with two different curves (a lower and an upper curve) for $\alpha \geq 0.95$.

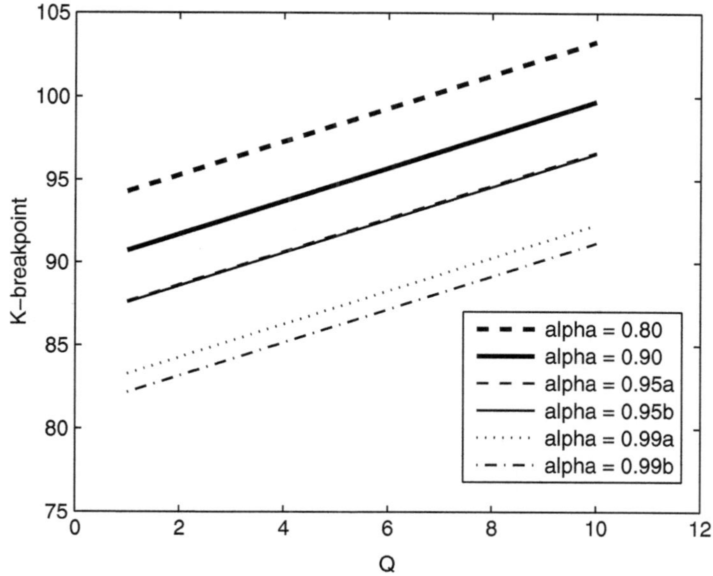

Fig. 10 Breakpoints of K against Q for 2,000 scenarios of price uncertainty

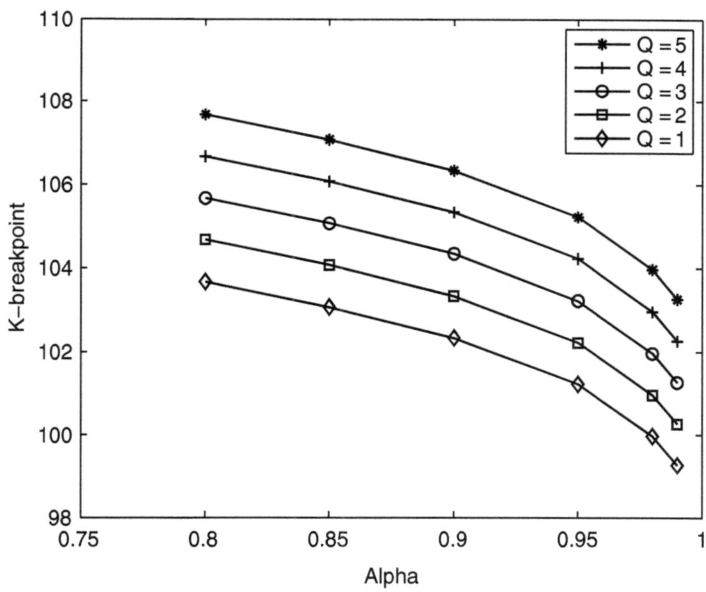

Fig. 11 Breakpoints of K against α for 3,000 scenarios of geological uncertainty

For $0.80 \leq \alpha < 0.95$ we have a single curve and we shall mine either 36 blocks or mine nothing, depending, respectively, on whether K is below or on the curve or K is strictly above the curve. For each of the two curves for $\alpha \geq 0.95$, when K is below or on each of the lower curves (labelled with the letter b in the legend), we shall

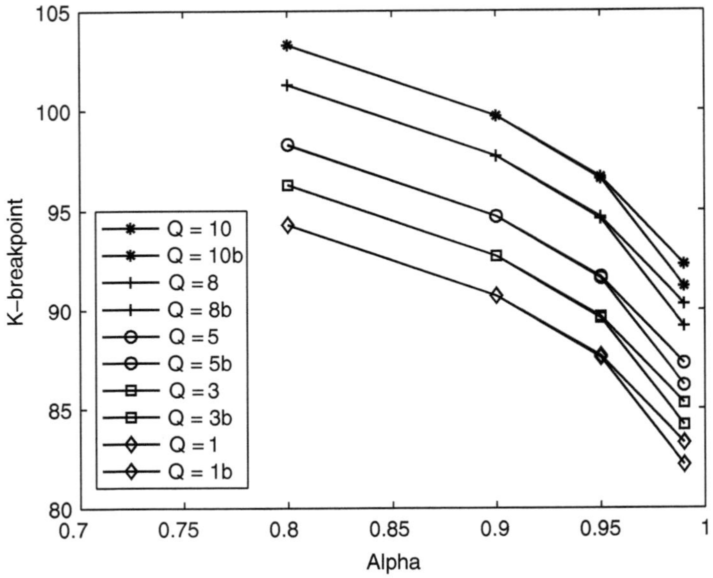

Fig. 12 Breakpoints of K against α for 2,000 scenarios of price uncertainty

mine all the 36 blocks, while we shall mine 25 blocks when K is above the lower curve, and for each of the upper curves we shall mine either 25 blocks or nothing, depending, respectively, on whether K is below or on the curve or K is strictly above the curve. These observations are clearly consistent with the behavior illustrated in Fig. 10.

3.4 Results for the Design Model

The 3,000 scenarios of the simulated geological uncertainty are used in Problem (11) for different values of α and varying Q. The results are given in Fig. 13. The same experiment was made for different α values for the 2,000 scenarios of the simulated price uncertainty and the results are shown in Fig. 14.

Figures 15 and 16 show Pareto optimal solutions of Problem (13) for different values of α. By using the 3,000 scenarios of the simulated geological uncertainty and $0.043 \leq \lambda \leq 0.2101$, we obtained the results in Fig. 15. Figure 16 is the results from the same problem by using the 2,000 scenarios of the simulated price uncertainty. In this case, $0.075 \leq \lambda \leq 0.6$. For the expected variable profits of 24, 44, 56, 72, 92, 104, and 108, the number of blocks mined is, respectively, 2, 4, 6, 9, 16, 25, and 36. As can be expected, the conditional value-at-risk depends on the confidence level—the higher the confidence level, the higher the conditional value-at-risk. In other words, the expected variable profit is higher for a lower confidence level, as one is then more willing to accept that a loss might occur.

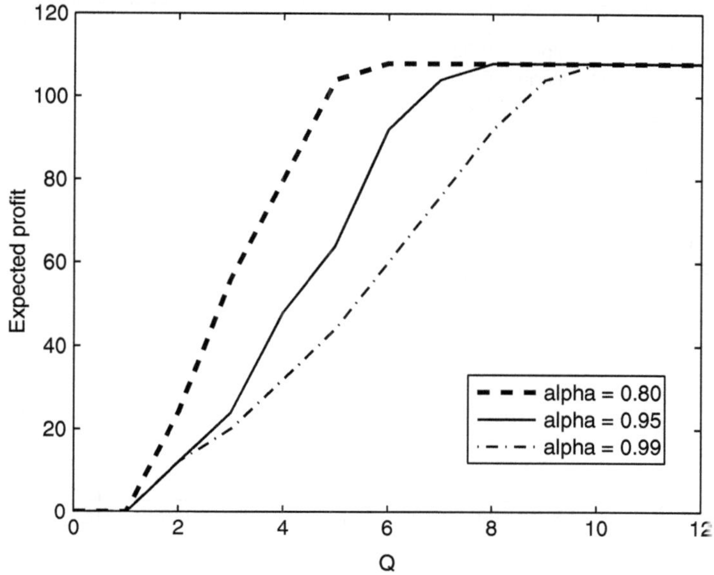

Fig. 13 Expected variable profit against Q for 3,000 scenarios of geological uncertainty

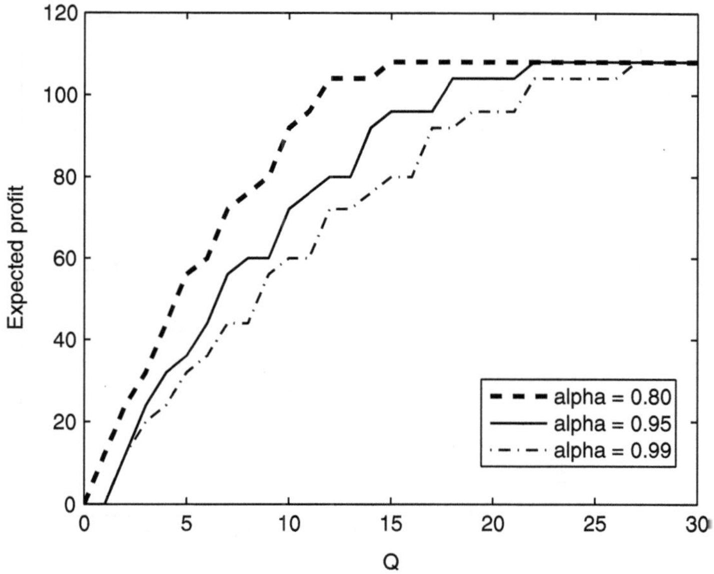

Fig. 14 Expected variable profit against Q for 2,000 scenarios of price uncertainty

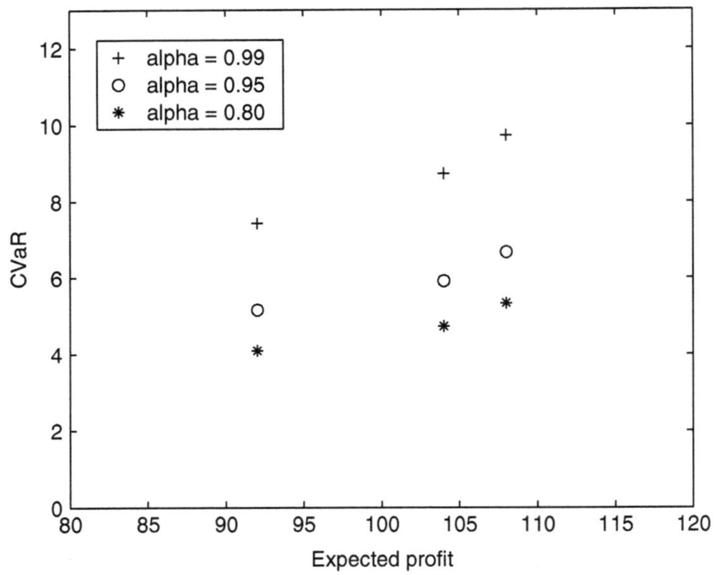

Fig. 15 CVaR against expected variable profit for 3,000 scenarios of geological uncertainty

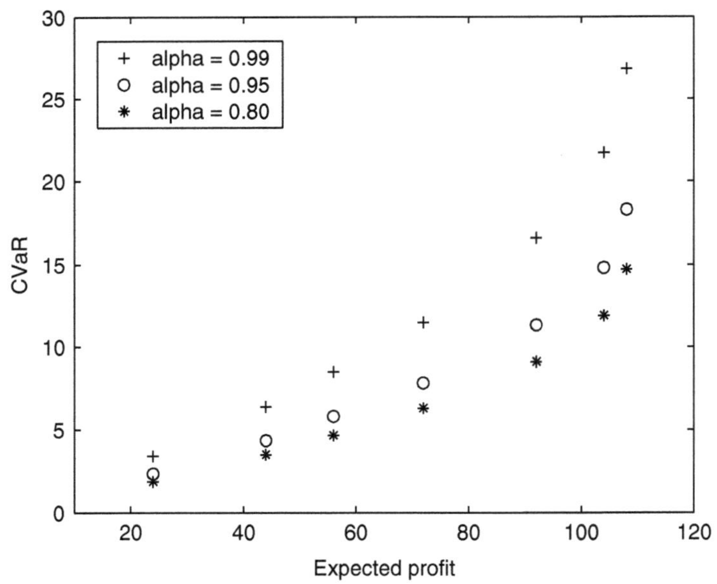

Fig. 16 CVaR against expected variable profit for 2,000 scenarios of price uncertainty

Table 1 Blocks mined for pit contours and corresponding expected variable profits

Contour number	Number of blocks mined	Additional blocks mined	Cumulative expected variable profit
0	0		0
1	2	77,76	24
2	4	75,59	44
3	9	78,79,60,61,45	72
4	16	80,74,62,58,46,44,32	92
5	25	81,82,63,64,47,48,33,34,22	104
6	36	83,73,65,57,49,43,35,31,23,21,13	108

Table 2 Interval of γ for the contours and corresponding CVaR for different values of α for 3,000 scenarios of the simulated geological uncertainty

	$\alpha = 0.99$		$\alpha = 0.95$		$\alpha = 0.80$	
Contour #	Interval of γ	CVaR	Interval of γ	CVaR	Interval of γ	CVaR
0	>11.980	0.00	>11.980	0.00	>11.990	0.00
1	9.993–11.980	3.20	9.996–11.980	2.81	9.998–11.990	1.90
2	6.996–9.993	4.72	6.997–9.996	3.70	6.998–9.998	2.50
3	4.998–6.996	6.59	4.998–6.997	5.08	4.999–6.998	3.39
4	2.997–4.998	7.43	2.998–4.998	5.97	2.999–4.999	4.09
5	0.998–2.997	8.72	0.998–2.998	6.90	0.999–2.999	4.72
6	0.000–0.998	9.73	0.000–0.998	7.78	0.000–0.999	5.33

Table 3 Interval of γ for the contours and corresponding CVaR for different values of α for 3,000 scenarios of the simulated price uncertainty

	$\alpha = 0.99$		$\alpha = 0.95$		$\alpha = 0.80$	
Contour #	Interval of γ	CVaR	Interval of γ	CVaR	Interval of γ	CVaR
0	>11.983	0.00	>11.986	0.00	>11.990	0.00
1	9.986–11.983	3.37	9.986–11.986	2.70	9.992–11.990	1.88
2	6.988–9.986	6.32	6.990–9.986	5.06	6.993–9.992	3.52
3	4.988–6.988	11.37	4.990–6.990	9.11	4.993–6.993	6.34
4	2.988–4.988	16.43	2.990–4.990	13.16	2.993–4.993	9.15
5	0.988–2.988	21.49	0.989–2.990	17.21	0.993–2.993	11.97
6	0.000–0.988	26.54	0.000–0.989	21.25	0.000–0.993	14.78

3.5 *Results for the Nested Pit Contours*

By varying the parameter γ, using a fixed value of $\varepsilon = 0.01$ and 3,000 scenarios each, of the simulated geological uncertainty and the simulated price uncertainty, Problem (16) is solved for different values of the confidence level α. In Table 1, we present the order in which the pit contours are generated, as well as the expected variable profit for each contour. Tables 2 and 3 show, respectively, the trend for the CVaR and the intervals of γ for corresponding contours.

The results from these tables enable the decision maker to know the order in which the blocks are to be mined, and they also show how the expected variable profit and CVaR change for this mining schedule.

4 Conclusion

We have presented optimization models for open-pit mining planning where uncertainty with respect to both geology and future market prices can be taken into account by means of the conditional value-at-risk measure. This measure was first applied in the field of financial planning, but has a rather natural and immediate application also to the planning situation under consideration here. The models presented comprise three stages of the mining planning process: the investment decision, the pit design, and the mining sequence.

The models have been verified through numerical experiments on a small-scale example problem for which uncertainty is simulated by means of randomly generated scenarios. The overall characteristics of the results are consistent with what could be expected, considering the properties of the optimization problems studied and the aims of the models. A somewhat unexpected result is that as much as thousands of scenarios are needed to simulate the uncertainty, in order to obtain reliable solutions to the models, even though the example problem is small scale.

An obvious direction for continued evaluation of the presented models and research would be applications to realistic instances of open-pit mining. Due to the large scales of such instances and the large numbers of scenarios needed to describe the uncertainty, this will most certainly necessitate the development of specialized solution methods, by exploiting the very special structures of our models.

References

1. Abdel Sabour, S.A., Dimitrakopoulos, R.G., Kumral, M.: Mine design selection under uncertainty. Mining Tech. **117**, 53–64 (2008)
2. Andersson, F., Mausser, H., Rosen, D., Uryasev, S.: Credit risk optimization with conditional value-at-risk criterion. Math. Program. Ser. B, **89**, 273–291 (2001)
3. Carneiro, M.C., Ribas, G.P., Hamacher, S.: Risk management in the oil supply chain: a CVaR approach. Ind. Eng. Chem. Res. **49**, 3286–3294 (2010)
4. Dimitrakopoulos, R., Farrelly, C., Godoy, M.C.: Moving forward from traditional optimisation: grade uncertainty and risk effects in open pit mine design. Trans. Inst. Min. Metall. Section A, Min. Ind. **111**, A82–A89 (2002)
5. Ferland, J.A., Amaya, J., Djuimo, M.S.: Application of a particle swarm algorithm to the capacitated open pit mining problem. In: Sen Gupta, G. (ed.) Autonomous Robots and Agents. Studies in Computational Intelligence, vol. 76, pp. 127–133. Springer, Berlin (2007)
6. Hochbaum, D., Chen, A.: Performance analysis and best implementations of old and new algorithms for the open-pit mining problem. Oper. Res. **48**, 894–914 (2000)
7. Krokhmal, P., Palmquist, J., Uryasev, S.: Portfolio optimization with conditional value-at-risk objective and constraints. J. Risk **4**, 43–68 (2002)
8. Lai, F.J., Bamford, W.E., Yuen, S.T.S., Li, T.: Implementing value at risk in slope risk evaluation. In: Proceedings of the International Symposium on Rock Slope Stability in Open Pit Mining and Civil Engineering, Santiago, Chile, pp. 1–10 (2009)
9. Lasdon, L.S.: Optimization Theory for Large Systems. Dover, New York (2002)
10. Lerchs, H., Grossmann, I.F.: Optimum design of open-pit mines. Trans. Can. Inst. Min. Metall. **LXVIII**, 17–24 (1965)

11. Lim, C., Sherali, H. D., Uryasev, S.: Portfolio optimization by minimizing conditional value-at-risk via nondifferentiable optimization. Comput. Optim. Appl. **46**, 391–415 (2010)
12. Mansini, R., Ogryczak, W., Speranza, M. G.: Conditional value at risk and related linear programming models for portfolio optimization. Ann. Oper. Res. **152**, 227–256 (2007)
13. Martinez, L.A.: Why accounting for uncertainty and risk can improve final decision-making in strategic open pit mine evaluation. In: Project Evaluation Conference, Melbourne, pp. 1–12 (2009)
14. Rockafellar, R.T., Uryasev, S.: Optimization of conditional value-at-risk. J. Risk **2**, 21–41 (2000)
15. Rockafellar, R.T., Uryasev, S.: Conditional value-at-risk for general loss distributions. J. Bank. Finance **26**, 1443–1471 (2002)
16. West, J.: Decreasing metal ore grades. J. Ind. Ecol. **15**, 165–168 (2011)

Incidence Graphs of Bipartite *G*-Graphs

Cerasela Tanasescu, Ruxandra Marinescu-Ghemeci, and Alain Bretto

Abstract Defining graphs from groups is a widely studied area motivated, for example, by communication networks. The most popular graphs defined by a group are Cayley graphs. G-graphs correspond to an alternative construction. After recalling the main properties of these graphs and their motivation, we propose a characterization result. With the help of this result, we show that the incidence graph of a symmetric bipartite G-graph is also a G-graph and we give a proof that, with some constraints, if the incidence graph of a symmetric bipartite graph is G-graph, the graph is also a G-graph. Using these results, we give an alternative proof for the fact that mesh of d-ary trees are G-graphs.

1 Introduction

A recent studied family of graphs constructed from groups are G-graphs. These graphs, introduced in [2], have also, like Cayley graphs, highly regular properties. In particular, because the algorithm for constructing the G-graphs is simple, it appears as a useful tool to construct new symmetric and semisymmetric graphs [3]. Moreover, thanks to G-graphs, some upper bounds in the cage graphs problem were improved ([4], see also [7]). One interesting direction is studying the G-graphs

C. Tanasescu (✉)
ESSEC Business School, 1 avenue Bernard Hirsch Cergy, France
e-mail: tanasescu@essec.edu

R. Marinescu-Ghemeci
University of Bucharest, Str. Academiei, 14, Bucharest, Romania
e-mail: verman@fmi.unibuc.ro

A. Bretto
Université de Caen, GREYC CNRS UMR-6072, Campus II, Bd Marechal Juin BP 5186, 14032 Caen cedex, France
e-mail: alain.bretto@info.unicaen.fr

A. Migdalas et al. (eds.), *Optimization Theory, Decision Making, and Operations Research Applications*, Springer Proceedings in Mathematics & Statistics 31,
DOI 10.1007/978-1-4614-5134-1_9,

properties and providing characterization theorems for G-graphs. In the following sections, we propose such characterization and study the incidence graphs of G-graphs. Based on the results obtained we present an alternative proof of the fact that mesh of d-ary trees are G-graphs.

2 Definitions

2.1 Group Definitions

Let $(G,\cdot,e)$ be a finite group [8], where e denotes the identity element of G. If no ambiguity occurs we denote the group by G.

For simplicity we denote $g \cdot g = g^2$ and we define g^k with $g \in G$ by $g^k = g^{k-1} \cdot g$.

"$\cdot$" denotes a canonical operation for a group. The operation can be sometimes denoted differently, as in $(\mathbb{Z}/n\mathbb{Z},+)$. Also, if the operation between two elements g_1, g_2 of a group can be easily revealed by the context, we will write $g_1 g_2$ instead of $g_1 \cdot g_2$.

We will use notation $G' \leq G$ if G' is a subgroup of G.

Cyclic group. For every g in G we define the order of g, denoted by $o(g)$, as the smallest positive integer k such that $g^k = e$. The set $(g) = \{e, g, g^2, \ldots, g^{o(g)-1}\}$ with the corresponding operation forms a subgroup of G, called the *cyclic group* of g.

Independent elements. Let $g_1, g_2, \ldots, g_k$ be k elements in G. $g_1, g_2, \ldots, g_k$ are called *independent elements* in G if and only if, for any two distinct elements g_i, g_j, if $g_i^p = g_j^k$ for some positive integer p, k, then $g_i^p = g_j^k = e$.

Left and Right Cosets. If H is a fixed subgroup of a group G and $x \in G$, the subset $Hx = \{hx | h \in H\}$ is called *right coset of H containing x*. The key property of cosets is that, for any $x, y \in G$, $Hx = Hy$ or $Hx \cap Hy = \emptyset$. The cosets of H yield thus a partition of G and we can find a subset T of G such that, for any $x \in G$ there is exactly one element $t \in T$ for which $t \in Hx$, that is,

$$G = \bigcup_{t \in T} Ht.$$

Such set T is called a *right transversal* for H in G.

The notions of *left coset* and *left transversal* are defined analogously.

Semi-direct-product of groups. We define a *(left) group action* of G on a set X as a binary function from $G \times X$ to X, $(g,x) \to g \cdot x$ satisfying conditions: $e \cdot x = x$ for every $x \in X$ and $g \cdot \left(g' \cdot x\right) = \left(g \cdot g'\right) \cdot x$ for all $g, g' \in G$ and $x \in X$. The action is *transitive* if for every $x, y \in X$, there exists $g \in G$ such that $g \cdot x = y$.

For $x \in X$ we define the *stabilizer subgroup* of x as the set of all elements in G that let x invariant: $Stab_G x = \{g : g \cdot x = x\}$.

Let $(H,\cdot)$ and $(Q,\cdot)$ be two groups and $\varphi : Q \times H \longrightarrow H$ an action of Q on H. The *semi-direct product* of H and Q by φ, denoted $H \rtimes_\varphi Q$, is defined as the group with underlying set $H \times Q$ and the operation: $(h,q)\cdot(h',q') = \left(h\cdot\varphi(q,h')\ q\cdot q'\right)$.

2.2 Graphs Definitions

In this paper, graphs are undirected, simple, without loops. Specifically, we define an undirected graph $\Gamma = (V;E)$ by the vertex set V and the edge set E. For a vertex $x \in V$ we call the *neighborhood* of x, the set $N(x)$ of vertices in V adjacent to x.

Intersection graph. An *Intersection graph* is an undirected graph obtained from a family of sets S_i, $i = 0,1,2,\ldots$ by creating a vertex v_i for each set S_i, and connecting two distinct vertices by an edge whenever the corresponding two sets S_i and S_j have a nonempty intersection: $|S_i \cap S_j| \geq 1$.

Equitable partition. Given a graph $\Gamma = (V;E)$ and a partition of its vertex set $\pi = \bigcup_{1\leq i\leq r} C_i$, we say that π is an *equitable partition* if and only if for all $1 \leq i \neq j \leq r$ there exists b_{ij} such that $\forall x \in C_i$ we have $|N(x) \cap C_j| = b_{ij}$. The edges between C_i and C_j induce a semiregular bipartite graph. All vertices from C_i have the same degree, as well as vertices from C_j.

Graph isomorphism. Let $\Gamma_1 = (V_1;E_1)$ and $\Gamma_2 = (V_2;E_2)$ be two graphs. An *isomorphism* from Γ_1 to Γ_2 is a bijection $f : V_1 \longrightarrow V_2$ such that $xy \in E_1$ if and only if $f(x)f(y) \in E_2$.

An isomorphism from Γ to itself is called *automorphism* of Γ. The identity automorphism will be denoted by *id*.
$Aut(\Gamma)$ denotes the group of automorphisms of a graph Γ under composition law.

Orbit partition. For a graph $\Gamma = (V;E)$, let $H \leq Aut(\Gamma)$ be a subgroup of its automorphisms group. We define an equivalence relation on V regarding H as follows: for any u and v in V, u is in relation with v if and only if there exists $h \in H$ such as $h(u) = v$. An *orbit* is an equivalence class. The *orbit partition* of V regarding H is the partition of V associated whith this relation.

Incidence Graph. The *Incidence graph* of a simple graph $\Gamma = (V;E)$ is the graph $I\Gamma = (IV = V \cup E; IE)$ where $IE = \{ab : a = xy \in E, b = x \text{ or } b = y, b \in V\}$.

3 G-Graphs Characterization

3.1 G-Graphs Definition

Definition. Consider G a finite group, with e the neutral element. Let S be a nonempty subset of G. The *right G-graph*, $\Phi(G;S)$, is the intersection graph of the right cosets of the cyclic groups (s), $s \in S$.

By this definition, we construct $\Phi(G;S) = (V;E)$ in the following way:

1. The vertices of $\Phi(G;S)$ are $V = \bigcup_{s \in S} V_s$ where $V_s = \{(s)x : x \in T_s\}$ and T_s is a right transversal for (s) in G.
2. For $(s)x, (t)y \in V$, there exists an edge between $(s)x$ and $(t)y$ if and only if $|(s)x \cap (t)y| \geq 1$.

Remark. If s and t are independents, we have $|(s)x \cap (t)y| \leq 1$.

We denote $S = \{s_1, s_2, \ldots, s_k\}$. Then $V = \bigcup_{i=1}^{k} V_{s_i}$.
We assume S is a set of independent elements.

Lemma 1. *$\Phi(G;S)$ is a $|S|$-partite graph and this partition is an equitable partition. Every vertex from a class V_{s_i} of the partition has the degree $(|S|-1)o(s_i)$.*

Proof. $V_{s_1}, V_{s_2}, \ldots, V_{s_k}$ is a $\Phi(G;S)$ partition. For any $i \in \{1,2,\ldots,k\}$, we have $V_{s_i} = \{(s_i)x | x \in T_{s_i}\}$, where T_{s_i} is a right transversal for (s_i), and cosets $(s_i)x$, for $x \in T_{s_i}$ form a partition of G.

Let i, j be two distinct indices from $\{1,2,\ldots,k\}$.

Let a be an element of a coset $(s_i)x$, $x \in T_{s_i}$. Then there exists an unique $y \in T_{s_j}$ such that $a \in (s_j)y$. Hence each element of coset $(s_i)x$ induces an edge in $\Phi(G;S)$ from vertex $(s_i)x$ to a vertex from V_j. But coset $(s_i)x$ has $o(s_i)$ elements, hence in the G-graph we have $|N((s_i)x) \cap V_j| = o(s_i)$. It follows that every two classes V_i and V_j induce a semiregular graph, hence $V = \bigcup_{i=1}^{k} V_{s_i}$ is an equitable partition. We immediately deduce that the degree of the vertex $(s_i)x$ is $o(s_i)(k-1)$. □

Remark. If $S = \{s_1, s_2\}$ and $o(s_1) = o(s_2) = p$, the G-graph $\Phi(G;S)$ is called *symmetric bipartite*, since it is bipartite and p-regular (all the vertices have degree p).

Short list of classical graphs which are identified as *G*-graphs [5]

1. Bipartite complete graphs $K_{n,k}$ $(G = \mathbb{Z}_n \times \mathbb{Z}_k, S = \{(1,0),(0,1)\})$
2. The cuboctahedral graph $\left(G = (\mathbb{Z}_2)^3, S = \{(1,0,0),(0,1,0),(0,0,1)\}\right)$
3. The square (G is the Klein's group, $G =< \{e,a,b,ab\} | a^2 = b^2 = e, ab = ba >$ and $S = \{a,b\}$)
4. The Heawood's graph $(\langle a,b \mid a^7 = b^3 = e, ab = baa \rangle, S = \{b, ba\})$
5. The Pappus graph $\left(G = \langle a,b,c \mid a^3 = b^3 = c^3 = e, ab = ba, ac = ca, bc = cba \rangle,\right.$ $\left.S = \{b,c\}\right)$

3.2 G-Graph Characterization

Lemma 2. *Let $\Phi_1(G;S) = (V_1;E_1)$ and $\Phi_2(G;S) = (V_2;E_2)$ be the right and left G-graphs of G. These two graphs are isomorphic.*

Proof. For every $s \in S$ the sets of left and right cosets of the cyclic group (s) have the same cardinality. If T_s is a left transversal of (s) in G, then $T_s^{-1} = \{t^{-1}, t \in T_s\}$ is a right transversal of (s) in G.

Consider the application $f : V_1 \to V_2$ such that $f((s)x) = x^{-1}(s)$. It is easy to see that f is well defined, since if $(s)x = (s)y$, then $x^{-1}(s) = y^{-1}(s)$ and f is a bijection. Moreover, there exists an edge between $(s_i)x$ and $(s_j)y$ if and only if there exist α and β such as $s_i^{\alpha}x = s_j^{\beta}y$ or, equivalent, $(s_i^{\alpha}x)^{-1} = (s_j^{\beta}y)^{-1}$, hence $x^{-1}s_i^{o(s_i)-\alpha} = y^{-1}s_j^{o(s_j)-\beta}$. But this relation holds if and only if there exists an edge between the vertices $x^{-1}(s_i)$ and $y^{-1}(s_j)$. It follows that $(s_i)x(s_j)y \in E_1$ if and only if $x^{-1}(s_i)y^{-1}(s_j) \in E_2$ hence f is an isomorphism. □

Let $\Phi(G;S) = (V;E)$ be a G-graph. For any $g \in G$ we can associate the morphism $\delta_g : V \longrightarrow V$, defined by $\delta_g((s)x) = (s)xg$. Using these notions we have the following Theorem.

Theorem 1. *Let $\Phi(G;S) = (V;E)$ be a G-graph.*

(1) δ_g is an automorphism of $\Phi(G;S)$ and $\delta_g(V_s) = V_s$ for every $s \in S$.
(2) $\delta_G = \{\delta_g, g \in G\}$ form a group under the composition law, and V_{s_i} for any fixed s_i is an orbit regarding δ_G.
(3) Settle s_i and s_j in S. For every $x,u \in V_{s_i}$, $y,v \in V_{s_j}$ with $xy, uv \in E$, there exists $g \in G$ such that $\delta_g(x) = u$ and $\delta_g(y) = v$.
(4) For every $(s)x \in V$ with $s \in S$, $Stab_{\delta_G}(s)x$ is a cyclic group of order $o(s)$, generated by $\delta_{x^{-1}sx}$.

Proof. (1) We will prove that δ_g is an automorphism of $\Phi(G;S)$.

- We have

$$\delta_g(V_s) = \{\delta_g((s)x) | x \in G\} = \{(s)xg | x \in G\} = \{(s)x | x \in G\} = V_s,$$

  since $G = \{x | x \in G\} = \{xg | x \in G\}$.
  It follows that $\delta_g(V_s) = V_s$ for every $s \in S$, hence δ_g is a bijection.
- δ_g is a morphism. Indeed, we have $(s)x(t)y \in E$ if and only if there exist i and j such as $s^i x = t^j y$, which is equivalent to $s^i xg = t^j yg$. Hence $(s)x(t)y$ is in E if and only if $(s)xg(t)yg$ is in E.

  It follows that δ_g is in $Aut\,(\Phi(G;S))$, for every $g \in G$.

(2) $\delta_G = \{\delta_g, g \in G\}$ forms a group under the composition law, since we have:

- $\delta_{g_1} \circ \delta_{g_2} = \delta_{g_1 g_2}, \forall g_1, g_2 \in G$
- $\delta_g \circ \delta_e = \delta_e \circ \delta_g = \delta_g, \forall g \in G$
- $(\delta_{g_1} \circ \delta_{g_2}) \circ \delta_{g_3} = \delta_{(g_1 g_2) g_3} = \delta_{g_1 (g_2 g_3)} = \delta_{g_1} \circ (\delta_{g_2} \circ \delta_{g_3}), \forall g_1, g_2, g_3 \in G$
- $\delta_g \circ \delta_{g^{-1}} = e$

At (1) we proved that $\delta_g(V_s) = V_s$ for every $s \in S$ and $g \in G$, hence V_s is an orbit regarding the group δ_G.

(3) Let $x = (s_i)a$ and $u = (s_i)b$. Since $xy \in E$, it follows that there exists q such that $y = (s_j)s_i^q a$. Similarly, since $uv \in E$, there exists k such that $v = (s_j)s_i^k b$. Consider now $g = a^{-1}s_i^{k-q}b \in G$. We have:

$$\delta_g(y) = (s_j)s_i^q ag = (s_j)s_i^k b = v$$

$$\delta_g(x) = (s_i)ag = (s_i)s_i^{k-q} b = (s_i)b = u.$$

(4) From definition, for every $(s)x \in V$ with $s \in S$,

$$Stab_{\delta_G}(s)x = \{\delta_g, \delta_g((s)x) = (s)x\}.$$

We have $\delta_g \in Stab_{\delta_G}(s)x$ if and only if $\delta_g((s)x) = (s)xg = (s)x$. Thus $\delta_g \in Stab_{\delta_G}(s)x$ if and only if there exist q,r such that $s^q xg = s^r x$, hence $g = x^{-1}s^{r-q}x$. It follows that $Stab_{\delta_G}(s)x = \{\delta_{x^{-1}s^p x}, p = 1,\ldots,o(s)\}$. Then $Stab_{\delta_G}(s)x$ is generated by $\delta_{x^{-1}sx}$, hence it is a cyclic group of order $o(s)$. □

Theorem 2. ***Characterization of a G-graph.*** *Let $\Gamma = (V;E)$ be a graph. Then Γ is a G-graph $\Phi(G;S)$, with S an independent set of elements, if and only if there exists H a subgroup of $Aut(\Gamma)$ such that, by denoting $V_1,\ldots,V_k$ the orbit partition of V regarding H, there exists a clique $x_1,\ldots,x_k$ in Γ, with $x_i \in V_i$, where*

(1) $Stab_H x_i$, $i \in \{1,\ldots,k\}$ are cyclic groups and $Stab_H x_i \cap Stab_H x_j = \{id\}$ for every $i \neq j$
(2) $|N(x_i) \cap V_j| = |Stab_H x_i|$, for every $i \neq j \in \{1,\ldots,k\}$.

Proof. "$\Longleftarrow$" Suppose there exist H and clique $\{x_1,\ldots,x_k\}$ satisfying the two properties. Denote by σ_i a generator for $Stab_H x_i$: $Stab_H x_i = (\sigma_i)$. Consider the G-graph $\Phi(H;\{\sigma_1,\ldots,\sigma_k\}) = (W;E)$. Since $Stab_H x_i \cap Stab_H x_j = \{id\}$ for every $i \neq j$, it follows that $\sigma_1,\ldots,\sigma_k$ are independent elements. We will prove that Γ is isomorphic to $\Phi(H;\{\sigma_1,\ldots,\sigma_k\})$.

We define the map $\varphi : W \longrightarrow V$, $\varphi((\sigma_i)a) = a^{-1}(x_i)$, for every $(\sigma_i)a \in W$.

φ is well defined since, if we have $(\sigma_i)a = (\sigma_i)b$, then there exists p such that $b = \sigma_i^p a$ or, equivalent, $b^{-1} = a^{-1}\sigma_i^{-p}$. Since $\sigma_i^{-p} \in Stab_H x_i$, we obtain $b^{-1}(x_i) = a^{-1}(x_i)$.

φ is surjective. Indeed, let $x \in V$. Then there exists $i \in \{1,\ldots,k\}$ such that $x \in V_i$. But, since V_i is an orbit of V regarding H and $x_i \in V_i$, there exists $a \in H$ such that $a(x_i) = x$. It follows that $x = \varphi((\sigma_i)a^{-1})$.

φ is injective: if $\varphi((\sigma_i)a) = \varphi((\sigma_j)b)$ then $a^{-1}(x_i) = b^{-1}(x_j)$. It follows that $i = j$, since $x_i \in V_i$ and $V_1,\ldots,V_k$ is an orbit partition regarding H. We obtain $a^{-1}(x_i) = b^{-1}(x_i)$, hence $ba^{-1} \in Stab_H x_i = (\sigma_i)$. Then there exists l such that $ba^{-1} = \sigma_i^l$ and thus $(\sigma_i)b = (\sigma_i)\sigma_i^l a = (\sigma_i)a$.

Suppose now that $(\sigma_i)a(\sigma_j)b \in F$. Then $i \neq j$ and there exist p,l such that $\sigma_i^p a = \sigma_j^l b$. It follows that $b = \sigma_j^{-l}\sigma_i^p a$, hence $b^{-1} = a^{-1}\sigma_i^{-p}\sigma_j^l$. But $x_i x_j \in E$ and $a^{-1}\sigma_i^{-p}$ is an automorphism, so $a^{-1}\sigma_i^{-p}(x_i)a^{-1}\sigma_i^{-p}(x_j) \in E$. Since $\sigma_j^l \in Stab_H x_j$ and $\sigma_i^{-p} \in Stab_H x_i$, $a^{-1}(x_i)a^{-1}\sigma_i^{-p}\sigma_j^l(x_j) \in E$, hence $a^{-1}(x_i)b^{-1}(x_j) \in E$.

Conversely, suppose $a^{-1}(x_i)b^{-1}(x_j) \in E$. From property (2) we have: $|N(x_i) \cap V_j| = |Stab_H x_i| = o(\sigma_i)$ and since σ_i and σ_j are independent, it follows that $N(x_i) \cap V_j = \{\sigma_i^l(x_j), l = 1,\ldots,o(\sigma_i)\}$ (and these elements are distinct). So there exists $l \in \{1,\ldots,o(\sigma_i)\}$ such that $ab^{-1}(x_j) = \sigma_i^l(x_j)$, hence $\sigma_i^{-l}ab^{-1} \in Stab_H x_j$. Then $\sigma_i^{-l}ab^{-1} = \sigma_j^p$ for some $p \in \{1,\ldots,o(\sigma_j)\}$. We obtain $\sigma_i^{-l}a = \sigma_j^t b$, hence $(\sigma_i)a(\sigma_j)b \in F$.

"$\Longrightarrow$" Suppose now that Γ is a G-graph $\Phi(G;\{s_1,\ldots,s_k\})$, with $V = \bigcup_{i=1}^k V_{s_i}$. Using Theorem 1, we consider $H = \delta_G \leq Aut(\Gamma)$. The orbit partition regarding H is $V = \bigcup_{i=1}^k V_{s_i}$. $Stab_H(s_i)e$ is a cyclic group of order $o(s_i)$. $\{(s_1)e,\ldots,(s_k)e\}$ is a clique, since $e \in (s_i)e \cap (s_j)e$ for every $i \neq j \in \{1,\ldots,k\}$. Moreover, $|N((s_i)e) \cap V_j| = o(s_i) = |Stab_H(s_i)e|$. □

4 Incidence Graphs of G-Graphs

Let $\Gamma = (V;E)$ be a simple graph with all vertices of degree at least 3, and $I\Gamma = (IV = V \cup E; IE)$ be its incidence graph [9]. Let $\Psi : Aut(\Gamma) \longrightarrow Aut(I\Gamma)$, $\Psi(f) \mapsto h$ such that: $h(x) = f(x)$ for all $x \in V$, $h(xy) = f(x)f(y)$ for all $xy \in E$.

Lemma 3. *Ψ is a group isomorphism.*

Proof. (a) Ψ is well defined, since $xy \in E$ if and only if $f(x)f(y) \in E$.

(b) *We prove that Ψ is an isomorphism.*

Let f_1 and $f_2 \in Aut(\Gamma)$. We have $\Psi(f_1) \circ \Psi(f_2) = h_1 \circ h_2$. It follows that $\Psi(f_1) \circ \Psi(f_2) = \Psi(f_1 \circ f_2)$, hence Ψ is a morphism.

Let f_1 and f_2 in $Aut(\Gamma)$ such as $\Psi(f_1) = \Psi(f_2)$. Then for every $x \in V$ we have $f_1(x) = f_2(x)$, hence $f_1 = f_2$, so Ψ is injective.

Settle h in $Aut(I\Gamma)$. Let $f = h|_V$. Since all vertices from V have degree at least 3 in $I\Gamma$ and all vertices from E have degree 2 in $I\Gamma$, it follows that $f(V) \subseteq V$ and $h(E) \subseteq E$. But, since f is injective, we obtain $f(V) = V$, so f is bijective. Also, application $h|_E : E \longrightarrow E$ is a bijection.

We will prove that f is a morphism.

Let $xy \in E$. Then $f(x)h(xy) \in IE$ and $f(y)h(xy) \in IE$. Since $f(x) \neq f(y)$, it means that $f(x)f(y) \in E$.

Suppose now that $f(x)f(y) \in E$. Then there exist $x'y' \in E$ such that $h(x'y') = f(x)f(y)$. But $x'(x'y') \in IE$ and $y'(x'y') \in IE$. It follows that $f(x')f(y') = h(x'y') = f(x)f(y)$, hence $\{x,y\} = \{x',y'\}$ and so $xy \in E$.

In conclusion $f \in Aut(\Gamma)$, hence Ψ is surjective. It follows that Ψ is an isomorphism. □

Let $\Gamma = \Phi(G;S) = (V;E)$ be a G-graph and $S = \{s_1, s_2\}$ a generating set for G with $o(s_1) = o(s_2) = p \geq 3$. Since $|S| = 2$, $\Phi(G;S)$ is bipartite; let $V = V_1 \cup V_2$.

Next we assume that s_1 and s_2 are independent elements and there exists f an automorphism of G that swaps the two elements contained in S: $f(s_1) = s_2$ and $f(s_2) = s_1$.

Proposition 1. *Automorphism f is of order 2.*

Proof. Let $a \in G$; then $a = \prod_{k=1}^{n} s_1^{\lambda_k} s_2^{\mu_k}$. We have

$$f^2(a) = f(f(a)) = f\left(f\left(\prod_{k=1}^{n} s_1^{\lambda_k} s_2^{\mu_k}\right)\right) = f\left(\prod_{k=1}^{n} s_2^{\lambda_k} s_1^{\mu_k}\right) = \prod_{k=1}^{n} s_1^{\lambda_k} s_2^{\mu_k} = a.$$

It follows that the order of f is two. □

In order to prove the main result of this section, we define an automorphism $\tau : V \longrightarrow V$ such that $\tau((s_i)x) = (f(s_i))f(x)$, for $s_i \in S, x \in G$. It is obvious that $\tau \in Aut(\Gamma)$. From the definition it is easy to see that the order of τ is 2.

Let $G^* = \{\delta_g, g \in G\} \cup \{\delta_g \circ \tau, g \in G\} = \delta_G \cup \delta_G \tau \subseteq Aut(\Gamma)$

Lemma 4. *G^* is a subgroup of $Aut(\Gamma)$.*

Proof. We will prove that G^* is a group. Since δ_G is a subgroup of $Aut(\Gamma)$, it suffices to prove that, for any $g_1, g_2 \in G$, automorphisms $(\delta_{g_1} \circ \tau)^{-1}$, $(\delta_{g_1} \circ \tau) \circ \delta_{g_2}$ and $(\delta_{g_1} \circ \tau) \circ (\delta_{g_2} \circ \tau)$ are in G^*.

$(\delta_{g_1} \circ \tau)^{-1} = \tau^{-1} \circ \delta_{g_1^{-1}} = \tau \circ \delta_{g_1^{-1}}$, since τ is of order 2

Let $(s)x \in V$. We have

$$(\tau \circ \delta_{g_1^{-1}})((s)x) = \tau((s)xg_1^{-1}) = (f(s))f(x)f(g_1^{-1}) = \delta_{f(g_1^{-1})}(\tau((s)x)),$$

hence $(\delta_{g_1} \circ \tau)^{-1} = \tau \circ \delta_{g_1^{-1}} = \delta_{f(g_1^{-1})} \circ \tau \in G^*$.

Then $(\delta_{g_1} \circ \tau) \circ \delta_{g_2} = \delta_{g_1} \circ (\tau \circ \delta_{g_2}) = \delta_{g_1} \circ \delta_{f(g_2)} \circ \tau = \delta_{f(g_2)g_1} \circ \tau \in G^*$

Also $(\delta_{g_1} \circ \tau) \circ (\delta_{g_2} \circ \tau) = \delta_{f(g_2)g_1} \circ \tau \circ \tau = \delta_{f(g_2)g_1} \in G^*$. □

Let H be the following group: $H = \Psi(G^*)$.

Theorem 3. *Incidence Graph.* *Let G be a group having a generating set $S = \{s_1, s_2\}$ such that s_1 and s_2 are independent elements of G with $o(s_1) = o(s_2) = p \geq 3$, and an automorphism f that swaps the two elements contained in S. Let $\Gamma = \Phi(G;S) = (V;E)$ be a G-graph and $I\Gamma = (IV = V \cup E; IE)$ be the incidence graph of Γ. Then $I\Gamma$ is a G-graph, $I\Gamma = \Phi(H;S')$ with $S' = \{s_1', s_2'\}$ and $o(s_1') = p$, $o(s_2') = 2$.*

Proof. We shall prove that the group $H = \Psi(G^*) \subseteq Aut(I\Gamma)$ satisfies all conditions from Theorem 2- *G-graph characterization.*

Step 1. *First we will prove that the partition $IV = V \cup E$ is the orbit partition of IV regarding H.*
Let $x, y \in V$ be two vertices of graph $\Gamma = \Phi(G;S)$. There are two possibilities:

- *Both x and y are in the same partition (for example) V_1. From Theorem 1 there exists $g \in V_1$ such as $\delta_g(x) = y$.*

- *x and y are in different partitions; assume $x \in V_1$ and $y \in V_2$. Since $x \in V_1$, then $x' = \tau(x)$ belongs to V_2 and there exists $\delta_{g_2} \in \delta_G$ such that $\delta_{g_2}(x') = y$. We obtain $(\delta_{g_2} \circ \tau)(x) = \delta_{g_2}(\tau(x)) = y$.*

Hence G^* acts transitively on all V. As Ψ is an isomorphism, $H = \Psi(G^*)$ acts transitively on V.

By Theorem 1 it follows that G^* acts transitively on E, and since G^* is isomorphic with $H = \Psi(G^*)$, H acts transitively on V and on E. In conclusion the orbit partition of IV regarding H is $IV = V \cup E$.

Step 2. *We consider a clique with* 2 *elements in $I\Gamma$ as follows.*

Let $x_1 = (s_1)e \in V$ and $x_2 = (s_1)e(s_2)e \in E$. Then $\{x_1, x_2\}$ is a clique with 2 elements in $I\Gamma$.

Step 3. *We will show that conditions* (1) *and* (2) *from the characterization theorem are verified.*

Since $\tau(V_1) = V_2$ and $\tau(V_2) = V_1$, it follows that $Stab_{G^*}x_1 = Stab_{\delta_G}x_1$. But, from Theorem 1, $Stab_{\delta_G}x_1$ is a cyclic group of order p. It follows that $Stab_H x_1 = \Psi(Stab_{G^*}x_1)$ is a cyclic group of order $p = o(s_1)$. But since p is the degree of x_1 in Γ, it follows that in $I\Gamma$ we have $|N(x_1) \cap E| = p = |Stab_H x_1|$.

Let $h \in Stab_H x_2 = Stab_H(s_1)e(s_2)e$. Then either $h \in Stab_{\delta_G}(s_1)e \cap Stab_{\delta_G}(s_2)e$ or ($h = \delta_g \circ \tau$ with $h((s_1)e) = (s_2)e$ and $h((s_2)e) = (s_1)e$).

But $Stab_{\delta_G}(s_1)e \cap Stab_{\delta_G}(s_2)e = \{id\}$ so it remains to consider only the second situation. We have $h = \delta_g \circ \tau$. Then

$$h((s_1)e) = \delta_g(\tau((s_1)e)) = (s_2)g = (s_2)e$$

and, similarly,

$$h((s_2)e) = \delta_g(\tau((s_2)e)) = (s_1)g = (s_1)e.$$

It follows that there exist p, l such that $g = s_1^p = s_2^q$. Since s_1 and s_2 are independent elements, it follows that $g = e$, hence $h = \tau$, which has order 2.

Thus $Stab_H(s_1)e(s_2)e$ is a cyclic group of order 2, and, since in $I\Gamma$ we have

$$N((s_1)e(s_2)e) = \{(s1)e, (s2)e\},$$

then

$$|N((s_1)e(s_2)e) \cap V| = 2 = |Stab_H(s_1)e(s_2)e|.$$

Moreover, since $\tau \notin Stab_{G^*}x_1$, it follows that $Stab_H x_1 \cap Stab_H x_2 = \{id\}$.

To conclude, we have shown that H satisfy for $I\Gamma$ all the conditions of Theorem 2- *G-graph characterization*. So $I\Gamma$ is a G-graph. □

5 Mesh of d-ary Tree

Next, we are only interested in the mesh of d-ary trees in one dimension, $MT(d,1)$.

Let B an alphabet of d letters. We denote by $|u|$ the length of the word u. The *mesh of d-ary trees* $MT(d,1)$ is the graph with the vertex set $V = \{(u,v); (|u| = 1$ and $|v| < 1)$ or $(|v| = 1$ and $|u| \leq 1)\}$, and $[(u,v),(u',v')] \in E(MT(d,1))$ if and only if $|u| = 1$,$u = u'$ and $v = v'\lambda$ or $|v| = 1$, $v = v'$ and $u = u'\lambda$ with $\lambda \in B$.

More precisely, if we consider $B = \mathbb{Z}_d = \{0,1,\ldots,d-1\}$ and denote by e the empty word, we can describe the vertex set of $MT(d,1)$ as

$$V = \{(e,i)|i \in B\} \cup \{(i,e)|i \in B\} \cup \{(i,j)|i,j \in B\}$$

and the edge set as

$$E = \{[(u,e),(u,v)]|u,v \in B\} \cup \{[(e,v),(u,v)]|u,v \in B\}.$$

The key properties of meshes of d-tree graphs are as follows:

- Number of vertices $N_v = d\,(d+2)$
- Number of edges $N_e = 2d(\frac{d^2-1}{d-1} - 1) = 2d^2$
- Diameter $D = 4$
- The mesh of d-ary trees is not a Cayley graph
- The mesh of d-ary trees is not vertex-transitive

These graphs are very important in interconnection networks [1].

The following diagrams show $MT(3,1)$ and $MT(4,1)$ (Fig. 1).

In the sequel we consider $n,k \geq 2$. Let $C_n \times C_k$ be the product of two cyclic groups. Such a product is generated by two elements, a and b, with $a^n = b^k = 1$. More precisely, $C_n \times C_k$ is the group with description $\langle a,b \mid a^n = 1, b^k = 1, ab = ba\rangle$. In [5] the following was shown:

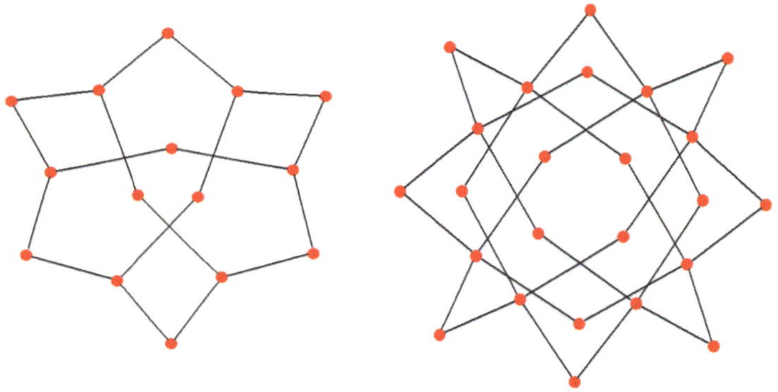

Fig. 1 $MT(3,1)$ and $MT(4,1)$

Proposition 2. *For $S = \{a,b\}$, the G-graph $\Phi(C_n \times C_k, S)$ of the product of two cyclic groups, is the complete bipartite graph $K_{n,k}$.*

It follows that for $n = k$ and $S = \{s_1 = (1,0), s_2 = (0,1)\}$ the graph $\Phi(\mathbb{Z}_n \times \mathbb{Z}_n, S)$ is a complete bipartite graph with degree equal to n.

From the definitions of $MT(n,1)$ and $K_{n,n}$ it is easy to see that the following result holds.

Corollary 1. *$MT(n,1)$ is the incidence graph of $K_{n,n}$.*

Corollary 2. *The graph $MT(n,1)$, for $n \geq 3$, is a G-graph.*

Remark. A direct proof for Corollary 2 can be found in [6].

References

1. Barth, D.: Bandwidth and cutwidth of the mesh of d-Ary trees. Euro-Par **I**, 243–246 (1996)
2. Bretto, A., Faisant, A.: Another way for associating a graph to a group. Math. Slovaca **55**(1), 1–8 (2005)
3. Bretto, A., Gilibert, L.: Symmetric and semisymmetric graphs construction using G-graphs. In: Kauers, M. (ed.) International Symposium on Symbolic and Algebraic Computation, (ISSAC'05), pp. 61–67, Beijing, China, 24–27 July. ACM, New York. ISBN:1-59593-095-7 (2005)
4. Bretto, A., Gilibert, L.: G-graphs for the cage problem: a new upper bound. In: International Symposium on Symbolic and Algebraic Computation, (ISSAC'07), Waterloo, Ontario, Canada, 29 July–1 August. ACM, New York. ISBN:978-1-59593-743-8 (2007)
5. Bretto, A., Faisant, A., Gillibert, L.: G-graphs: A new representation of groups. J. Symbolic Comput. **42**(5), 549–560 (2007). ISSN:0747-7171
6. Bretto, A., Jaulin, C., Gillibert, L., Laget, B.: Hamming graphs and mesh of *d*-ary trees are *G*-graphs. In: Asian Symposium in Computer Mathematics, 15–18 December 2007, Singapore. LNCS, vol. 5081, pp. 139–150. Springer, Berlin (2008)
7. Lu, J., Moura, J.M.: Structured LDPC codes for high-density recording: large girth and low error floor. IEEE Trans. Magn. **42**, 208–213 (2006)
8. Robinson, D.J.S.: A course in the theory of groups. Graduate Texts in Mathematics, Springer, Berlin, **80** (1982)
9. Zhong-fu, Z., Bing, Y., Jing-wen, L., Lin-zhong, L., Jian-fang, W., Bao-gen, X.: On Incidence Graphs. Ars Combinatoria **87**, 213–223 (2008)

A Tight Bound on the Worst-Case Number of Comparisons for Floyd's Heap Construction Algorithm

Ioannis Paparrizos

Abstract In this paper a tight bound on the worst-case number of comparisons for Floyd's well-known heap construction algorithm is derived.[1] It is shown that at most $2n - 2\mu(n) - \sigma(n)$ comparisons are executed in the worst case, where $\mu(n)$ is the number of ones and $\sigma(n)$ is the number of zeros after the last one in the binary representation of the number of keys n.

Key words Algorithm analysis • Worst case complexity • Data structures • Heaps

1 Introduction

Floyd's heap construction algorithm [3] proposed in 1964 as an improvement of the construction phase of the classical heapsort algorithm introduced earlier that year by Williams [9] in order to develop an efficient in-place general sorting algorithm. The importance of heaps in representing priority queues and speeding up an amazing variety of algorithms is well documented in the literature. Moreover, the classical heapsort algorithm and, hence, Floyd's heap construction algorithm as part of it is contained and analyzed in each textbook discussing algorithm analysis, see [1] and [2] for example.

[1]This paper was also presented at Student Research Forum of SOFSEM'11 [Paparrizos, A tight bound on the worst-case number of comparisons for Floyd's heap construction algorithm (2011)].

I. Paparrizos (✉)
Computer Science Department, Columbia University, New York, NY, USA
e-mail: jopa@cs.columbia.edu

A. Migdalas et al. (eds.), *Optimization Theory, Decision Making, and Operations Research Applications*, Springer Proceedings in Mathematics & Statistics 31,
DOI 10.1007/978-1-4614-5134-1_10,

Floyd's algorithm is optimal as long as complexity is expressed in terms of sets of functions described via the asymptotic symbols O, Θ, and Ω. Indeed, its linear complexity $\Theta(n)$, both in the worst and best case, cannot be improved as each object must be examined at least once. However, it is an established tradition to analyze algorithms solving comparison-based problem by counting mainly comparisons, see, for example, Knuth [5] who states that the theoretical study of comparison counting gives us a good deal of useful insight into the nature of sorting processes.

Despite the overwhelming attention received by the computer community in the more than 45 years of its life, a tight bound on the worst-case number of comparisons holding for all values of n, is, to our knowledge, still unknown. Kruskal et al. [6] showed that $2n - 2\lceil \log(n+1) \rceil$ is a tight bound on the worst-case number of comparisons, if $n = 2^k - 1$, where k is a positive integer, and, to our knowledge, this is the only value of n for which a tight upper bound has been reported in the literature.

Schaffer [8] showed that $n - \lceil \log(n+1) \rceil + \lambda(n)$, where $\lambda(n)$ is the number of zeros in the binary representation of n, is the sum of heights of sub-trees rooted at internal nodes of a complete binary tree, see also [4] for an interesting geometric approach to the same problem. Using this result we show that $2n - 2\mu(n) - \sigma(n)$ is a tight bound on the worst-case number of comparisons for Floyd's heap construction algorithm. Here, $\mu(n)$ is the number of ones and $\sigma(n)$ is the number of zeros after the last (right most) one in the binary representation of the number of keys n.

2 Floyd's Heap Construction Algorithm

A *maximum heap* is an array H the elements of which satisfy the property:

$$H(\lfloor i/2 \rfloor) \geq H(i), i = 2, 3, \ldots, n. \tag{1}$$

Relation (1) will be referred to as the heap property. A *minimum heap* is similarly defined; just reverse the inequality sign in (1) from $\geq$ to $\leq$. When we simply say a heap we will always mean a maximum heap. A nice property of heaps is that they can be represented by a *complete binary tree*. Recall that a complete binary tree is a binary tree in which the root lies in level zero and all the levels except the last one contain the maximum possible number of nodes. In addition, the nodes at the last level are positioned as far to the left as possible. If $n = 2^k - 1$, the last level $\lfloor \log n \rfloor = k - 1$ contains the maximum possible number $2^k - 1$ of nodes. In this case the complete binary tree is called *perfect*. The *distinguished path*, introduced in [5], of a complete binary tree that connects the root node 1 with the last leaf node n, will play an important role in deriving our results. It is well known, see for example [5], that the nodes of the distinguished path correspond to the digits of the binary expression of n. Figure 1 illustrates a complete binary tree, its distinguished path

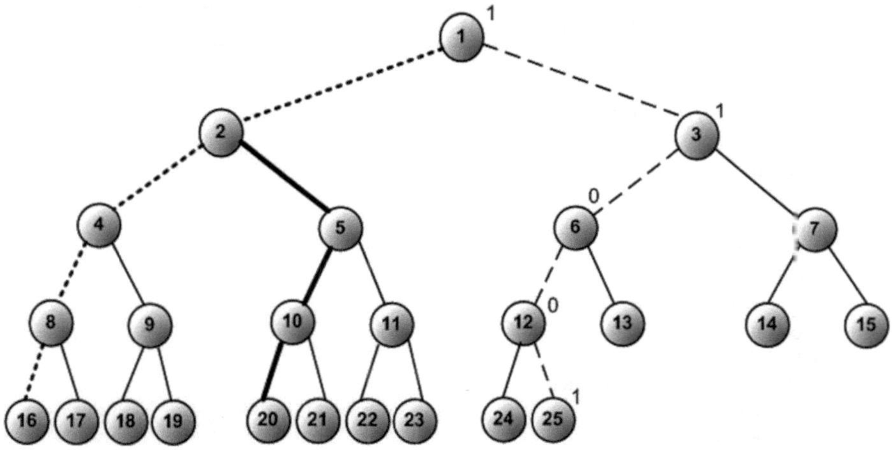

Fig. 1 A complete binary tree, its distinguished path (*dashed edges*), a special path (*thick edges*) and the leftmost path (*dotted edges*). The numbers by the nodes of the distinguished path are the digits of the binary expression (11001) of $n = 25$

and the corresponding binary expression of n. In terms of binary trees the heap property is stated as follows:

The value of a child is smaller than or equal to the value of its parent.

It is easily verified that the value of the root is the largest value. Also, each subtree T_j of the complete binary tree representing a heap is also a heap and, hence, the value $H(j)$ is the largest value among those that correspond to the nodes of T_j. A sub-array $H(i:n)$ for which the heap property is satisfied by each node j the parent of which is an element of $H(i:n)$, is also called a heap. Here the expression $j:n$ denotes the sequence of indices $j, j+1, j+2, \ldots, n$.

An *almost heap* is a sub-array $H(i:n)$ all nodes of which satisfy the heap property except possibly node i. If key $H(i)$ violates the heap property, then $H(i) < \max\{H(2i), H(2i+1)\}$.

The main procedure of Floyd's heap construction algorithm, called in this paper heapdown, works as follows. It is applied to an almost heap $H(j:n)$ and converts it into a heap. In particular, if $m = H(j)$ satisfies the heap property $H(j) \geq \max\{H(2j), H(2j+1)\}$ and, hence, $H(j:n)$ is a heap, the algorithm does nothing. Otherwise, it swaps key $m = H(j)$ with the maximum child key $H(j_{\max})$. Then, it considers the child $j_{\max}$ which currently contains key m, and repeats the procedure until the heap property is restored. Algorithm 1 shows a formal description of the algorithm.

Algorithm 1: HEAPDOWN(H(i...n))

```
while 2i+1 ≤ n do
   k = 2i
   if H(k) < H(k+1) then
      k = k+1
   end if
   if H(i) < H(k) then
      swap(H(i),H(k))
      i = k
   else
      return H(i...n)
   end if
end while
if 2i = n and H(i) < H(n) then
   swap(H(i),H(n))
   return H(i...n)
end if
```

Algorithm 2: FLOYD-BUILDHEAP(H)

```
for i = ⌊n/2⌋ to 1 step -1 do
   heapdown(H(i...n))
end for
return H
```

Floyd's heap construction algorithm, called Floyd—buildheap procedure in this paper, applies procedure heapdown to the sequence of almost heaps

$$H(\lfloor n/2 \rfloor : n), H(\lfloor n/2 \rfloor - 1 : n), \ldots, H(1 : n). \tag{2}$$

As the sub-array $H(\lfloor n/2 \rfloor + 1 : n)$ consists of leafs and, therefore, it is a heap and procedure heapdown converts an almost heap to a heap, the correctness of procedure Floyd—buildheap is easily shown.

When procedure heapdown is applied to the almost heap $H(j : n)$ key $m = H(j)$ moves down one level per iteration. In general, two comparisons are executed per level, one comparison to find the maximum child and one to determine whether key m should be interchanged with the maximum child key. However, there is a case in which just one comparison is executed. This happens when key m is positioned at node $\lfloor n/2 \rfloor$ and n is even. Then, internal node $n/2$ has just one child, the last node n, and therefore no comparison is needed to find the maximum child. We will see in the next section, when we will investigate the worst case of procedure Floyd-buildheap, that this situation happens quite often, if n is even. Procedure FLOYD-BUILDHEAP describes formally Floyd's algorithm.

3 A Tight Bound on the Worst-Case Number of Comparisons

It is well known that the number of interchanges performed by Floyd's heap construction algorithm is bounded above by the sum $t(n)$ of heights of sub-trees rooted at the internal nodes of a complete binary tree. Schaffer [8] showed that:

$$t(n) = n - \lceil \log(n+1) \rceil + \lambda(n), \tag{3}$$

where $\lambda(n)$ is the number of zeros in the binary representation of n. For the sake of completeness of the presentation we provide a short proof based on the geometric idea described in [4]. We associate a *special path* with each internal node of the binary tree. The special path connects a node, say j, with a leaf of the subtree T_j rooted at node j. The first edge of the special path is a *right edge* and all the remaining edges are *left edges*, see Fig. 1. In particular, the nodes of the special path are $j, 2j+1, 2^2j+2, 2^3j+2^2, \ldots, 2^kj+2^{k-1}$. Observe now that the edges of all special paths cover all the edges of the binary tree exactly ones except the $\lfloor \log n \rfloor$ edges of the *leftmost path*, see Fig. 1. As no two special paths contain a common edge, the number of edges of all special paths is $n-1-\lfloor \log n \rfloor$.

The lengths of special paths are closely related to the heights of the sub-trees. Recall that the *length* of a path is the number of edges it contains. Denote by $sp(j)$ the special path corresponding to node j. If internal node j does not belong in the distinguished path, then $length(sp(j)) = h(T_j)$. If internal node j belongs in the distinguished path and the right edge $(j, 2j+1)$ is an edge of the distinguished path, then $length(sp(j)) = h(T_j)$. In that case the first edge of $sp(j)$ belongs in the distinguished path and the digit of the binary expression of n corresponding to node j is 1. If internal node j belongs in the distinguished path and the left edge $(j, 2j)$ is an edge of the distinguished path, then $h(T_j) = length(sp(j)) + 1$. In that case the first edge of $sp(j)$ does not belong to the distinguished path and the digit of the binary expression of n corresponding to node j is 0. Summing up all heights of internal nodes we get

$$t(n) = n - 1 - \lfloor \log n \rfloor + \lambda(n) = n - \lceil \log(n+1) \rceil + \lambda(n). \tag{4}$$

In computing, our tight bound on the worst-case number of comparisons, two cases must be considered, n even and n odd. We first take care of the case n is odd.

Lemma 1. *Let n be odd. Then the maximum number of comparisons executed by Floyd's heap construction algorithm is*

$$2t(n) = 2(n - \lceil \log(n+1) \rceil + \lambda(n)). \tag{5}$$

Proof. If n is odd, each internal node has exactly two children and, hence, each key swap corresponds to two key comparisons. Therefore $2t(n)$ is an upper bound on the number of comparisons. □

We show now that this bound is tight. To this end we construct a special worst-case array H. In particular H satisfies the following properties

1. The elements of H are the n distinct keys $1,2,\ldots,n$.
2. The nodes in the distinguished path are assigned the $\lceil \log(n+1) \rceil$ largest keys. In particular, the nodes in levels $0,1,2,\ldots,\log(n)$ are assigned the keys $n - \lceil \log (n+1) \rceil + 1, n - \lceil \log(n+1) \rceil + 2, \ldots, n$, respectively.
3. If j is a node not belonging in the distinguished path, sub-tree T_j is a minimum heap.

Apply now procedure Floyd-buildheap to the array H described previously. When procedure heapdown is called on the almost heap $H(j:n)$ and j is not a node of the distinguished path, key $m = H(j)$ will move all the way down to the bottom level of sub-tree T_j. This is so because key m is the smallest among the keys corresponding to nodes of the sub-tree rooted at node j, see property 3. Also, two comparisons are executed per level. When procedure heapdown is applied to an almost heap $H(j:n)$, where j is a node of the distinguished path, key $m = h(j)$ will follow the distinguished path all the way down to the bottom level taking the position of leaf node n, see property 2. Again, two comparisons are executed per level and, hence, the number $2n - 2\lceil \log(n+1) \rceil + 2\lambda(n)$ is a tight bound on the worst-case number of comparisons. □

Next lemma takes care of the case n even.

Lemma 2. *If n is even the exact worst case number of comparisons for Floyd's heap construction algorithm is*

$$2(n - \lceil \log(n+1) \rceil + \lambda(n)) - \sigma(n), \tag{6}$$

where $\sigma(n)$ is the number of zeros after the last one in the binary representation of n.

Proof. Let $(b_m b_{m-1} \ldots b_2 b_1 b_0)$ be the binary representation of n. Let also $b_k b_{k-1} \ldots b_2 b_1 b_0$ be the last $k+1$ digits of the binary representation of n such that $b_k = 1$ and $b_{k-1} = b_{k-2} = \cdots = b_1 = b_0 = 0$. □

As n is even $b_0 = 0$ and, hence, $k \geq 1$. Consider now an internal node of height $j \leq k$ lying at the distinguished path. It is easily verified, using inductively the well-known property $\lfloor \lfloor n/2 \rfloor / a \rfloor = \lfloor n/a^2 \rfloor$ of the floor function, that the index at that node is $\lfloor n/2^j \rfloor$. When procedure Floyd-buildheap calls procedure heapdown on the almost heap $H(\lfloor n/2^j \rfloor : n)$ key $m = H(\lfloor n/2^j \rfloor)$ will move down the levels either following the distinguished path or moving to the right of it at some point. This is so because all the edges $(n, \lfloor n/2 \rfloor), (\lfloor n/2 \rfloor, \lfloor n/2^2 \rfloor), \ldots, (\lfloor n/2^{j-1} \rfloor, \lfloor n/2^j \rfloor)$ of the distinguished path are left edges. In the former case at most $2j-1$ comparisons are executed and this happens when key $H(\lfloor n/2^j \rfloor)$ is placed either at the bottom level or at the level next to bottom. In the latter case at most $2(j-1)$ comparisons are executed. Hence for each node of the distinguished path at height $j = 1,2,\ldots,k$ the maximum number of comparisons is one less than 2 times the height of the sub-tree rooted at that node. For all the remaining internal nodes i the maximum

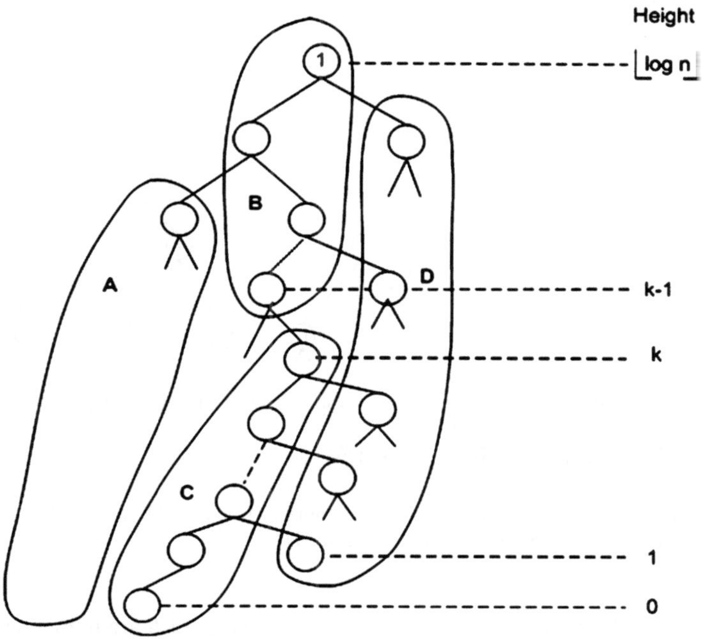

Fig. 2 Partition of the nodes of a complete binary tree into sets A, B, C, D

number of comparisons is $2h(T_i)$, where $h(T_i)$ is the height of the sub-tree rooted at node i. As the number of internal nodes of the distinguished path at heights $1, 2, \ldots, k$ is $\sigma(n)$, the previous arguments show that the number

$$2(n - \lceil \log(n+1) \rceil + \lambda(n)) - \sigma(n) \tag{7}$$

is an upper bound on the number of comparisons for procedure Floyd-buildheap.

We describe now an array H on which procedure Floyd-buildheap executes exactly $2(n - \lceil \log(n+1) \rceil + \lambda(n)) - \sigma(n)$ comparisons, thus showing that this number is a tight upper bound for n even. In order to describe the structure of the worst-case example H we partition the nodes of the complete binary tree into 4 sets A, B, C, D. Set A contains all the nodes on the left side of the distinguished path. Set D contains all nodes lying on the right side of the distinguished path. Set C contains the nodes of the distinguished path of height $j = 0, 1, 2, \ldots, k$ and set B contains all the remaining nodes of the distinguished path. Figure 2 illustrates a complete binary tree and the sets of nodes A, B, C, D.

The structure of array H is described in the following properties:

1. The elements of H are the n distinct keys $1, 2, \ldots, n$.
2. If i, j, k, l are nodes belonging to the sets A, B, C, D, respectively, then

$$H(i) > H(j) > H(k) > H(l). \tag{8}$$

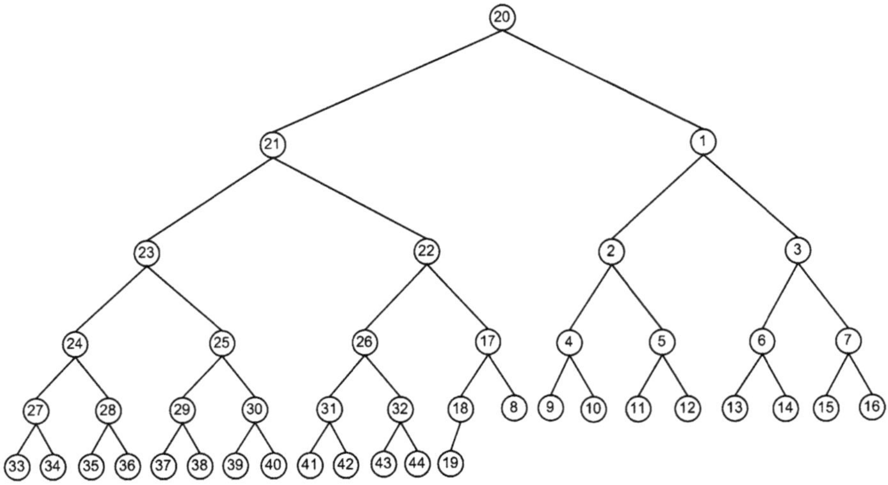

Fig. 3 A worst-case complete binary tree for Lemma 2. It is $n = 44$, $k = 2$, $|A| = 23, |B| = 3$, $|C| = 3, |D| = 15$. The number inside node j is the key $H(j)$

3. The keys in the distinguished path that belong to the set B are in increasing order from the top to the bottom. The keys in the distinguished path that belong to the set C are in increasing order from the top to the bottom.
4. If j is a node not belonging to the distinguished path, the sub-tree T_j is a minimum heap.

Although there are more than one way to assign the keys $1,2,\ldots,n$ to the elements of H so that properties (2)–(4) are satisfied, an easy way to do that is as follows. Place the $|A|$ largest keys to the sub-trees on the left of the distinguished path so that each sub-tree is a minimum heap. The symbol $|A|$ denotes the number of elements of set A. Obviously $|A|$ is the number of nodes on the left of the distinguished path. Place in increasing order from top to bottom levels the next $|B|$ largest elements at the nodes of the distinguished path that belong to the set B. Also, place in increasing order from top to down levels the next $|C|$ largest elements at the nodes of the distinguished path that belong to the set C. Obviously $|B|+|C| = 1+\lfloor\log(n)\rfloor = \lceil\log(n+1)\rceil$. Finally, place the remaining $|D|$ smallest keys, i.e, the keys $1,2,\ldots,|B|$ at the sub-trees right to the distinguished path so that each sub-tree is a minimum heap. Figure 3 illustrates such a worst-case example for $n = 44$.

Apply now procedure Floyd-buildheap on the array H described previously. Let $H(j:n), j = \lfloor n/2\rfloor, \lfloor n/2\rfloor-1,\ldots,1$ be the almost heap on which procedure heapdown is applied to after it is called by procedure Floyd-buildheap. If j is not a node at the distinguished path, key $H(j)$, because of property (4), will move all the way down to the bottom level of sub-tree T_j and $2h(T_j)$ comparisons will be executed. If j is a node of the distinguished path belonging to set C, key $H(j)$, because of properties (2)–(4), will follow the distinguished path never making a right turn. In this case, key $H(j)$ will be placed at node n executing

$2h(T_j) - 1$ comparisons. Finally, if node j belongs in set B, key $H(j)$ will definitely make a left turn before reaching the node $\lfloor n/2^k \rfloor$ of the distinguished path, see properties (2) and (3). Then it will move all the way down to bottom level executing 2 comparisons per level. Again $2h(T_j)$ comparisons are executed.

Summing up the comparisons for all the calls of procedure heapdown we see that the total number of comparisons is as stated in the Lemma. □

Observe that the array H described in the previous Lemma is not a minimum heap. In particular the sub-trees rooted at nodes of the distinguished path are not minimum heaps.

Theorem 1. *The number $2n - 2\mu(n) - \sigma(n)$, where $\mu(n)$ is the number of ones and $\sigma(n)$ is the number of zeros after the last one in the binary representation of n, is a tight bound on the worst-case number of comparisons for Floyd's heap construction algorithm.*

Proof. If n is odd, then $b_0 = 1$ and, hence, $\sigma(n) = 0$. Combining Lemmas 1 and 2 we see that a tight bound on the worst-case number of comparisons is the number $2[n - \lceil \log(n+1) \rceil + \lambda(n)] - \sigma(n) = 2[n - (\lambda(n) + \mu(n)) + \lambda(n)] - \sigma(n) = 2n - 2\mu(n) - \sigma(n)$. □

4 Conclusion

Deriving worst case tight upper bound examples for an algorithm implies that the worst-case complexity of the algorithm cannot be improved. We derived our worst-case examples by the use of simple geometric ideas. As the binary trees and heaps are involved in many other algorithms for which worst-case tight examples are not known, we hope that our results will contribute in solving those problems.

References

1. Aho, A.V., Hopcroft, J.E., Ullman, J.D.: The Design and Analysis of Computer Algorithms. Addison-Wesley Series in Computer Science and Information Processing, Addison-Wesley Publishing Company (1974)
2. Cormen, T., Leiserson, C., Rivest, R., Stein, C.: Introduction to Algorithms, 2nd edn. MIT, Cambridge (2001)
3. Floyd, R.: Algorithm 245: treesort 3. Comm. ACM **7**, 701 (1964)
4. Goodrich, M.T., Tamassia, R.: Algorithm Design, Foundations, Analysis, and Internet Examples. Wiley, New York (2002)
5. Knuth, D.: The art of computer programming, vol. 3, 2nd edn. Searching and Sorting. Addison Wesley, Redwood city (1998)
6. Kruskal, C.P., Weixelbaum, D.: The worst case analysis of heapsort. Technical Report no 018a, Department of Computer Science, New York University (1979)

7. Paparrizos, I.: A tight bound on the worst-case number of comparisons for Floyd's heap construction algorithm. In: Proceedings of the 37th International Conference on Current Trends in Theory and Practice of Computer Science (2011)
8. Schaffer, R.: Analysis of heapsort. Ph.D. Thesis, Department of computer science, Princeton University (1992)
9. Williams, J.W.J.: Algorithm 232: heapsort. Comm. ACM **6**, 347–348 (1964)

A Parallel Implementation of the Revised Simplex Algorithm Using OpenMP: Some Preliminary Results

Nikolaos Ploskas, Nikolaos Samaras, and Konstantinos Margaritis

Abstract Linear Programming (LP) is a significant research area in the field of operations research. The simplex algorithm is the most widely used method for solving Linear Programming problems (LPs). The aim of this paper is to present a parallel implementation of the revised simplex algorithm. Our parallel implementation focuses on the reduction of the time taken to perform the basis inverse, due to the fact that the total computational effort of an iteration of simplex type algorithms is dominated by this computation. This inverse does not have to be computed from scratch at any iteration. In this paper, we compute the basis inverse with two well-known updating schemes: (1) The Product Form of the Inverse (PFI) and (2) A Modification of the Product Form of the Inverse (MPFI); and incorporate them with revised simplex algorithm. Apart from the parallel implementation, this paper presents a computational study that shows the speedup among the serial and the parallel implementations in large-scale LPs. Computational results with a set of benchmark problems from Netlib, including some infeasible ones, are also presented. The parallelism is achieved using OpenMP in a shared memory multiprocessor architecture.

Key words Linear programming • Revised simplex method • Basis inverse • Parallel computing • OpenMP

N. Ploskas • N. Samaras (✉) • K. Margaritis
Department of Applied Informatics, University of Macedonia, 156 Egnatia Str., 54006 Thessaloniki, Greece
e-mail: ploskas@uom.gr; samaras@uom.gr; kmarg@uom.gr

A. Migdalas et al. (eds.), *Optimization Theory, Decision Making, and Operations Research Applications*, Springer Proceedings in Mathematics & Statistics 31,
DOI 10.1007/978-1-4614-5134-1_11,

1 Introduction

Linear Programming (LP) is the process of minimizing or maximizing a linear objective function $z = \sum_{i=1}^{n} c_i \cdot x_i$ subject to a number of linear equality and inequality constraints. Several methods are available for solving LPs, among which the simplex algorithm is the most widely used. We assume that the problem is in its general form. Formulating the linear problem, we can describe it as shown in (LP1).

$$
\begin{aligned}
\min \quad & c^{\mathrm{T}}x \\
\text{subject to} \quad & Ax = b \\
& x \geq 0
\end{aligned}
\qquad (LP1)
$$

where $A \in R^{m \times n}$, $(c,x) \in R^n$, $b \in R^m$, and T denotes transposition. Without loss of generality we assume that A has full rank, $rank(A) = m$, where $(m < n)$. The simplex method searches for an optimal solution by moving from one feasible solution to another, along the edges of the feasible region. The dual problem associated with the linear problem in (LP1) is shown in (DP).

$$
\begin{aligned}
\min \quad & b^{\mathrm{T}}w \\
\text{subject to} \quad & A^{\mathrm{T}}w + s = c \\
& s \geq 0
\end{aligned}
\qquad (DP)
$$

where $w \in R^m$ and $s \in R^n$. As in the solution of any large-scale mathematical system, the computational time for large LPs is a major concern. Parallel programming is a good practice for solving computationally intensive problems in operations research. The application of parallel processing for LP has been introduced in the early 1970s. However, only since the beginning of the 1980s attempts have been made to develop parallel implementations. A lot of architectural features have been used in practice. Preliminary parallel approaches were developed for network optimization, direct search methods, and global optimization. A growing number of optimization problems demand parallel computing capabilities. Any performance improvement in the parallelization of the revised simplex would be of great interest.

One of the earliest parallel tableau simplex methods on a small-scale distributed memory Multiple-Instruction Multiple-Data (MIMD) machines is the one introduced by Finkel [8]. Stunkel [28] implemented both the tableau and the revised simplex method on a 16-processor Intel hypercube computer, achieving a speedup of between 8 and 12 for small problems from the Netlib set [9]. Helgason, Kennington, and Zaki [16] proposed an algorithm to implement the revised simplex using sparse matrices methods on shared memory MIMD computer. Furthermore, Shu and Wu [26] and Shu [25] parallelized the explicit inverse and the LU decomposition of the basis simplex algorithms. Hall and McKinnon [12, 15] have implemented two parallel schemes for the revised simplex method. The first of

Hall and McKinnon's parallel revised simplex implementations was ASYNPLEX [12]. In this implementation one processor is devoted to the basis inversion and the remaining processors perform simplex iterations. ASYNPLEX was implemented on a Cray T3D, achieving a speedup of between 2.5 and 4.8 for four modest Netlib problems. The second of Hall and McKinnon's parallel revised simplex implementations was PARSMI [15]. PARSMI was tested on modest problems from the Netlib set, resulting in a speedup of between 1.7 and 1.9. Hall [13] implemented a variant of PARSMI on a 8-processor shared memory Sun Fire E15k, leading in a speedup of between 1.8 and 3.

Simplex algorithms for general LPs on Single Instruction Multiple Data (SIMD) have been reported by [1]. Luo and Reijns [20] presented an implementation of the revised simplex method, achieving a speedup of more than 12, when solving modest Netlib problems on 16 transputers. Eckstein et al. [7] implemented a parallelization of standard and revised simplex method in a CM2 machine. Lentini et al. [19] worked on the standard simplex method with the tableau stored as a sparse matrix, resulting in a speedup of between 0.5 and 2.7, when solving medium-sized Netlib problems on four transputers. Thomadakis and Liu [29] worked on the standard simplex method on MasPar MP-1 and MP-2 machines, achieving a speedup of up to three, when solving large randomly generated problems. Badr et al. [2] implemented a dense standard simplex method on eight computers, leading in a speedup of five when solving small random dense LPs. Yarmish and Slyke [30] presented a distributed implementation of the standard simplex method that is not affected by problem density and their implementation outperformed revised method at density slightly above 10% when using 7 processors. Mamalis et al. [21] proposed a parallel implementation framework of the standard full tableau simplex using a column and a row distribution scheme. Their implementation tested on a linux-cluster of eight Xeon processors and the results showed that the column distribution scheme performs quite better than the row distribution scheme. Previous attempts to develop simplex implementations with the aim of exploiting high performance computing architectures are reviewed by Hall [14]. Finally, computational results for parallelizing the network simplex method are reported in [3, 6, 24].

The use of GPUs for general purpose computations is a quite recent topic, which was applied to linear programming. Greeff [10] implemented the revised simplex method on a GPU using OpenGL and Cg and was able to achieve a speedup of up to 11.4 over an identical CPU implementation. Jung and O'Leary [18] and Owens et al. [23] also presented an implementation using Cg and OpenGL. Spampinato and Elster [27] proposed a GPU implementation of the revised simplex method, based on the CUDA architecture and achieved a speedup of up to 2.5. Recently, Bieling, Peschlow and Martini [5] also presented an implementation of the revised simplex algorithm and achieved a speedup of up to 10.

This paper presents a parallelization of the revised simplex algorithm on a shared memory multiprocessor architecture. The focus of this parallelization is on the basis inverse. The structure of the paper is as follows. In Sect. 2, the revised simplex algorithm is described and presented. In Sect. 3, two methods that have been widely used for basis inversion are analyzed. Section 4 presents the parallel revised simplex algorithm and Sect. 5 gives the computational results. Finally, the conclusions of this paper are outlined in Sect. 6.

2 Revised Simplex Algorithm

The linear problem in (LP1) can be written as shown in (LP2).

$$\begin{aligned} \min \quad & c_B^{\mathrm{T}} x_B + c_N^{\mathrm{T}} x_N \\ \text{subject to} \quad & A_B x_B + A_N x_N = b \\ & x_B, x_N \geq 0 \end{aligned} \tag{LP2}$$

In (LP2), A_B is a $m \times m$ non-singular sub-matrix of A, called basic matrix or basis. The columns of A which belong to subset B are called basic and those which belong to N are called non-basic. The solution $x_B = (A_B)^{-1} b, x_N = 0$ is called a basic solution. A solution $x = (x_B, x_N)$ is feasible iff $x > 0$. Otherwise, (LP2) is infeasible. The solution of (DP) is computed by the relation $s = c - A^{\mathrm{T}} w$, where $w = (c_B)^{\mathrm{T}} (A_B)^{-1}$ are the simplex multipliers and s are the dual slack variables. The basis A_B is dual feasible iff $s \geq 0$.

In each iteration, simplex algorithm interchanges a column of matrix A_B with a column of matrix A_N and constructs a new basis $A_{\overline{B}}$. Any iteration of simplex type algorithms is relatively expensive. The total work of an iteration of simplex type algorithms is dominated by the determination of the basis inverse. This inverse, however, does not have to be computed from scratch during each iteration. Simplex type algorithms maintain a factorization of basis and update this factorization in each iteration. There are several schemes for updating basis inverse. Two well-known schemes are (1) the Product Form of the Inverse (PFI) and (2) a Modification of the Product Form of the Inverse, developed by Benhamadou [4]. These methods, in order to compute the new basis, use only information about the entering and leaving variables along with the current basis. A formal description of the revised simplex algorithm is given in Table 1.

3 Methods Used for Basis Inversion

The revised simplex algorithm differs from the original method. The former uses the same recursion relations to transform only the inverse of the basis in each iteration. It has been implemented to reduce the computation time of the basis inversion and is particularly effective for sparse linear problems. In this section, we will review two methods that have been widely used for basis inversion: (1) the Product Form of the Inverse and (2) a Modification of the Product Form of the Inverse.

Table 1 Revised simplex algorithm

Step 0. *(Initialization).*
Start with a feasible partition (A_B, A_N). Compute $(A_B)^{-1}$ and vectors x_B, w and s_N.
Step 1. *(Test of optimality).*
if $s_N \geq 0$ then STOP. The linear problem is optimal.
else
  Choose the index l of the entering variable using a pivoting rule.
  Variable x_l enters the basis.
Step 2. *(Minimum ratio test).*
Compute the pivot column $h_l = (A_B)^{-1} A_l$.
if $h_l \leq 0$ then STOP. The linear problem is unbounded.
else
  Choose the leaving variable $x_{B[r]} = x_k$ using the following relation:

$$x_{B[r]} = \frac{x_{B[r]}}{h_{il}} = min\left\{\frac{x_{B[i]}}{h_{il}} : h_{il} < 0\right\}$$

Step 3. *(Pivoting).*
Swap indices k and l. Update the new basis inverse $(A_{\overline{B}})^{-1}$, using PFI or MPFI.
Go to Step 1.

3.1 Product Form of the Inverse

The PFI scheme, in order to compute the new basis, uses information only about the entering and leaving variables along with the current basis. The new basis inverse can be updated at any iteration using the (1).

$$(A_{\overline{B}})^{-1} = (A_B E)^{-1} = E^{-1} (A_B)^{-1} \tag{1}$$

where E^{-1} is the inverse of the eta-matrix and can be computed by (2).

$$E^{-1} = I - \frac{1}{h_{rl}} (h_l - e_l) e_l^{\mathrm{T}} = \begin{bmatrix} 1 & & -h_{1l} & & \\ & \ddots & \vdots & & \\ & & 1/h_{rl} & & \\ & & \vdots & \ddots & \\ & & -h_{ml}/h_{rl} & & 1 \end{bmatrix} \tag{2}$$

If the current basis inverse is computed using regular multiplication, then the complexity of the PFI is $\Theta\left(m^3\right)$.

3.2 A Modification of Product Form of the Inverse

MPFI updating scheme has been presented by Benhamadou [4]. The key idea is that the current basis inverse $(A_{\overline{B}})^{-1}$ can be computed from the previous inverse $(A_B)^{-1}$ using a simple outer product of two vectors and one matrix addition, as shown in (3).

$$(A_{\overline{B}})^{-1} = \left(A_{\overline{B}_{r.}}\right)^{-1} + v \otimes \left(A_{B_{r.}}\right)^{-1} \tag{3}$$

The updating scheme of the inverse is shown in (4).

$$\begin{array}{c} (A_B)^{-1} : \begin{vmatrix} b_{r1} & \cdots & b_{rr} & \cdots & b_{rm} \end{vmatrix} \\ (A_{\overline{B}})^{-1} = \begin{vmatrix} b_{11} & \cdots & b_{1m} \\ \vdots & \ddots & \vdots \\ 0 & 0 & 0 \\ \vdots & \ddots & \vdots \\ b_{m1} & \cdots & b_{mm} \end{vmatrix} + \begin{vmatrix} -\frac{h_{1l}}{h_{rl}} \\ \vdots \\ -\frac{1}{h_{rl}} \\ \vdots \\ -\frac{h_{ml}}{h_{rl}} \end{vmatrix} \end{array} \tag{4}$$

The outer product requires m^2 multiplications and the addition of two matrices requires m^2 additions. The total cost of the above method is $2m^2$ operations (multiplications and additions). Hence, the complexity is $\Theta(m^2)$.

4 Parallel Revised Simplex Algorithm

The parallelization of all the individual steps of the revised simplex algorithm is limited and very hard to achieve. However, it is also essential for any algorithm to perform basis inverse in parallel with simplex iterations, otherwise basis inverse will become the dominant step and limit the possible speedup. Our parallel implementation focuses on the reduction of the time taken to perform the basis inverse. The basis inversion is done with the Product Form of the Inverse and a Modification of the Product Form of the Inverse, as described in the previous section.

Both methods take as input the previous basis inverse $(A_B)^{-1}$, the pivot column (h_l), the index of the leaving variable (k) and the number of the constraints (m).

The most time-consuming step of PFI scheme is the matrix multiplication of (1). Our parallel algorithm uses the block matrix multiplication algorithm for this step. This algorithm suggests a recursive divide-and-conquer solution, as described in [11, 17]. This method has significant potential for parallel implementations, especially on shared memory implementations. Block size plays important role on the total performance of the block matrix multiplication. In order to choose the suitable value in our implementation, we have experimented with this parameter and found that the appropriate block size is 16 when using 4 cores.

Let us assume that we have p processors. Table 2 shows the steps that we used to compute the new basis inverse $(A_{\overline{B}})^{-1}$ with the PFI scheme. Table 3 shows the steps that we used to compute the new basis inverse $(A_{\overline{B}})^{-1}$ with the MPFI scheme.

Table 2 Parallel PFI

Step 0.
Compute the column vector:

$$v = \left[-\frac{h_{1l}}{h_{rl}} \cdots \frac{1}{h_{rl}} \cdots -\frac{h_{ml}}{h_{rl}} \right]^{\mathrm{T}}$$

Each processor computes in parallel m/p elements of v.

Step 1.
Replace the rth column of an identity matrix with the column vector v. Each processor assigns in parallel m/p elements to the identity matrix. This matrix is the inverse of the Eta-matrix.

Step 2.
Compute the new basis inverse using (1) with block matrix multiplication. Each processor will compute m/p rows of the new basis.

Table 3 Parallel MPFI

Step 0.
Compute the column vector:

$$v = \left[-\frac{h_{1l}}{h_{rl}} \cdots \frac{1}{h_{rl}} \cdots -\frac{h_{ml}}{h_{rl}} \right]^{\mathrm{T}}$$

Each processor computes in parallel m/p elements of v.

Step 1. (The following steps are computed in parallel)

Step 1.1. Compute the outer product $v \otimes (A_{B_r})^{-1}$ with block matrix multiplication.

Step 1.2. Copy matrix $(A_B)^{-1}$ to matrix $(A_{\overline{B}})^{-1}$. Set the rth row of $(A_{\overline{B}})^{-1}$ equal to zero. Each processor computes in parallel m/p rows of $(A_{\overline{B}})^{-1}$.

Step 2.
Compute the new basis inverse using relation (3). Each processor computes in parallel m/p rows of the new basis.

5 Computational Experiments

In this section we report the computational results of running our implementations on a set of LPs available through Netlib. The three most usual approaches to analyzing algorithms are (1) worst-case analysis, (2) average-case analysis and (3) experimental analysis. Computational studies have been proven useful tools in order to examine the practical efficiency of an algorithm or even compare algorithms by using the same problem sets. The computational comparison has been performed on a quad-processor Intel Xeon 3.2 GHz with 2 Gbyte of main memory running under Ubuntu 10.10 64-bit and performed on GCC 4.5.2. In the following computational results all reported CPU times were measured in seconds. The algorithms have been implemented using C++ and OpenMP using the techniques discussed in Sect. 4. In all LPs from the Netlib collection, the parallel versions of the simplex algorithm converge to the same solution.

Table 4 Statistics of the benchmarks

Problem	Constraints	Variables	Non-zeros A	Sparsity A (%)
agg	488	163	2,410	3.03
agg2	516	302	4,284	2.75
agg3	516	302	4,300	2.76
bandm	305	472	2,494	1.73
brandy	220	249	2,148	3.92
e226	223	282	2,578	4.10
fffff800	524	854	6,227	1.39
israel	174	142	2,269	9.18
lotfi	153	308	1,078	2.29
sc105	105	103	280	2.59
sc205	205	203	551	1.32
scfxm1	330	457	2,589	1.72
scfxm2	660	914	5,183	0.86
scfxm3	990	1,371	7,777	0.57
scrs8	490	1,169	3,182	0.56
share1b	117	225	1,151	4.37
share2b	96	79	694	9.15
ship04l	402	2,118	6,332	0.74
ship04s	402	1,458	4,352	0.74
ship08l	778	4,283	12,802	0.38
ship08s	778	2,387	7,114	0.38
ship12l	1,151	5,427	16,170	0.26
ship12s	1,151	2,763	8,178	0.26
stocfor1	117	111	447	3.44
klein2	477	54	4,585	17.80
klein3	994	88	12,107	13.84
Average	474.96	1,044.84	4,754.88	

5.1 *Problem Instances*

The test set used in our experiments were the Netlib set of LPs. The Netlib library is a well-known suite containing many real-world LPs. Ordóñez and Freund [22] have shown that 71% of the Netlib LPs are ill-conditioned.

Below there are some useful information about the data set, which was used in the computational study. The first column of Table 4 includes the name of the problem, the second the number of constraints, the third the number of variables, the fourth the nonzero elements of matrix A and the fifth the density of the coefficient matrix A. Let $nnz(A)$ denote the number of nonzeros in the matrix A. The density of matrix A is defined as the ratio of the $nnz(A)$ to the total number of its elements.

All LPs have been presolved. The purpose of the presolve analysis is to improve linear problem's numerical properties and computational characteristics. The last row of each table shows the average value of each column.

Table 5 Basis inverse and total time of the serial and parallel implementations

	Serial implementations				Parallel implementations			
	PFI		MPFI		PFI		MPFI	
Problem	Time of basis inverse	Total time	Time of basis inverse	Total time	Time of basis inverse	Total time	Time of basis inverse	Total time
agg	2.83	4.58	2.28	4.05	1.52	3.32	1.13	2.95
agg2	3.54	5.65	2.78	4.97	1.81	3.96	1.52	3.69
agg3	3.28	5.61	2.62	5.03	1.85	4.05	1.48	3.78
bandm	1.01	1.62	0.74	1.41	0.84	1.52	0.61	1.29
brandy	1.30	2.76	1.06	2.55	1.04	2.48	0.76	2.22
e226	1.66	3.34	1.38	3.09	1.22	2.82	0.85	2.50
fffff800	6.10	12.77	4.86	11.64	5.18	11.58	4.31	10.89
israel	0.82	1.73	0.63	1.65	0.53	1.48	0.45	1.31
lotfi	0.33	0.85	0.29	0.80	0.25	0.78	0.21	0.68
sc105	0.08	0.09	0.03	0.07	0.01	0.07	0.02	0.06
sc205	0.55	0.91	0.51	0.85	0.40	0.74	0.20	0.56
scfxm1	3.72	7.11	2.96	6.38	3.17	6.31	2.49	5.80
scfxm2	30.76	62.34	24.26	56.40	26.34	58.56	21.22	52.54
scfxm3	109.06	244.48	83.97	219.22	91.67	224.76	72.93	209.23
scrs8	11.17	22.20	8.69	19.81	9.56	20.20	7.34	17.90
share1b	0.15	0.29	0.09	0.26	0.11	0.25	0.08	0.23
share2b	0.05	0.11	0.04	0.10	0.01	0.10	0.03	0.09
ship04l	5.20	14.53	4.18	13.65	4.43	13.65	3.56	12.90
ship04s	1.76	4.55	1.52	4.31	1.54	4.22	1.21	4.10
ship08l	33.78	94.70	26.30	86.45	30.02	91.90	22.10	82.50
ship08s	7.43	19.07	5.99	17.49	6.50	18.09	5.01	16.95
ship12l	120.21	335.98	89.94	305.95	100.05	317.80	75.64	292.00
ship12s	17.20	43.52	13.45	39.16	12.99	42.84	10.77	36.60
stocfor1	0.04	0.06	0.03	0.05	0.02	0.06	0.02	0.04
klein2	17.36	33.01	13.53	28.80	9.85	25.30	7.55	22.74
klein3	192.50	413.33	148.69	368.50	106.92	326.70	75.30	295.01
Average	22.00	51.35	16.95	46.26	16.07	45.52	12.18	41.48

Table 6 Basis inverse and total time of the serial and parallel implementations

	Speedup			
	PFI		MPFI	
Problem	Basis inverse	Total	Basis inverse	Total
agg	1.86	1.38	2.02	1.37
agg2	1.96	1.43	1.83	1.35
agg3	1.77	1.39	1.77	1.33
bandm	1.20	1.07	1.21	1.09
brandy	1.25	1.11	1.39	1.15
e226	1.36	1.18	1.62	1.24
fffff800	1.18	1.10	1.13	1.07
israel	1.55	1.17	1.40	1.26
lotfi	1.32	1.09	1.38	1.18
sc105	8.00	1.29	1.50	1.17
sc205	1.38	1.23	2.55	1.52
scfxm1	1.17	1.13	1.19	1.10
scfxm2	1.17	1.06	1.14	1.07
scfxm3	1.19	1.09	1.15	1.05
scrs8	1.17	1.10	1.18	1.11
share1b	1.36	1.16	1.13	1.13
share2b	5.00	1.10	1.33	1.11
ship04l	1.17	1.06	1.17	1.06
ship04s	1.14	1.08	1.26	1.05
ship08l	1.13	1.03	1.19	1.05
ship08s	1.14	1.05	1.20	1.03
ship12l	1.20	1.06	1.19	1.05
ship12s	1.32	1.02	1.25	1.07
stocfor1	2.00	1.00	1.50	1.25
klein2	1.76	1.30	1.79	1.27
klein3	1.80	1.27	1.97	1.25
Average	1.79	1.15	1.44	1.17

5.2 *Computational Results*

The algorithms described in Sect. 4 have been experimentally implemented. Table 5 presents the results from the execution of the serial and parallel implementations of the above-mentioned updating schemes. For each implementation, the table shows the CPU time for the basis inverse and the total CPU time.

In order to show more clearly the superiority of parallel implementations over the serial ones, we provide Table 6. Table 6 presents the speedup obtained by the parallel implementations regarding the CPU time for the basis inverse and the total CPU time, for both PFI and MPFI schemes. We now plot the ratios taken from Table 6 in Fig. 1. The total time is in logarithmic scale.

From the above results, we observe: (1) the MPFI scheme is in most problems faster than PFI both in serial and in parallel implementation, (2) using PFI scheme,

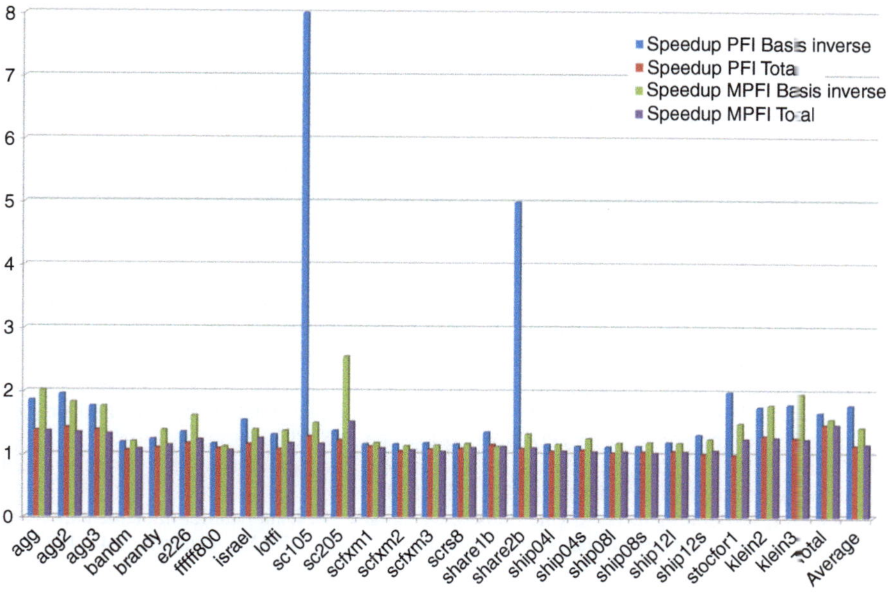

Fig. 1 Basis inverse and total time of the serial and parallel implementations

the speedup gained from the parallelization is of average 1.79 for the time of basis inverse and 1.15 for total time, and (3) using MPFI scheme, the speedup is of average 1.44 for the time of basis inverse and 1.17 for total time. Super-linear speedup for problems sc105 and share2b sometimes occurs due to a reduction in processor idle time when using multiple threads.

Notice that the speedup is above the ideal one due to the control flow. Revised simplex algorithm includes many steps that present data-dependency relations, which can affect the speedup. Two of these steps are the choice of the entering and leaving variables.

6 Conclusions

A parallel implementation for the revised simplex algorithm has been described in this paper. Some preliminary computational results on Netlib problems have reported a speedup of average 1.79 and 1.44 regarding the basis inverse procedure, using PFI and MPFI updating schemes, respectively. These results could be further improved by performance optimization. In future work, we plan to implement our parallel algorithm combining the Message Passing Interface (MPI) and OpenMP programming models to exploit parallelism beyond a single level. Furthermore, we intend to port our algorithm to a GPU implementation based on the CUDA architecture.

References

1. Agrawal, A., Blelloch, G.E., Krawitz, R.L., Phillips, C.A.: Four vector-matrix primitives. In: Proceedings of the ACM Symposium on Parallel Algorithms and Architectures, pp. 292–302 (1989)
2. Badr, E.S., Moussa, M., Papparrizos, K., Samaras, N., Sifaleras, A.: Some computational results on MPI parallel implementations of dense simplex method. In: Proceedings of World Academy of Science, Engineering and Technology. Presented in the 17th International Conference on Computer & Information Science and Engineering (CISE 2006), 8–10 December, Cairo, Egypt, vol. 23, pp. 39–42 (2006)
3. Barr, R.S., Hickman, B.L.: Parallel simplex for large pure network problems: computational testing and sources of speedup. Oper. Res. **42(1)**, 65–80 (1994)
4. Benhamadou, M.: On the simplex algorithm "revised form". Adv. Eng. Softw. **33**, 769–777 (2002)
5. Bieling, J., Peschlow, P., Martini, P.: An efficient GPU implementation of the revised simplex method. In: Proceedings of the IPDPS Workshops, pp. 1–8 (2010)
6. Chang, M.D., Engquist, M., Finkel, R., Meyer, R.R.: A parallel algorithm for generalized networks. Ann. Oper. Res. **14**(1–4), 125–145 (1988)
7. Eckstein, J., Boduroglu, I., Polymenakos, L., Goldfarb, D.: Data-parallel implementations of dense simplex methods on the connection machine CM-2. ORSA J. Comput. **7(4)**, 402–416 (1995)
8. Finkel, R.A.: Large-grain parallelism: three case studies. In: Jamieson, H. (ed.) Proceedings of Characteristics of Parallel Algorithms. MIT, Cambridge (1987)
9. Gay, D.M.: Electronic mail distribution of linear programming test problems. Math. Program. Soc. COAL Newslett. **13**, 10–12 (1985)
10. Greeff, G.: The revised simplex method on a GPU. Stellenbosch University, South Africa, Honours Year Project (2004)
11. Hake, J.F.: Parallel algorithms for matrix operations and their performance in multiprocessor systems. In: Kronsjo, L., Shumsheruddin, D. (edS.) Advances in Parallel Algorithms. Halsted Press, New York (1993)
12. Hall, J.A.J., McKinnon, K.I.M.: PARSMI, a parallel revised simplex algorithm incorporating minor iterations and Devex pricing. In: Wasniewski, J., Dongarra, J., Madsen, K., Olesen, D. (eds.) Applied Parallel Computing. LNCS, vol. 1184. Springer, Berlin (1996)
13. Hall, J.A.J.: SYNPLEX: a task-parallel scheme for the revised simplex method. Contributed talk at the 2nd International Workshop on Combinatorial Scientific Computing (2005)
14. Hall, J.A.J.: Towards a practical parallelisation of the simplex method. Comput. Manag. Sci. **7**, 139–170 (2010)
15. Hall, J.A.J., McKinnon, K.I.M.: ASYNPLEX an asynchronous parallel revised simplex algorithm. Ann. Oper. Res. **81**, 27–50 (1998)
16. Helgason, R.V., Kennington, J.L., Zaki, H.A.: A parallelization of the simplex method. Ann. Oper. Res. **14**(1–4), 17–40 (1988)
17. Horowitz, E., Zorat, A.: Divide-and-conquer for parallel processing. IEEE Trans. Comput. **C-32**(6), 582–585 (1983)
18. Jung, J.H., O'Leary, D.P.: Implementing an interior point method for linear programs on a CPU-GPU system. Electr. Trans. Numer. Anal. **28**, 174–189 (2008)
19. Lentini, M., Reinoza, A., Teruel, A., Guillen, A.: SIMPAR: a parallel sparse simplex. Comput. Appl. Math. **14(1)**, 49–58 (1995)
20. Luo, J., Reijns, G.L.: Linear programming on transputers. In: van Leeuwen, J. (ed.) Algorithms, Software, Architecture. IFIP Transactions A (Computer Science and Technology). Elsevier, Amsterdam (1992)
21. Mamalis, B., Pantziou, G., Kremmydas D., Dimitropoulos, G.: Reexamining the parallelization schemes for standard full tableau simplex method on distributed memory environments. In: Proceedings of the 10th IASTED PDCN Conference, pp. 115–123 (2011)

22. Ordóñez, F., Freund, R.: Computational experience and the explanatory value of condition measures for linear optimization. SIAM J. Optim. **14(2)**, 307–333 (2003)
23. Owens, J.D., Houston, M., Luebke, D., Green, S., Stone, J.E.: GPU Computing. Proc. IEEE **96**(5), 879–899 (2008)
24. Peters, J.: The network simplex method on a multiprocessor. Networks **20(7)**, 845–859 (1990)
25. Shu, W.: Parallel implementation of a sparse simplex algorithm on MIMD distributed memory computers. J. Parallel Distr. Comput. **31**(1), 25–40 (1995)
26. Shu, W., Wu, M.: Sparse implementation of revised simplex algorithms on parallel computers. In: Proceedings of the 6th SIAM Conference on Parallel Processing for Scientific Computing, Norfolk (1993)
27. Spampinato, D.G., Elster, A.C.: Linear optimization on modern GPUs. In: Proceedings of the 2009 IEEE International Symposium on Parallel and Distributed Processing (2009)
28. Stunkel, C.B.: Linear optimization via message-based parallel processing. In: Proceedings of the International Conference on Parallel Processing, vol. 3, pp. 264–271 (1988)
29. Thomadakis, M.E., Liu, J.C.: An efficient steepest-edge simplex algorithm for SIMD computers. In: Proceedings of the International conference on Super-Computing (1996)
30. Yarmish, G., Slyke, R.V.: A distributed scaleable simplex method. J. Supercomput. **49**(3), 373–381 (2009)

Maximum Induced Matchings in Grids

Ruxandra Marinescu-Ghemeci

Abstract An induced matching in a graph is a matching such that no two edges are joined by an edge of G. For a connected graph G, denote by $i\mu(G)$ the maximum cardinality of an induced matching in G. In this paper we study the proble of finding a maximum induced matching in grid graphs with n lines and m columns—$G_{n,m}$, and determine the exact value for $i\mu(G_{n,m})$ when n or m are even.

1 Introduction

Let $G = (V,E)$ be a connected graph. A *matching* in G is a subset of edges $M \subseteq E$ such that no two edges of M are adjacent. A matching M in G is called *induced* if no two edges of M are joined by an edge of G. Denote by $i\mu(G)$ the maximum cardinality of an induced matching in G. A *maximum induced matching* in G is an induced matching with $i\mu(G)$ edges [2, 11]. The problem of finding a maximum induced matching of a graph (MIM) was introduced as a variation of the maximum matching problem, motivated by the "risk-free" marriage problem. Induced matchings have many applications, for example in networking [1, 7] and discrete mathematics [6]. MIM is proved to be NP-hard [2, 11] and it remains NP-hard even for bipartite graphs of maximum degree 3 [9, 10], line graphs [8], and for k-regular bipartite graph for any $k \geq 3$ [5].

Denote by $G_{n,m}$ the grid graph with n lines and m columns. It is known that MIM in subgrids is NP-hard [1]. In this paper we study the maximum induced problem in grids and determine the exact value for $i\mu(G_{n,m})$ when n or m are even. In the terms of covering a chessboard with domino pieces, the problem of finding $i\mu(G_{n,m})$ can be formulated as follows: given a $n \times m$ chessboard, how many domino pieces can

R. Marinescu-Ghemeci (✉)
University of Bucharest, Str. Academiei, 14, Bucharest, Romania
e-mail: verman@fmi.unibuc.ro

A. Migdalas et al. (eds.), *Optimization Theory, Decision Making, and Operations Research Applications*, Springer Proceedings in Mathematics & Statistics 31,
DOI 10.1007/978-1-4614-5134-1_12,

be placed on the chessboard such that each domino piece covers exactly two squares and two pieces placed on the chessboard can touch only in corners (not on edges).

2 Maximum Induced Matching in Grids

Let $G_{n,m}$ be the grid graph with n lines and m columns and M a matching in $G_{n,m}$. A vertex $v \in V$ is called *M-saturated* if $v \in V(M)$ and *M-unsaturated* otherwise. If H is an induced subgraph in $G_{n,m}$, we denote by $s_M(H) = |V(M) \cap V(H)|$ the number of M-saturated vertices of H. Also, we denote by $s\mu(G_{n,m})$ the maximum number of vertices that can be saturated in a subgrid isomorphic to $G_{n,m}$ of a grid by an induced matching of the grid:

$$s\mu(G_{n,m}) = \max\{s_M(H)|\ H \cong G_{n,m}\, subgrid\, in\, a\, grid G_{p,q},\\ M\, induced\, matching\, in\, G_{p,q}\}.$$

Then $s\mu(G_{n,m}) = s\mu(G_{m,n})$ and $i\mu(G_{m,n}) = i\mu(G_{n,m}) \leq \left\lfloor \frac{s\mu(G_{n,m})}{2} \right\rfloor$.

Let $H_1,\ldots,H_p$ be subgraphs in $G_{n,m}$. We say that $(H_1,\ldots,H_p)$ is a partition of $G_{n,m}$ if the subgraphs are vertex-disjoint and $V(G_{n,m}) = \bigcup_{i=1}^{p} V(H_i)$.

Partition $(H_1,\ldots,H_p)$ is called a $(k_1G_{n_1,m_1}, k_2G_{n_2,m_2},\ldots,k_qG_{n_q,m_q})$-partition of $G_{n,m}$ if k_i subgraphs from partition are isomorphic to G_{n_i,m_i}, for $1 \leq i \leq q$.

Lemma 1. *The following properties hold:*

(a) $s\mu(G_{3,3}) = 5$ *and equality holds only for two configurations (modulo a rotation), shown in Fig. 1a.*

(b) $s\mu(G_{3,1}) = 2$; *(c)* $s\mu(G_{3,2}) = 4$; *(d)* $s\mu(G_{3,4}) = 6$; *(e)* $s\mu(G_{2,2}) = 2$.

Proof. As illustrated in Fig. 1, we have $s\mu(G_{3,3}) \geq 5$, $s\mu(G_{3,1}) \geq 2$, $s\mu(G_{3,2}) \geq 4$, $s\mu(G_{3,4}) \geq 6$, $s\mu(G_{2,2}) \geq 2$. The reverse inequalities can be easily proved by carefully analyzing all the situations that can occur, based on the remark that if u,v,w are 3 distinct vertices such that uv and vw are edges, then at most 2 of them can be saturated by an induced matching. □

Theorem 1. *Let $m,n \geq 2$ be two positive integers, with m even. Then*

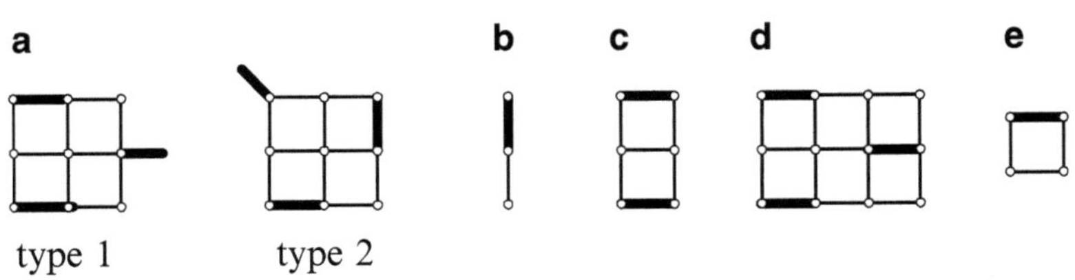

Fig. 1 All possible configurations for $s\mu(G_{n,m})$, with $G_{n,m} \in \{G_{3,3}, G_{3,1}, G_{3,2}, G_{3,4}, G_{2,2}\}$

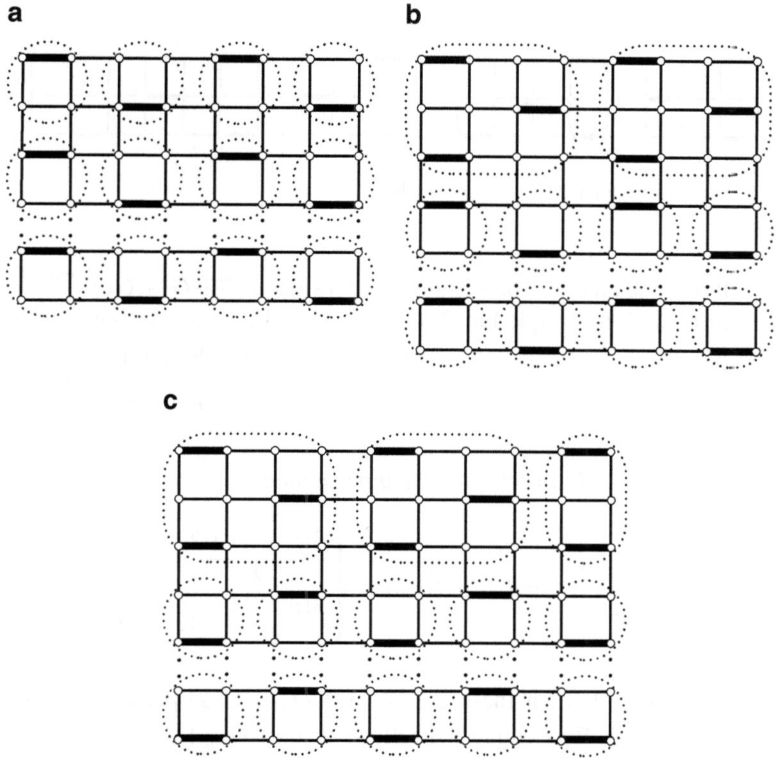

Fig. 2 Configurations for $s\mu(G_{n,m})$, with m even

$$s\mu(G_{n,m}) = \begin{cases} \dfrac{mn+2}{2}, & if\, n\, is\, odd\, and\, m = 4k+2 \\ \dfrac{mn}{2}, & otherwise \end{cases} \quad and \quad i\mu(G_{n,m}) = \left\lceil \frac{mn}{4} \right\rceil.$$

Proof. We have $i\mu(G_{n,m}) \leq \frac{s\mu(G_{n,m})}{2}$.

Case 1. If n is even, then we consider a $(\frac{mn}{4}G_{2,2})$-partition of $G_{n,m}$ (for example see Fig. 2a). Using Lemma 1 for every subgraph $H \cong G_{2,2}$ from the partition we obtain $s\mu(G_{n,m}) \leq 2 \cdot \frac{mn}{4} = \frac{mn}{2}$. The matching M' described in Fig. 2a is an induced matching in $G_{n,m}$ which saturates $\frac{mn}{2}$ vertices, so $i\mu(G_{n,m}) = \frac{s\mu(G_{n,m})}{2} = \frac{mn}{4}$.

Case 2. If n is odd and $m = 4k$ there exists a $(\frac{m}{4}G_{3,4}, \frac{m(n-3)}{4}G_{2,2})$-partition of $G_{n,m}$ (for example Fig. 2b). Then from Lemma 1 for subgraphs isomorphic to $G_{2,2}$ and $G_{3,4}$, we have $s\mu(G_{n,m}) \leq 6 \cdot \frac{m}{4} + 2 \cdot \frac{m(n-3)}{4} = \frac{mn}{2}$. The matching M' described in Fig. 2b is an induced matching which saturates $\frac{mn}{2}$ vertices, hence again we obtain $i\mu(G_{n,m}) = \frac{s\mu(G_{n,m})}{2} = \frac{mn}{4}$.

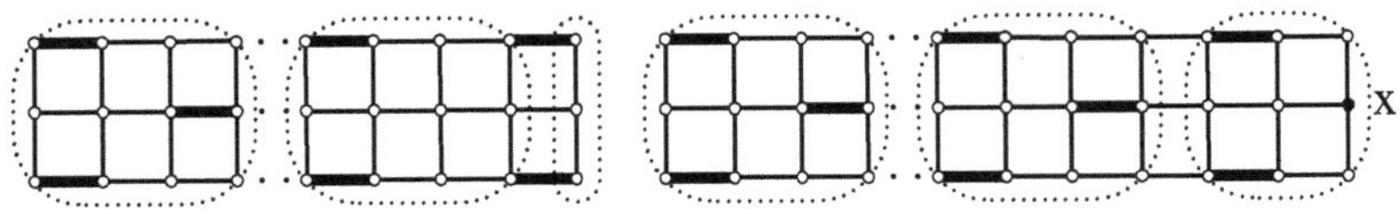

Fig. 3 Configurations for $s\mu(G_{3,m})$, with m odd

Case 3. If n is odd and $m = 4k+2$, then we consider a $(\frac{m-2}{4}G_{3,4}, G_{3,2}, \frac{m(n-3)}{4}G_{2,2})$-partition of $G_{n,m}$ (for example Fig. 2c). Using again Lemma 1 we have $s\mu(G_{n,m}) \leq 6\cdot\frac{m-2}{4} + 4 + 2\cdot\frac{m(n-3)}{4} = \frac{mn+2}{2}$. An induced matching which saturates $\frac{mn+2}{2}$ vertices is described in Fig. 2c, hence in this case we have $i\mu(G_{n,m}) = \frac{s\mu(G_{n,m})}{2} = \frac{mn+2}{4} = \lceil\frac{mn}{4}\rceil$. □

Theorem 2. *If $m \geq 3$ is an odd positive integer, then*

$$s\mu(G_{3,m}) = \frac{3m+1}{2} \quad \textit{and} \quad i\mu(G_{3,m}) = \begin{cases} \frac{3(m-1)+2}{4}, & \textit{if} \quad m = 4k+3 \\ \frac{3(m-1)}{4} + 1, & \textit{otherwise.} \end{cases}$$

Proof. If $m = 4k+1$, then there exists a $(\frac{m-1}{4}G_{3,4}, G_{3,1})$-partition of $G_{3,m}$ (for example see Fig. 3a). By Lemma 1 it follows that $s\mu(G_{3,m}) \leq 6\cdot\frac{m-1}{4} + 2 = \frac{3m+1}{2}$. An induced matching in $G_{3,m}$ that saturates $\frac{3m+1}{2}$ vertices is described in Fig. 3a. We obtain $i\mu(G_{3,m}) = \frac{s\mu(G_{3,m})}{2} = \frac{3m+1}{4}$.

If $m = 4k+3$, then we consider a $(\frac{m-3}{4}G_{3,4}, G_{3,3})$-partition of $G_{3,m}$ (for example see Fig. 3b). Again, by Lemma 1, it follows that an induced matching in a grid can saturate at most $6\cdot\frac{m-3}{4} + 5 = \frac{3m+1}{2}$ vertices of a subgrid isomorphic to $G_{3,m}$. Since this bound is odd and a matching in a graph saturates an even number of vertices, it follows that $i\mu(G_{3,m}) \leq \frac{s\mu(G_{3,m})-1}{2} \leq \frac{3m-1}{4}$. An induced matching in $G_{3,m}$ that saturates $\frac{3m-1}{2}$ vertices is described in Fig. 3b. Moreover, if we consider $G_{3,m}$ as subgrid in a grid, this matching can be easily extended such that vertex x is also saturated. Hence we obtain $s\mu(G_{3,m}) = \frac{3m+1}{2}$ and $i\mu(G_{3,m}) = \frac{3m-1}{4}$. □

Theorem 3. *Let n, m be two odd integers with $m \geq n \geq 5$. Then*

$$i\mu(G_{n,m}) \geq \begin{cases} \frac{n(m-1)}{4} + 1 & \textit{if}\ m = 4k+1 \\ \frac{n(m-1)+2}{4} & \textit{if}\ m = 4k+3\ \textit{and}\ n < \frac{m+5}{2} \\ \frac{n(m-1)+2}{4} + 1 & \textit{if}\ m = 4k+3\ \textit{and}\ \frac{m+5}{2} \leq n. \end{cases}$$

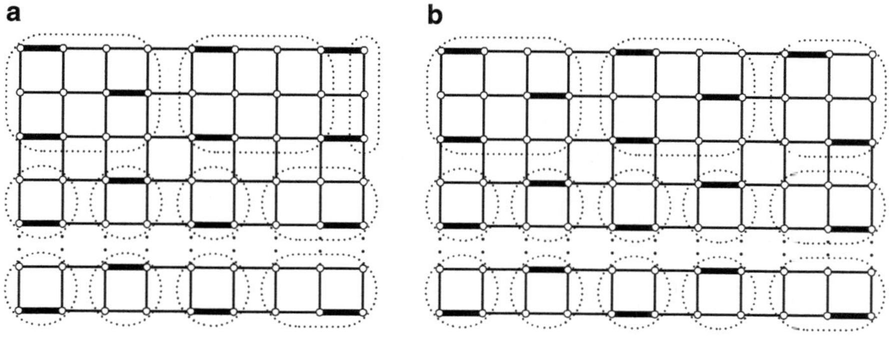

Fig. 4 Maximum induced matchings in $G_{n,m}$, with m,n odd

Proof. If $m=4k+1$, let M be the matching described in Fig. 4a. By considering the $(\frac{m-1}{4}G_{3,4},G_{3,1},\frac{(n-3)(m-3)}{4}G_{2,2},\frac{n-3}{2}G_{2,3})$-partition of $G_{n,m}$ suggested in the figure, it is easy to see that the number of M-saturated vertices is

$$s_M(G_{n,m})=6\cdot\frac{m-1}{4}+2+2\cdot\frac{m-3}{2}\cdot\frac{n-3}{2}+2\cdot\frac{n-3}{2}=\frac{n(m-1)+4}{2}$$

Hence $i\mu(G_{n,m})\geq|M|=\frac{s_M(G_{n,m})}{2}=\frac{n(m-1)}{4}+1.$

Assume now that $m=4k+3$. If $n<\frac{m+5}{2}$, let M be the matching described in Fig. 4b. By considering the $(\frac{m-3}{4}G_{3,4},G_{3,2},\frac{(n-3)(m-3)}{4}G_{2,2},\frac{n-3}{2}G_{2,3})$-partition of $G_{n,m}$ suggested in the figure, we obtain that the number of M-saturated vertices is

$$s_M(G_{n,m})=6\cdot\frac{m-3}{4}+4+2\cdot\frac{n-3}{2}+2\cdot\frac{m-3}{2}\cdot\frac{n-3}{2}=\frac{n(m-1)+2}{2}$$

Hence $i\mu(G_{n,m})\geq|M|=\frac{s_M(G_{n,m})}{2}=\frac{n(m-1)+2}{4}.$

If $n\geq\lceil\frac{m+5}{2}\rceil$ let M be the matching described in Fig. 5. By considering the $(G_{\frac{m+5}{2}},(n-\frac{m+5}{2}-1)G_{2,m},G_{1,m})$-partition of $G_{n,m}$ shown in figure and counting the number of saturated vertices from each subgraph of partition we obtain

$$\begin{aligned}|M|&=\frac{s_M(G_{n,m})}{2}=\frac{(m-1)(m+5)}{8}+1+\frac{1}{2}\left(n-\frac{m+5}{2}-1\right)\frac{m-1}{2}+\frac{m-1}{4}\\&=\frac{n(m-1)+6}{4}.\end{aligned}$$

□

Theorem 4. *Let $m\geq 5$ be an odd positive integer. Then*

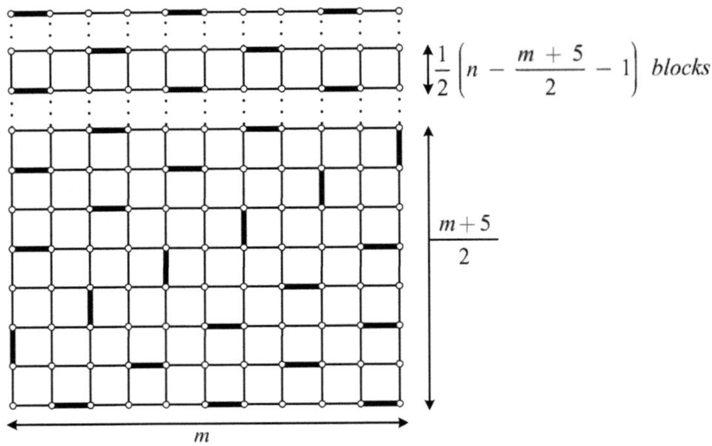

Fig. 5 Maximum induced matchings in $G_{n,m}$, with $m = 4k+3$ and $n \geq (m+5)/2$

(a) $s\mu(G_{5,m}) = \dfrac{5m+1}{2}$ *and* $i\mu(G_{5,m}) = \begin{cases} \frac{5(m-1)+2}{4}, & if \quad m = 4k+3 \\ \frac{5(m-1)}{4}+1, & otherwise. \end{cases}$

(b) *If* $m = 4k+1$ *then* $s\mu(G_{7,m}) = \dfrac{7m-1}{2}$ *and* $i\mu(G_{7,m}) = \dfrac{7(m-1)}{4}+1.$

(c) *If* $m = 4k+3$ *then* $s\mu(G_{7,m}) = \dfrac{7m+3}{2}$, $i\mu(G_{7,7}) = 12$ *and* $\dfrac{7(m-1)+2}{4} \leq i\mu(G_{7,m}) \leq \dfrac{7(m-1)+2}{4}+1.$

Proof. Let $G = G_{p,q}$ be a grid and M an induced matching in G. Let H be a subgraph isomorphic to $G_{n,m}$ in G. Denote by y^i_j the vertex from line i and column j in H and by $G^{i:s}_{j:k}$ the subgrid induced in H by vertices y^a_b with $i \leq a \leq s$, $j \leq b \leq k$.

(a) Let $n = 5$. We consider two $(G_{2,m}, G_{3,m})$-partitions of H: first partition determined by subgraphs $H_1 = G^{1:2}_{1:m} \cong G_{2,m}$ induced by the first two lines, and $H_2 = G^{3:5}_{1:m} \cong G_{3,m}$ induced by the remaining 3 lines, and second determined by subgraphs $H'_1 = G^{4:5}_{1:m} \cong G_{2,m}$ induced by the last two lines, and $H'_2 = G^{1:3}_{1:m} \cong G_{3,m}$. We have $s_M(H) \leq s_M(H_1) + s_M(H_2)$. By Theorems 1 and 2, $s_M(H_1) \leq m+1$ and $s_M(H_2) \leq \frac{3m+1}{2}$.

Case 1. $m = 4k+1$

Consider first $m = 5$.

Assume that $s_M(H_1) = m+1 = 6$ and $s_M(H_2) = \frac{3m+1}{2} = 8$. We consider the $(G_{2,2}, G_{2,3})$-partition of H_1 determined by $G^{1:2}_{1:2}$ and $G^{1:2}_{3:5}$. In order to have $s_M(H_1) = 6 = s\mu(G_{2,5})$, we must have $s_M(G^{1:2}_{3:5}) = s\mu(G_{2,3}) = 4$ and $s_M(G^{1:2}_{1:2}) = s\mu(G_{2,2}) = 2$. It follows that $y^1_3y^2_3, y^1_5y^2_5 \in M$ and then $y^1_1y^2_1 \in M$. Moreover, since $s_M(G^{3:5}_{1:4}) \leq s\mu(G_{3,4}) = 6$ and $s_M(G^{3:5}_{5:5}) \leq s\mu(G_{3,1}) = 2$, in order to have $s_M(H_2) = 8$ we must have $s_M(G^{3:5}_{1:4}) = 6$ and $s_M(G^{3:5}_{5:5}) = 2$. It follows that $y^4_5y^5_5 \in M$. But then vertices $y^3_1, y^3_3, y^4_4, y^5_4$ are unsaturated, and it is easy to see that in this situation $s_M(G^{3:5}_{1:4}) < 6$, contradiction.

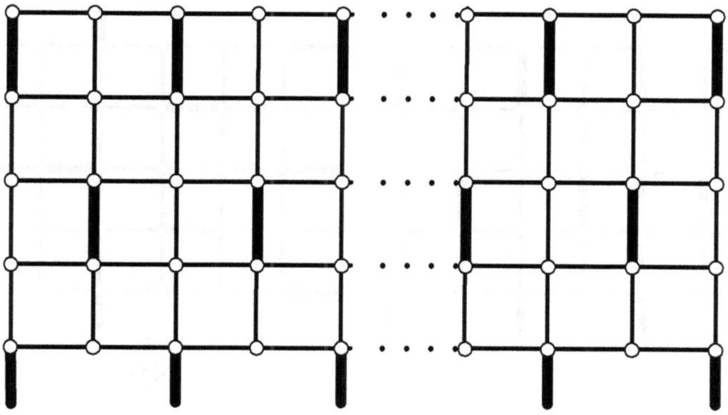

Fig. 6 A configuration for $s\mu(G_{5,m})$, with $m = 4k+1$

It follows that $s_M(H) \le s_M(H_1) + s_M(H_2) - 1 = 13$ and then $s\mu(G_{5,5}) \le 13$. The matching illustrated in Fig. 6 saturates 13 vertices in a subgrid isomorphic to $G_{5,5}$, hence $s\mu(G_{5,5}) = 13$.

If $m > 5$, since $m-5$ is even, by Theorem 1 we have

$$s\mu(G_{5,m}) \le s\mu(G_{5,5}) + s\mu(G_{5,m-5}) = 13 + \frac{5(m-5)}{2} = \frac{5m+1}{2}.$$

As illustrated in Fig. 6 there exists an induced matching which saturates $\frac{5m+1}{2}$ vertices (having only vertical edges), hence $s\mu(G_{5,m}) = \frac{5m+1}{2}$.

Then we also have $i\mu(G_{5,m}) \le \left\lfloor \frac{s\mu(G_{5,m})}{2} \right\rfloor = \left\lfloor \frac{5m+1}{4} \right\rfloor = \frac{5m-1}{4}$. By Theorem 3 we obtain $i\mu(G_{5,m}) \ge \frac{5(m-1)}{4} + 1 = \frac{5m-1}{4}$, hence $i\mu(G_{5,m}) = \frac{5m-1}{4}$.

Case 2. $m = 4k+3$

Consider first $m = 7$.

We will prove that $s_M(H) \le 18$ and determine the configurations for which equality holds.

If $s_M(H_1) < 7$ or ($s_M(H_1) = 7$ and $s_M(H_2) < 11$), then $s_M(H) \le 17$. Then it suffices to consider the following situations.

2.a. $s_M(H_1) = m+1 = 8$ or $s_M(H_1') = 8$.

By symmetry, it suffices to assume $s_M(H_1) = 8$. Considering a $(2G_{2,2}, G_{2,3})$-partition of H_1, as in previous case, it follows that we must have $y_1^1y_1^2, y_3^1y_3^2, y_5^1y_5^2, y_7^1y_7^2 \in M$.

Assume $s_M(H_2) = \frac{3m+1}{2} = 11$. Since $s_M(G^{3:5}_{1:3}) \le s\mu(G_{3,3}) = 5$ and $s_M(G^{3:5}_{4:7}) \le s\mu(G_{3,4}) = 6$, in order to have $s_M(H_2) = 11$ we must have $s_M(G^{3:5}_{1:3}) = 5$. That is impossible since vertices y_1^3, y_3^3 are unsaturated (Lemma 1).

It follows that $s_M(H_2) \le 10$, hence $s_M(H) \le 18$. Moreover, if we have $s_M(H_2) = 10$, then we must have $s_M(G^{3:5}_{1:3}) = 4$ and $s_M(G^{3:5}_{4:7}) = 6$ and this is possible only in situation illustrated in Fig. 7 (c_1).

2.b. $s_M(H_1) = s_M(H_1') = 7$, $s_M(H_2) = s_M(H_2') = 11$.

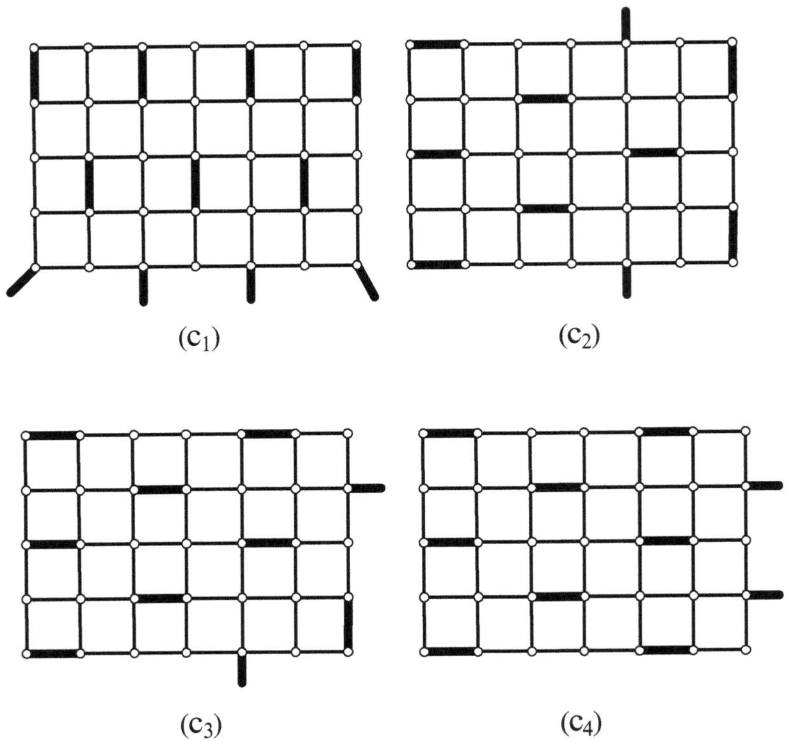

Fig. 7 Configurations for $s\mu(G_{5,7})$

Considering $(G_{3,3}, G_{3,4})$ partitions of H_2 and H_2', by Lemma 1 and Theorem 2 it follows that $s_M(G^{3:5}_{1:3}) = 5$ and $s_M(G^{3:5}_{4:7}) = 6$. By symmetry, using similar arguments, we obtain $s_M(G^{3:5}_{5:7}) = s_M(G^{1:3}_{1:3}) = s_M(G^{1:3}_{5:7}) = 5$ and $s_M(G^{3:5}_{1:4}) = s_M(G^{1:3}_{4:7}) = s_M(G^{1:3}_{1:4}) = 6$.

Moreover, by Theorem 2 we have $s_M(G^{1:5}_{1:3}) \leq 8$, hence $s_M(G^{1:2}_{1:3}) = 3$. Using the same arguments we obtain that $s_M(G^{1:2}_{5:7}) = s_M(G^{4:5}_{1:3}) = s_M(G^{4:5}_{5:7}) = 3$. It follows that $s_M(G^{4:5}_{3:3}) = s_M(G^{1:2}_{3:3}) = s_M(H_1) - 6 = 1$ and $s_M(G^{3:3}_{1:3}) = s_M(G^{3:3}_{5:7}) = 2$.

Then $y_3^1 y_3^2, y_3^4 y_3^5, y_5^1 y_5^2, y_5^4 y_5^5 \notin M$. Moreover, since subgrids $G^{3:5}_{1:3}, G^{3:5}_{5:7}, G^{1:3}_{1:3}, G^{1:3}_{5:7}$ are isomorphic to $G_{3,3}$, by Lemma 1, there exist only two types of configurations possible for each of these subgrids (modulo a rotation), as shown in Fig. 1a. It is easy to see that not all configurations can be of type 2 (since $y_3^1 y_3^2, y_3^4 y_3^5, y_5^1 y_5^2, y_5^4 y_5^5 \notin M$). Then, by symmetry, it suffices to consider the situations when $G^{3:5}_{1:3}$ has a configuration of type 1. By a simple analysis of these situations, it follows that there are only 3 possible configurations for edges of M such that the determined number of saturated vertices are reached, shown in Fig. 7 (c_2), (c_3), (c_4).

In all situations we obtain that $s_M(H) \leq 18$, hence $s\mu(G_{5,7}) \leq 18$.

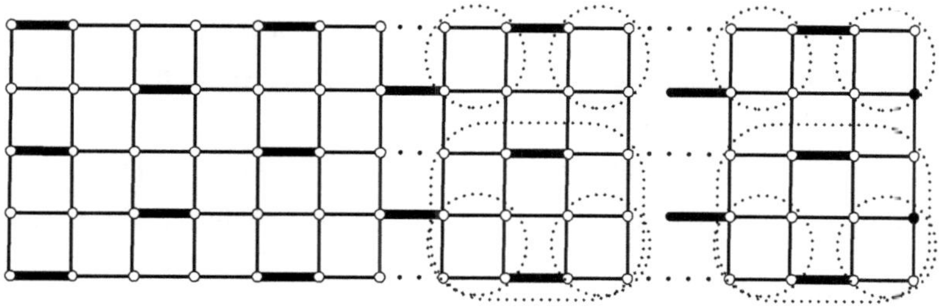

Fig. 8 A configuration for $s\mu(G_{5,m})$, with $m = 4k+3$

Assume now that $m > 7$. Then, since $m-5$ is even, using Theorem 1 we obtain

$$s\mu(G_{5,m}) \leq s\mu(G_{5,7}) + s\mu(G_{5,m-7}) = 18 + \frac{5(m-7)}{2} = \frac{5m+1}{2}.$$

As illustrated in Fig. 6 there exists an induced matching which saturates $\frac{5m+1}{2}$ vertices, hence $s\mu(G_{5,m}) = \frac{5m+1}{2}$.

If $H = G$ (i.e. $p = 5$ and $q = m$), we have

$$s_M(H) \leq s_M(G_{1:7}^{1:5}) + s_M(G_{8:m}^{1:5}) \leq 18 + s\mu(G_{5,m-7}) = \frac{5m+1}{2},$$

and equality can hold only if configuration (c_4) occurs for $G_{1:7}^{1:5}$ and $s_M(G_{8:m}^{1:5}) = \frac{5(m-7)}{2}$.

But, if we consider partitions of $G_{8:m}^{1:5}$ into subgrids isomorphic to $G_{2,2}$ and $G_{3,4}$, as shown in Fig. 8, then each subgrid isomorphic to $G_{2,2}$ must have 2 saturated vertices and each subgrid isomorphic to $G_{3,4}$ must have 6 saturated vertices. It follows that all edges of M must be horizontal and then $s_M(G_{m-1:m}^{1:2}) < 2$, contradiction. Hence $s_M(H) \leq \frac{5m+1}{2} - 1 = \frac{5m-1}{2}$. If M is a maximum induced matching in G we obtain $i\mu(G_{5,m}) \leq \left\lfloor \frac{s_M(H)}{2} \right\rfloor \leq \left\lfloor \frac{5m-1}{4} \right\rfloor = \frac{5m-3}{4}$. From Theorem 3 we have that $i\mu(G_{5,m}) \geq \frac{5m-3}{4}$, hence equality holds.

(b) If $n = 7$ and $m = 4k+1 \geq 7$, then consider $G_1 \cong G_{1:5}^{1:7}$ and $G_2 \cong G_{6:m}^{1:7}$ a $(G_{7,5}, G_{7,m-5})$-partition of $G_{7,m}$. Since $m-5$ is a multiple of 4, $s_M(G) \leq s\mu(G_{7,5}) + s\mu(G_{7,m-5}) = 18 + \frac{7(m-5)}{2}$ and it is easy to see that equality holds for a matching with all edges horizontal, as suggested for $m = 5$ (Fig. 8). If $H = G$ equality can hold only if configuration (c_1) (rotated) occurs for G_1 and $s_M(G_2) = \frac{7(m-5)}{2}$. Using the same arguments as for $n = p = 5$, if we consider a partition of G_2 into subgrids isomorphic to $G_{2,2}$ and $G_{3,4}$, then each subgrid isomorphic to $G_{2,2}$ must have 2 saturated vertices and each subgrid isomorphic to $G_{3,4}$ must have 6 saturated vertices. It follows that all edges of M must be horizontal and then $s_M(G_{m-1:m}^{1:2}) < 2$, contradiction. Hence $s_M(G) \le 17 + \frac{7(m-5)}{2} = \frac{7m-1}{2}$. Then $i\mu(G_{7,m}) \le \lfloor \frac{7m-1}{4} \rfloor = \frac{7m-3}{4}$. By Theorem 3 we have $i\mu(G_{7,m}) \ge \frac{7(m-1)}{4} + 1 = \frac{7m-3}{4}$, hence $i\mu(G_{7,m}) = \frac{7m-3}{4}$, for $m = 4k+1$.

If $m = 4k+3$, using similar arguments, we obtain $s\mu(G_{7,m}) = 18 + \frac{7(m-5)+2}{2} = \frac{7m+3}{2}$ and $i\mu(G_{7,m}) \le \left\lfloor \frac{s\mu(G_{7,m})-1}{2} \right\rfloor = \frac{7m-1}{4}$. By Theorem 3, the result follows. □

Theorem 5. *Let $n,m \ge 2$ be two odd integers. Then*

$$i\mu(G_{n,m}) \le \left\lfloor \frac{nm+1}{4} \right\rfloor.$$

Proof. If $n = 4k+3$, consider a $(G_{3,m}, \frac{n-3}{4} G_{4,m})$-partition of $G_{n,m}$.

Then $i\mu(G_{n,m}) \le \frac{s\mu(G_{3,m})}{2} + \frac{n-3}{4} \cdot \frac{s\mu(G_{4,m})}{2}$. By Theorems 2 and 1 we obtain $i\mu(G_{n,m}) \le \frac{3m+1}{4} + \frac{n-3}{4} \cdot \frac{4m}{4} = \frac{nm+1}{4}$.

If $n = 4k+1$, consider a $(G_{5,m}, \frac{n-5}{4} G_{4,m})$-partition of $G_{n,m}$. Then $i\mu(G_{n,m}) \le \frac{s\mu(G_{5,m})}{2} + \frac{n-5}{4} \cdot \frac{4m}{4}$, and by Theorems 4 and 1 we obtain $i\mu(G_{n,m}) \le \frac{5m+1}{4} + \frac{n-5}{4} \cdot \frac{4m}{4} = \frac{nm+1}{4}$. □

Conjecture 1. The lower bounds from Theorem 3 are actually the exact values for $i\mu(G_{n,m})$.

As I reminded at the beginning of these paper, MIM in NP-hard in general and also for classes with some given properties. But there are graphs for which polynomial-time algorithms were developed [3, 7, 9], such as interval graphs, trees [7], weakly chordal graphs [4], graphs of bounded clique-width [8], hypercubes [5]. It is interesting then to refine the boundary line between hard and easy cases for MIM problem.

This is one motivation to study whether if MIM in grids is polynomial and the existence of polynomial algorithms for other families of graphs, such as Cayley graph (hypercubes and grids on a tours are Cayley graphs) or classes of bipartite graphs. Thus, a more particular conjecture is as well interesting.

Conjecture 2. There exists a polynomial time algorithm for the problem of finding a maximum induced matching in grids.

Also, it is important to find good approximation algorithms for MIM in the cases of graphs for which this problem is known to be NP-hard, such as subgrids. For example, it is not known if there exist PTAS in subgrids.

3 Conclusion

The main results obtained for $i\mu(G_{n,m})$, where $2 \leq n \leq m$ are summarized in the table below.

	m even	$m=4k+1$	$m=4k+3$
n even	$\frac{mn}{4}$	$\lceil \frac{mn}{4} \rceil$	$\lceil \frac{mn}{4} \rceil$
$n=3$	$\lceil \frac{mn}{4} \rceil$	$\frac{3(m-1)}{4}+1$	$\frac{3(m-1)+2}{4}$
$n=5$	$\lceil \frac{mn}{4} \rceil$	$\frac{5(m-1)}{4}+1$	$\frac{5(m-1)+2}{4}$
$n=7$	$\lceil \frac{mn}{4} \rceil$	$\frac{7(m-1)}{4}+1$	$\in \left\{ \frac{7(m-1)+2}{4}, \frac{7(m-1)+2}{4}+1 \right\}$
$n \geq 9$ odd, $n < \frac{m+5}{2}$	$\lceil \frac{mn}{4} \rceil$	$\geq \frac{n(m-1)}{4}+1$	$\geq \frac{n(m-1)+2}{4}$
$n \geq 9$ odd, $n \geq \frac{m+5}{2}$	$\lceil \frac{mn}{4} \rceil$	$\geq \frac{n(m-1)}{4}+1$	$\geq \frac{n(m-1)+2}{4}+1$

Acknowledgements I would like to thank Marc Demange from ESSEC Business School for suggesting me this problem and for his support.

References

1. Bonifaci, V., Korteweg, P., Marchetti-Spaccamela, A., and Stougie, L.: Minimizing flow time in the wireless gathering problem. *ACM Transactions on Algorithms* 7(3), (2011) http://arxiv.org/abs/0802.2836.
2. Cameron, K.: Induced matchings. Discrete Appl. Math. **24**, 97–102 (1989)
3. Cameron, K.: Induced matchings in intersection graphs. Discrete Math. **278**, 1–9 (2004)
4. Cameron, K., Sritharan, R., Tang, Y.: Finding a maximum induced matching in weakly chordal graphs. Discrete Math. **266**, 133–142 (2003)
5. Dabrowski, K., Demange, M., and Lozin, V.V.: New results on maximum induced matchings in bipartite graphs and beyond. *submitted*, http://homepages.warwick.ac.uk/~mariac/
6. Faudree, R.J., Gyárfas, A., Schelp, R.H., Tuza, Z.: Induced matchings in bipartite graphs. Discrete Math. **78**, 83–87 (1989)
7. Golumbic, M.C., Lewenstein, M.: New results on induced matchings. Discrete Appl. Math. **101**, 157–165 (2000)
8. Kobler, D., Rotics, U.: Finding maximum induced matchings in subclasses of claw-free and P5-free graphs, and in graphs with matching and induced matching of equal maximum size. Algorithmica **37**, 327–346 (2003)
9. Lozin, V.V.: On maximum induced matchings in bipartite graphs. Inform. Process. Lett. **81** 7–11 (2002)
10. Rusu, I.: Maximum weight edge-constrained matchings. Discrete Appl. Math. **156**, 662–672 (2008)
11. Stockmeyer, L.J., Vazirani, V.V.: Np-completeness of some generalizations of the maximum matching problem. I.P.L. **15**, 14–19 (1982)

Determining the Minimum Number of Warehouses and their Space-Size for Storing Compatible Items

Dimitra Alexiou and Stefanos Katsavounis

Abstract We present an exact procedure for determining the smallest number of necessary warehouses and their space-size for storing compatible items. The required floor space housing of every item is known. The method developed here refers to store compatible items in the same warehouse in order to diminish the maximum necessary space-size of every warehouse and consequently to the determination of the minimum number of needed warehouses. The problem is formulated in the context of graph theory. Compatible items stored in the same warehouse are the elements of a color class of a specific coloring of a weighted conflict graph $G = (V, E, W)$, where the vertices of V represent the items to be stored and all couples of non-compatible items define the edge set E. The elements of W are the numbers assigned to the vertices of V that express the required storing space of every corresponding item. That is the problem is reduced to find a coloring of G that correspond to an optimal solution.

1 Introduction

The problem of avoiding the storing of non-compatible items in the same warehouse occurs in diverse practical situations related to the inventory.

This eventuality arises when the warehousing of a particular set of items can cause their deterioration, or be a fire hazard, as well as for reasons of organizational

D. Alexiou
Department of Spatial Planning and Development Engineering, School of Engineering, Aristotle University of Thessaloniki, Veria, Greece
e-mail: alexd@tee.gr

S. Katsavounis (✉)
Department of Production and Management Engineering, School of Engineering, Demokritos University of Thrace, Xanthi, Greece
e-mail: skatsav@pme.duth.gr

A. Migdalas et al. (eds.), *Optimization Theory, Decision Making, and Operations Research Applications*, Springer Proceedings in Mathematics & Statistics 31,
DOI 10.1007/978-1-4614-5134-1_13,

scheduling as it is the case, for example, in the construction industry where the time interval for the removal or for the storing of each item is closely estimated.

The warehouse storing policies are directly related to the well-known bin-packing problem, see [2,9]. Nevertheless, limited studies appear in the literature that take into account the simultaneous housing of only compatible items, the methods in [1, 5–7] are approximation procedures, while an exact approach is given in [8] based on a set covering formulation. In this paper we develop an exact method that finds the smallest size as well as the minimum number of needed warehouses.

The problem is formulated in the context of graph theory. A conflict weighted graph $G = (V, E, W)$ is stated, the vertices of which represent the items to be stored, while all couples of non-compatible items are the elements of edge set E. To every vertex $v_i \in V = \{v_1, v_2, \ldots, v_n\}$ corresponds to a number $w_i \in W = \{w_1, w_2, \ldots, w_n\}$ that expresses the storing space required for the item associated with v_i.

An *independent set* of G is a subset of V so that no two elements of V are adjacent. Conclusively, a subset of compatible items allowable to be stored simultaneously in the same warehouse corresponds to a subset $S \subset V$ of the conflict graph G for which $\Gamma(S) \cap S = \emptyset$ where $\Gamma(S) = \bigcup_{v \in S} \Gamma(v), \Gamma(v)$ denote the set of adjacent vertices of vertex v. Thereby our problem turns out to partition the elements of V into the smallest number L of independent sets $S_i \subset V$, that is, to determine the sets of family

$$\mathscr{P} = \{S_1, S_2, \ldots, S_L\} \text{ such that:}$$

$$L = \min\{K\} \text{ and for } i, j \in \{1, 2, \ldots, K\} \text{ it holds that}$$

$$S_i \cap S_j = \emptyset, i \neq j \quad \wedge \quad \Gamma(S_i) \cap S_i = \emptyset \quad \wedge \quad \bigcup_{i=1}^{K} S_i = V \tag{1}$$

An assignment of k colors to the vertices of a graph G so that no two adjacent nodes share the same color is a *k-coloring* of G. The smallest number of colors needed to color G is its *chromatic number* $x(G)$. The vertices assigned the same color evidently form an independent set and constitute a *color class*, thus the compatible items stored in the same warehouse at the same time correspond to a color class of the conflict graph G.

The proposed problem is dealt within the framework of coloring graph G taking into account the relationships that implicate the parameters involved.

A coloring can be represented by the set family $\mathscr{P} = \{S_1, S_2, \ldots, S_L\}$ where every set $S_i, i = 1, 2, \ldots, L$ denotes a color class. Following, a natural number is assigned to every vertex, so the elements of a color class are expressed by the corresponding numbers.

The next section deals with the development of the proposed algorithm AGC (Alternative Graph Coloring) that determines the families $\mathscr{P}$ with the smallest number L of color classes $S_i, i = 1, 2, \ldots, L$ that concurrently satisfy relation 1.

2 Algorithm AGC

Algorithm AGC is an implicit enumeration backtracking algorithm that generates the entire alternative non-isomorphic k-colorings of a graph. The inclusion of an optimality test produces the non-isomorphic $x(G)$ colorings of G. Two colorings of G are said to be isomorphic if they contain exactly the same color classes.

Example:

The $x(G)$ =3-colorings $\mathscr{P}_1, \mathscr{P}_2$, and $\mathscr{P}_3$ of graph G shown in Fig. 1 are non-isomorphic, while coloring $\mathscr{P}_4$ is isomorphic to $\mathscr{P}_1$.

2.1 Algorithm: General Description

Before outlining the basic operations of AGC some necessary notations are introduced.

$S_L = \{s_{L_1}, s_{L_2}, \ldots, s_{L_{n_L}}\}$ with $s_{L_j} < s_{L_{j+1}}, j \in \{1,2,\ldots,n_L - 1\}$ is a generated independent set representing a color class in a lexicographic order, $L = \{1,2,\ldots,m\}$ where L is the number of distinct colors used and s_{L_1} is the smallest vertex of class S_L. Clearly L is the number of color classes and $S_i \cap S_j = \emptyset, i \neq j, i, j \in \{1,2,\ldots,L\}$.

F: contains the colored vertices i.e., $F = \bigcup_{i=1}^{m} S_i$ and $\bar{F} = V - F$

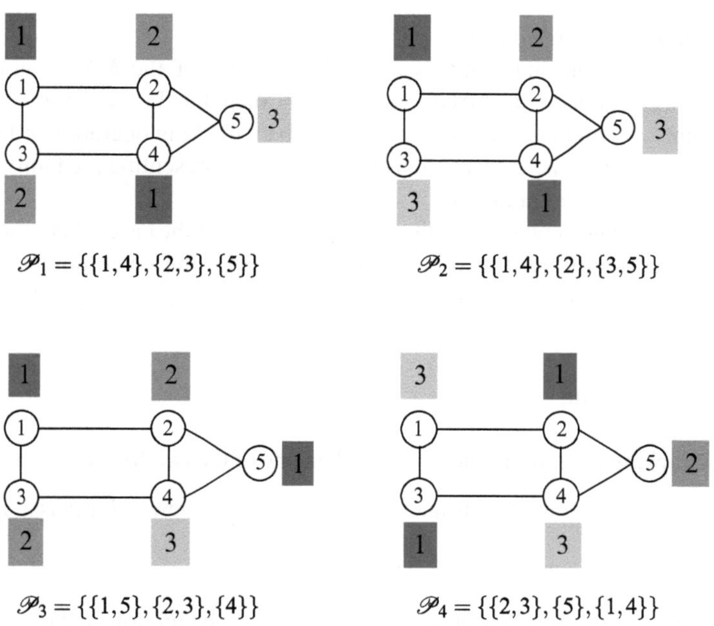

Fig. 1 3-colorings example

$\bar{\Gamma}(y) = \{y_1, y_2, \ldots, y_{n_y}\}, y_i < y_{i+1}, i \in \{1, 2, .., n_y - 1\}$ represents the nonadjacent vertices of vertex y that are greater than y, i.e. $y_i \notin \Gamma(y), y_i > y, i \in \{1, 2, \ldots, n_y\}$

n_L: The number of colored vertices that belong to the same color class L.

MNL: The smallest number of used colors in a coloring found so far by the algorithmic process.

k: Number of distinct colorings $\mathscr{P}_k$ generated so far.

2.1.1 Branching

The starting stage in order to generate a color class S_L consists of placing in S_L the smallest uncolored vertex. Let v be the latest vertex inserted in S_L. The process augments S_L repeatedly by a vertex y greater than v, where y is the smallest uncolored non-adjacent vertex to v as well as to any element of S_i and which has not been used to extend S_L in a previous branching phase, namely, $S_i = \{i_1, i_2, \ldots, i_k\} \subseteq S_L$ is augmented by vertex i_{k+1}, where

$$i_{k+1} = \min\{y \in (\bar{\Gamma}(i_k) \cap S_i \cap \bar{F}), \text{ and } y \text{ not used in a previous phase to augment } S_L\}$$

At that particular instant the vertices used in an earlier stage to extend S_L and not currently in S_L are uncolored due to a backtracking operation, these vertices will be contained in a subsequent generated color class. The process of extending S_L ends when one of the two following cases appears.

1. *All vertices of G are colored*
   If the current coloring contains a smaller number of the MNL color classes previously retained, then AGC sets $k \leftarrow 1$, then the last generated coloring represents $\mathscr{P}_1$ and the process passes to the backtracking operation, otherwise it sets $k \leftarrow k+1$ and the current coloring is retained representing the family set $\mathscr{P}_k$.
2. *If such a vertex does not exist*
   The algorithm goes on to form the next color class. In the case where the number L of the currently generated color classes exceeds the corresponding number MNL retained so far, the procedure passes to the backtracking operations.

2.1.2 Backtracking

The backtracking operation is performed when one of the two following cases arises.

- The procedure attains the instance in order to form a new color class, S_{L+1} with $L+1 > MNL$.
- A new L-coloring $\mathscr{P}_K$ with $L = MNL$ is detected.

In the occurrence of either of the two cases above, the un-coloring of all vertices in the last color class S_L is actuated and the procedure backtracks from the last vertex $s_{L_{n_L}}$ inserted in $L \leftarrow L-1$, if $L < 1$ the algorithm ends.

2.2 Formal Statement of Algorithm AGC

FV: The number of colored vertices
$h(s_{L_i})$: The position in $\bar{\Gamma}(s_{L_{i-1}})$ of vertex $s_{L_i} \in S_L$ for $i > 1$, and $h(s_{L_1}) = 0$. The vertices in the preceding positions $h(s_{L_j}) < h(s_{L_i})$ in $\Gamma(s_{L_{i-1}})$ have been used in a previous stage so as to extend S_L.
$\bar{\Gamma}_{h(y)}(y)$: Contains the vertices of $\bar{\Gamma}(y)$ placed in positions $h(y) + 1, h(y) + 2, \ldots, h(y) + n_y$

$$\bar{\Gamma}_{h(y)}(y) = \{y_{h(y)+1}, y_{h(y)+2}, \ldots, y_{h(y)+n_y}\} \subset \bar{\Gamma}(y)$$

```
Bool ← 1
While Bool = 1
    Let v be the smallest uncolored vertex v ∉ F
    L ← L+1{ next color class }
    If L ≤ MNL then { coloring number optimality test }
        n_L ← 1, s_L1 ← v, S_L = {s_L1}, F ← F+{v}, FV ← FV+1
        If FV = N then
            CALL COL, CALL BACK
        else
            DO
                L ← L−1: CALL BACK
            LOOP
        endif
    endif
    DO
        CALL CHECK { extend S_L if possible }
        if Bool = 1 then
            EXIT DO
        endif
        FV ← FV+1, n_L ← n_L+1, S_L ← S_L+{y}, F ← F+{y}
        if FV = N then
            CALL COL, CALL BACK
        endif
    LOOP
END While
```

```
BACK {Backtracking}
    DO i = 1,n
        F ← F − {s_Li}
    LOOP
    FV ← FV − n_L
    DO
```

```
        L ← L − 1
        if L < 1 then
            CALL WRT {yield result and STOP}
        endif
        y ← s_{L n_L}, F ← F − {y}, FV ← FV − 1, n_L ← n_L − 1, y ← s_{L n_L}
    LOOP While n_L = 0
Return
```

```
CHECK {independency test}
r ← s_{n_L − 1}
    if S_L ∩ {y} ≠ ∅, ∀y ∈ {Γ̄_{h(r)}(r) − F̄}
        Bool ← 1
    else
        let q be the smallest index for which S_L ∩ {y} = ∅, y_q ∈ Γ̄_{h(r)}(r)
        y ← y_{h_q}, h(y) ← q
    endif
Return
```

```
COL { store a new coloring 𝒫_k }
    If L < MNL then
        k ← 0, MNL ← L
    endif
    k ← k + 1, 𝒫_k ← {S_1, S_2, .., S_l}
Return
```

3 Number and Size of Warehouses

Without loss of the generality it is assumed that all used warehouses are the same size. The housing of compatible items in the minimum number of warehouses generally leads to numerous disparate housing systems, since every distinct housing method corresponds to a discrete k-coloring $\mathscr{P}$ of the conflict graph G, where k is the smallest possible number of warehouses needed. Apparently k is the chromatic number $x(G)$ of G when no restriction is given to warehouse size. To a specific coloring class $S_L = \{s_{L_1}, s_{L_2}, \ldots, s_{L_{n_L}}\}$ we associate the *class-capacity* $WS(L) = \bigcup_{i=1}^{n_L} w_{L_i}, w_{L_i} \in W$ is the required storing space of a specific item, as mentioned in the introduction. We state the *coloring-capacity* $WC(\mathscr{P}_q)$ to be the greatest *class-capacity* of a color class $\{S_1^q, S_2^q, \ldots, S_L^q\} = \mathscr{P}_q$. That is $WC(\mathscr{P}_q) = \max\{WS(S_i^q)\}, i \in [1, L]$.

Since it is assumed that all warehouses are the same size, the selected warehouse with the minimum size is derived from coloring $\mathscr{P}_{\text{opt}}$ with the smallest coloring-capacity among the non-isomorphic colorings, therefore,

$$WC(\mathscr{P}_{\text{opt}}) = \min\{WC(\mathscr{P}_i)\}, i \in [1, k]$$

Table 1 Coloring classes and coloring-capacities

$\mathscr{P}$	L	S_L	$w_{L_i}, i \in [1, n_L]$	$WS(L)$	$WC(\mathscr{P})$	$WC(\mathscr{P}_{\text{opt}})$
	1	$S_1^1 = \{1,4\}$	$\{12,10\}$	22	22	22
$\mathscr{P}_1$	2	$S_2^1 = \{2,3\}$	$\{15,6\}$	21		
	3	$S_3^1 = \{5\}$	$\{20\}$	20		
	1	$S_1^2 = \{1,4\}$	$\{12,10\}$	22		
$\mathscr{P}_2$	2	$S_2^2 = \{2\}$	$\{15\}$	15		
	3	$S_3^2 = \{3,5\}$	$\{6,20\}$	26	26	
	1	$S_1^3 = \{1,5\}$	$\{12,20\}$	32	32	
$\mathscr{P}_3$	2	$S_2^3 = \{2,3\}$	$\{15,6\}$	21		
	3	$S_3^3 = \{4\}$	$\{10\}$	10		

Next a small example of the above conceptions on the graph in Fig. 1 is shown in Table 1, where the sizes of the storing surface of the associated items are the elements of W.

$$W = \{12, 15, 6, 10, 20\} \text{ that is } w_1 = 12, w_2 = 15, w_3 = 6, w_4 = 10, w_5 = 20$$

The optimal coloring is $\mathscr{P}_{\text{opt}} = \mathscr{P}_1 = \{\{1,4\},\{2,3\},\{5\}\}$ that corresponds to a minimum warehouse size of $WC(\mathscr{P}_{\text{opt}}) = 22$.

It is observed that the warehouse sizes *(class-capacity)* that correspond to an optimal coloring lead to a uniform distribution; this is an added advantage that concerns the decrement of the unused warehouse space $U(\mathscr{P}_i)$ of a coloring $\mathscr{P}_i$. In the short example, the unused space for each generated distinct coloring can easily be deduced from Table 1 as follows

$$U(\mathscr{P}_1) = (22 - 22) + (22 - 21) + (22 - 20) = \mathbf{3},$$
$$U(\mathscr{P}_2) = (26 - 22) + (26 - 15) + (26 - 26) = \mathbf{15}, \quad \text{and}$$
$$U(\mathscr{P}_3) = (32 - 32) + (32 - 21) + (32 - 10) = \mathbf{33}$$

In all the above it was assumed that there were no restrictions concerning the size of the warehouses. In real life situations, however, the case may arise that, for the given conflict graph G, the size of the available warehouses is smaller than those obtained by the corresponding $WC(\mathscr{P}_{\text{opt}})$. In this eventuality, clearly the smallest number of warehouses needed is greater than the chromatic number $x(G)$.

A size optimality test concerning the magnitude of *class-capacity* $WS(L)$ is incorporated in AGC during the extension phase of a color class S_L preventing the inclusion of an item in a coloring class S_L. The procedure, thus, advances to the backtracking operations whenever its associated capacity $WS(L)$ reaches the situation of being greater than a given surface space restriction, say *SPR*. A more illustrative example in given in graph G of Fig. 2, where the numbers in italic near the vertices represent the space needed for storing the corresponding items.

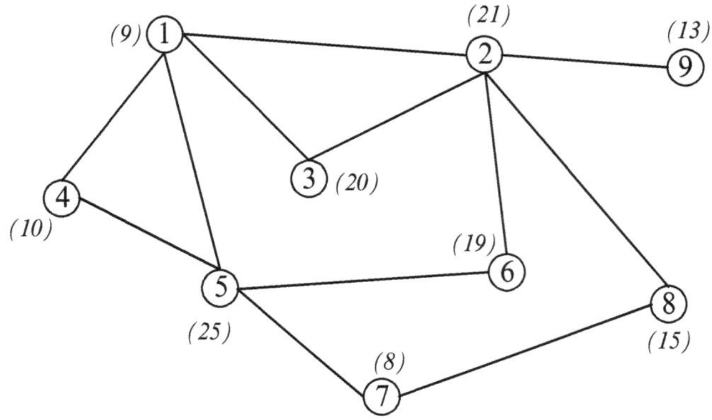

Fig. 2 4-colorings example

The chromatic number of G is $x(G) = 3$ and there are 14 distinct 3-colorings of G, therefore there exist 14 different ways for storing all items without any restrictions on the sizes of the warehouses, under this assumption the optimal solution concerning the warehouses size is the coloring

$$\mathscr{P}_{\text{opt}} = \{S_1 = \{1,6,7,9\}, S_2 = \{2,5\}, S_3 = \{3,4,8\}\} \Rightarrow$$
$$WC(\mathscr{P}_{\text{opt}}) = \max\{WS(S_1) = 49, WS(S_2) = 46, WS(S_3) = 45\} = \mathbf{49}$$

The worst solution is the coloring $\mathscr{P}_w = \{S_1 = \{1,7\}, S_2 = \{2,5\}, S_3 = \{3,4,6,8,9\}\}$ with

$$WC(\mathscr{P}_w) = \max\{S_1 = 17, S_2 = 46, S_3 = 77\} = \mathbf{77}$$

Let $SPR = 48$ be the maximum warehouse space capacity. In this case there exist 114 distinct 4 colorings of G for which $WC(\mathscr{P}_i) \leq 48, i = 1,2,\ldots,114$ and

$$\mathscr{P}_{\text{opt}} = \{S_1 = \{1,6,7\}, S_2 = \{2,4\}, S_3 = \{3,8\}, S_4 = \{5,9\}\}$$
$$WC(\mathscr{P}_{\text{opt}}) = \max\{WS(S_1) = 36, WS(S_2) = 31, WS(S_3) = 35, WS(S_4) = 38\}\} = \mathbf{38}$$

The worst solution is the coloring

$$\mathscr{P}_w = \{S_1 = \{1\}, S_2 = \{2,5\}, S_3 = \{3,8,9\}, S_4 = \{4,6,7\}\}$$
$$WC(\mathscr{P}_w) = \max\{WS(S_1) = 9, WS(S_2) = 46, WS(S_3) = 48, WS(S_4) = 37\}\} = \mathbf{48}$$

It is important to observe the uniform storing distribution in the optimal solution of the items in the warehouses. Also to note that the algorithm works exclusively with the colorings that may lead to an optimal solution due to the size optimality test on *class-capacity* $WS(L)$ and of course does not generate all 114 distinct 4-colorings.

4 Conclusions

Inventory management is needed in diverse types of enterprises. The problem of concurrently determining the minimum number and size of the warehouses necessary for storing distinct items is evidently a meaningful task.

In the previous sections, a new method was presented in order to confront the problem where a subset of couples of non-compatible items must not be stored simultaneously in the same warehouse.

The problem is modeled in the context of graph theory, more specifically, the procedure developed is a depth-first branch and bound technique controlled by feasibility and optimality tests seeking an optimal solution among the distinct coloring of a conflict weighted graph G.

Although the problem is classified as $\mathbb{NP}$-*Hard*, see [3,4] the density of a conflict graph related to compatible items is in general low enough permitting the successful application of the proposed algorithm to problems of which the size are met in real-world situations.

References

1. Basnet, C., Wilson, J.: Heuristics for determining the number of warehouses for storing non-compatible products. Int. Trans. Oper. Res. **12**, 527–538 (2005)
2. Coffman, E.G., Jr., Garey, M.R., Johnson, D.S.: Approximation algorithms for bin-packing - an updated survey. In: Ausiello, G., Lucertini, M., Serafini, P. (eds.) Algorithm Design for Computer System Design, pp. 49–106. Springer, New York (1984)
3. Coffman, E.G., Garey, M.R, Johnson, D.S.: Approximation Algorithms for NP-Hard Problems. PWS Publishing, Boston (1996)
4. Garey, M.R., Johnson D.S.: Computers and Intractability: A Guide to the Theory of NP-Completeness. W.H. Freeman & Co, New York (1979)
5. Gendreau M., Laporte G., Semet F.: Heuristics and lower bounds for the bin packing problem with conflicts. Comp. Oper. Res. **31**, 347–358 (2004)
6. Kalfakakou, R., Katsavounis, S., Tsouros, K.: Minimum number of warehouses for storing simultaneously compatible products. Int. J. Prod. Econ. **81–82**, 559–564 (2003)
7. Khanafer, A., Clautiaux, F., Talbi, E.G: New lower bounds for bin packing problems with conflicts. Eur. J. Oper. Res. **206**, 281–288 (2010)
8. Malaguti, E., Toth, P.: A survey on vertex coloring problems. Int. Trans. Oper. Res **17**, 1–34 (2010)
9. Muritiba, A.E.F, Iori, M., Malaguti, E., Toth, P.: Algorithms for the bin packing problem with conflicts. INFORMS J. Comp. **22**, 401–415 (2010)

Duality for Multiple Objective Fractional Programming with Generalized Type-I Univexity

Ioan M. Stancu-Minasian and Andreea Mădălina Stancu

Abstract In this paper, a multiobjective fractional subset programming problem (Problem (P)) is considered. A new class of $(\mathscr{F}, b, \phi, \rho, \theta)$-type-I univex function is introduced and a general dual model for (P) is presented. Based on these functions, weak, strong and converse duality theorems are derived. Almost all results presented in the literature were obtained under the assumption that the function $\mathscr{F}$ is sublinear in the third argument. Here, our results hold assuming only the convexity of this one.

1 Introduction

In this paper, we shall present a semiparametric dual model for the following multiobjective fractional subset programming problem

$$\min \ \varphi(S) = \left(\frac{F_1(S)}{G_1(S)}, \frac{F_2(S)}{G_2(S)}, \ldots, \frac{F_p(S)}{G_p(S)} \right) \quad \text{(P)}$$

subject to

$$H_j(S) \leqq 0, \ \ j \in \underline{q} = \{1, 2, \ldots, q\}, \ S \in A^n,$$

where A^n is the n-fold product of the σ-algebra A of subsets of a given set X, $F_i : A^n \to \mathbb{R}$, $G_i : A^n \to \mathbb{R}$, $i \in \underline{p} = \{1, 2, \ldots, p\}$, and $H_j : A^n \to \mathbb{R}$, $j \in \underline{q}$, such that for each $i \in \underline{p}$, $G_i(S) > 0$, for all $S \in \mathscr{P}$. We denoted by $\mathscr{P} = \{S \in A^n : H_j(S) \leqq 0, \ j \in \underline{q}\}$ the set of all feasible solutions for (P).

I.M. Stancu-Minasian • A.M. Stancu (✉)
Institute of Mathematical Statistics and Applied Mathematics, of the Romanian Academy, Calea 13 Septembrie Nr. 13, Sector 5, Bucharest, Romania
e-mail: stancu_minasian@yahoo.com; andreea_madalina_s@yahoo.com

A. Migdalas et al. (eds.), *Optimization Theory, Decision Making, and Operations Research Applications*, Springer Proceedings in Mathematics & Statistics 31,
DOI 10.1007/978-1-4614-5134-1_14,

For any vectors $x,y \in \mathbb{R}^n$, we use the following conventions: $x<y$ iff $x_i<y_i, i\in \underline{n}=\{1,2,\dots n\}$; $x\leqq y$ iff $x_i \leqq y_i, i\in\underline{n}=\{1,2,\dots n\}$; $x\leq y$ iff $x\leqq y$, but $x\neq y$; $x\nleq y$ means the negation of $x\leq y$.

Definition 1. A feasible solution $S^0 \in \mathscr{P}$ is said to be an efficient solution to (P) if there is no other $S\in\mathscr{P}$ such that

$$\left(\frac{F_1(S)}{G_1(S)},\frac{F_2(S)}{G_2(S)},\dots,\frac{F_p(S)}{G_p(S)}\right)\leq\left(\frac{F_1(S^0)}{G_1(S^0)},\frac{F_2(S^0)}{G_2(S^0)},\dots,\frac{F_p(S^0)}{G_p(S^0)}\right).$$

Following the introduction of the notion of convexity for set functions by Morris [8] and its extension for n-set functions by Corley [4], various generalizations of convexity for set and n-set functions were proposed in Lee [5], Lin [6], Preda [9,11], Stancu-Minasian and Preda [15], Zalmai [17, 18]. More precisely, quasiconvexity and pseudoconvexity for set functions were defined in Lee [5], and in Lin [6] for n-set functions, $(\mathscr{F},\rho)$-convexity in Preda [9, 10], (ρ,b)-vexity in Preda [11], $(\mathscr{F},\alpha,\rho,\theta)$-$V$-convexity in Zalmai [17], $(\mathscr{F},b,\phi,\rho,\theta)$-univexity in Zalmai [18] and $(\mathscr{F},\rho,\sigma,\theta)$-$V$-type-I in Mishra [7]. Also, in Bector et al. [2], some types of generalized convexity and optimality and duality results for a multiobjective programming problem involving n-set functions were given. For more information about fractional programming problems, the reader may consult the recently research bibliography compiled by Stancu-Minasian [14].

For formulating and proving various duality results, we shall introduce the class of generalized convex n-set functions called $(\mathscr{F},b,\phi,\rho,\theta)$-type-I univex functions. Until now, $\mathscr{F}$ was assumed to be a sublinear function in the third argument. In our approach, we suppose that $\mathscr{F}$ is a convex function in the third argument. The idea of replacing a sublinear function with a convex function in certain types of sufficiency and duality results is not new (see, for instance Bătătorescu et al. [1], Chinchuluun et al. [3], Preda et al. [12, 13] and Yuan et al. [16]).

2 Definitions and Preliminaries

Let (X,A,μ) be a finite atomless measure space with $L_1(X,A,\mu)$ separable, and let d be the pseudometric on A^n defined by

$$d(R,S) := \left[\sum_{k=1}^{n}\mu^2\left(R_k\Delta S_k\right)\right]^{1/2},$$

where $R=(R_1,\dots,R_n)$, $S=(S_1,\dots,S_n)\in A^n$ and Δ stands for the symmetric difference. Thus, (A^n,d) is a pseudometric space.

For $h\in L_1(X,A,\mu)$ and $T\in A$, the integral $\int_T h\mathrm{d}\mu$ is denoted by $\langle h,\chi_T\rangle$, where $\chi_T\in L_\infty(X,A,\mu)$ is the indicator (characteristic) function of T.

Definition 2. [8] A function $F : A \to \mathbb{R}$ is said to be differentiable at $S^* \in A$ if there exist $DF(S^*) \in L_1(X, A, \mu)$, called the derivative of F at S^*, and $V_F : A \times A \to \mathbb{R}$ such that

$$F(S) = F(S^*) + \langle DF(S^*), \chi_S - \chi_{S^*} \rangle + V_F(S, S^*)$$

for each $S \in A$, where $V_F(S, S^*)$ is $o(d(S, S^*))$, that is, $\lim_{d(S,S^*) \to 0} \dfrac{V_F(S, S^*)}{d(S, S^*)} = 0$.

Definition 3. [4] A function $G : A^n \to \mathbb{R}$ is said to have a partial derivative at $S^* = (S_1^*, \ldots, S_n^*) \in A^n$ with respect to its ith argument if the function $F(S_i) = G(S_1^*, \ldots, S_{i-1}^*, S_i, S_{i+1}^*, \ldots, S_n^*)$ has derivative $DF(S_i^*)$, $i \in \underline{n} = \{1, 2, \ldots, n\}$.

We define $D_i G(S^*) = DF(S_i^*)$ and write $DG(S^*) = (D_1 G(S^*), \ldots, D_n G(S^*))$.

Definition 4. [4] A function $G : A^n \to \mathbb{R}$ is said to be differentiable at S^*, if there exist $DG(S^*)$ and $W_G : A^n \times A^n \to \mathbb{R}$ such that

$G(S) = G(S^*) + \sum_{i=1}^n \left\langle D_i G(S^*), \chi_{S_i} - \chi_{S_i^*} \right\rangle + W_G(S, S^*)$, where $W_G(S, S^*)$ is $o(d(S, S^*))$ for all $S \in A^n$.

In the following we consider $\mathscr{F} : A^n \times A^n \times \mathbb{R} \to \mathbb{R}$. Also consider $F : A^n \to \mathbb{R}$ and $G : A^n \to \mathbb{R}$ to type-I the differentiable functions. The definitions below generalizes two definitions of Zalmai [18].

Let ρ_i and ρ_j real numbers and let $\rho = (\rho_i, \rho_j)$, $i \in I$, $j \in J$.

Definition 5. A pair $(F_i, G_j)\,(i \in I, j \in J)$ is said to be $(\mathscr{F}, b, \phi, \rho, \theta)$-pseudo quasi univex type-I at $S^* \in A^n$ according to the partition $\{I, J\}$ if there exist a function $b : A^n \times A^n \to \mathbb{R}_+$, a function $\theta : A^n \times A^n \to A^n \times A^n$ such that $S \neq S^* \Longrightarrow \theta(S, S^*) \neq (0, 0)$, a function $\phi : \mathbb{R} \to \mathbb{R}$, and real numbers ρ such that for all $S \in A^n$ the implications

$$\mathscr{F}(S, S^*; b(S, S^*) DF_i(S^*)) \geqq -\rho_i d^2(\theta(S, S^*)) \Longrightarrow \phi\,(F_i(S) - F_i(S^*)) \geqq 0, i \in I \tag{1}$$

and

$$\phi(-G_j(S^*)) \leqq 0 \Rightarrow \mathscr{F}(S, S^*; b(S, S^*) DG_j(S^*)) \leqq -\rho_j d^2(\theta(S, S^*)),\ j \in J \tag{2}$$

do hold.

If the second (implied) inequality in (1) is strict $(S \neq S^*)$, then we say that (F_i, G_j) is $(\mathscr{F}, b, \phi, \rho, \theta)$-strictly pseudo quasi univex type-I at $S^* \in A^n$ according to the partition $\{I, J\}$.

Definition 6. A pair $(F_i, G_j)\,(i \in I, j \in J)$ is said to be $(\mathscr{F}, b, \phi, \rho, \theta)$-quasi pseudo univex type-I at $S^* \in A^n$ according to the partition $\{I, J\}$ if there exist a function $b : A^n \times A^n \to \mathbb{R}_+$, a function $\theta : A^n \times A^n \to A^n \times A^n$ such that $S \neq S^* \Longrightarrow \theta(S, S^*) \neq (0, 0)$, a function $\phi : \mathbb{R} \to \mathbb{R}$, and real numbers ρ such that for all $S \in A^n$ the implications

$$\phi\,(F_i(S) - F_i(S^*)) \leqq 0 \Rightarrow \mathscr{F}(S, S^*; b(S, S^*) DF_i(S^*)) \leqq -\rho_i d^2(\theta(S, S^*)),\ i \in I \tag{3}$$

and

$$\mathscr{F}(S,S^*;b(S,S^*)DG_j(S^*)) \geqq -\rho_j d^2(\theta(S,S^*)) \Longrightarrow \phi(-G_j(S^*)) \geqq 0, j \in J \quad (4)$$

do hold.

If the first (implied) inequality in (3) is strict ($S \neq S^*$), and the second (implied) inequality in (4) is strict ($S \neq S^*$), then we say that (F_i, G_j) is $(\mathscr{F}, b, \phi, \rho, \theta)$-prestrictly quasi strictly pseudo univex type-I at $S^* \in A^n$ according to the partition $\{I,J\}$.

Definition 7. A pair $(F_i, G_j)\,(i \in I, j \in J)$ is said to be $(\mathscr{F}, b, \phi, \rho, \theta)$-quasi quasi univex type-I at $S^* \in A^n$ according to the partition $\{I,J\}$ if there exist a function $b: A^n \times A^n \to \mathbb{R}_+$, a function $\theta: A^n \times A^n \to A^n \times A^n$ such that $S \neq S^* \Longrightarrow \theta(S,S^*) \neq (0,0)$, a function $\phi: \mathbb{R} \to \mathbb{R}$, and real numbers ρ such that for all $S \in A^n$ the implications

$$\phi(F_i(S) - F_i(S^*)) \leqq 0 \Rightarrow \mathscr{F}(S,S^*;b(S,S^*)DF_i(S^*)) \leqq -\rho_i d^2(\theta(S,S^*)),\ i \in I \quad (5)$$

and

$$\phi(-G_j(S^*)) \leqq 0 \Rightarrow \mathscr{F}(S,S^*;b(S,S^*)DG_j(S^*)) \leqq -\rho_j d^2(\theta(S,S^*)),\ j \in J \quad (6)$$

do hold.

If the first (implied) inequality in (5) is strict ($S \neq S^*$), then we say that (F_i, G_j) is $(\mathscr{F}, b, \phi, \rho, \theta)$-prestrictly quasi quasi univex type-I at $S^* \in A^n$ according to the partition $\{I,J\}$.

Definition 8. A pair $(F_i, G_j)\,(i \in I, j \in J)$ is said to be $(\mathscr{F}, b, \phi, \rho, \theta)$-pseudo pseudo univex type-I at $S^* \in A^n$ according to the partition $\{I,J\}$ if there exist a function $b: A^n \times A^n \to \mathbb{R}_+$, a function $\theta: A^n \times A^n \to A^n \times A^n$ such that $S \neq S^* \Longrightarrow \theta(S,S^*) \neq (0,0)$, a function $\phi: \mathbb{R} \to \mathbb{R}$, and real numbers ρ such that for all $S \in A^n$ the implications

$$\mathscr{F}(S,S^*;b(S,S^*)DF_i(S^*)) \geqq -\rho_i d^2(\theta(S,S^*)) \Longrightarrow \phi(F_i(S) - F_i(S^*)) \geqq 0, i \in I \quad (7)$$

and

$$\mathscr{F}(S,S^*;b(S,S^*)DG_j(S^*)) \geqq -\rho_j d^2(\theta(S,S^*)) \Longrightarrow \phi(-G_j(S^*)) \geqq 0, j \in J \quad (8)$$

do hold.

If the second (implied) inequality in (7) and (8) is strict ($S \neq S^*$), then we say that (F_i, G_j) is $(\mathscr{F}, b, \phi, \rho, \theta)$-strictly pseudo strictly pseudo univex type-I at $S^* \in A^n$ according to the partition $\{I,J\}$.

In [17], Zalmai state for Problem (P) the following necessary conditions for efficiency.

Theorem 1. *Assume that F_i, G_i, $i \in \underline{p}$, and H_j, $j \in \underline{q}$, are differentiable at $S^* \in A^n$, and that for each $i \in \underline{p}$, there exists $\widehat{S_i} \in A^n$ such that*

$$H_j(S^*) + \sum_{k=1}^{n} \left\langle D_k H_j(S^*), \chi_{\widehat{S_k}} - \chi_{S_k^*} \right\rangle < 0,\ j \in \underline{q},$$

and for each $l \in \underline{p} \backslash \{i\}$,

$$\sum_{k=1}^{n} \left\langle G_i(S^*) D_k F_l(S^*) - F_i(S^*) D_k G_l(S^*), \chi_{\widehat{S_k}} - \chi_{S_k^*} \right\rangle < 0.$$

If S^* is an efficient solution to (P), then there exist $u^* \in U = \{u \in \mathbb{R}^p : u > 0, \sum_{i=1}^{p} u_i = 1\}$ and $v^* \in \mathbb{R}_+^q$ such that

$$\sum_{k=1}^{n} \left\langle \sum_{i=1}^{p} u_i^* \left[G_i(S^*) D_k F_i(S^*) - F_i(S^*) D_k G_i(S^*) \right] + \sum_{j=1}^{q} v_j^* D_k H_j(S^*), \chi_{S_k} - \chi_{S_k^*} \right\rangle \geqq 0,$$

We *for all $S \in A^n$, $v_j^* H_j(S^*) = 0$, $j \in \underline{q}$. We denoted by $\mathbb{R}_+^q$ the positive orthant of the q-dimensional space $\mathbb{R}^q$, i.e. $\mathbb{R}_+^q = \{x = (x_{1,\ldots,}x_q) \in \mathbb{R}^q : x_j \geqq 0, j \in \underline{q}\}$.*

We shall call an efficient solution S^* to (P) satisfying the first two conditions for some $\widehat{S_i}$, $i \in \underline{p}$, a regular efficient solution.

3 A Zalmai's Semiparametric Duality Model

In this section we present a general duality model for (P). Here we use two partitions of the index sets $\underline{q}$ and, respectively, $\underline{p}$.

Let $\{I_0, I_1, \ldots, I_k\}$ be a partition of the index set $\underline{p}$ and $\{J_0, J_1, \ldots, J_m\}$ be a partition of the index set $\underline{q}$ such that $K = \{0, 1, \ldots, k\} \subset M = \{0, 1, \ldots, m\}$, and let the function $\Omega_t(S, \cdot, u, v) : A^n \to \mathbb{R}$ be defined for fixed S, u and v, and $t \in K$ by

$$\Omega_t(S, T, u, v) = \sum_{i \in I_t} u_i \left[F_i(S) G_i(T) - F_i(T) G_i(S) \right] + \sum_{j \in J_t} v_j H_j(T)$$

We associate with the problem (P) the dual problem

$$\max \delta(T, u, v) = \left(\frac{F_1(T)}{G_1(T)}, \frac{F_2(T)}{G_2(T)}, \ldots, \frac{F_p(T)}{G_p(T)} \right) \quad \text{(D)}$$

subject to

$$\mathscr{F}\left(S,T;b(S,T)\left\{\sum_{i=1}^{p}u_i\left[G_i(T)DF_i(T)-F_i(T)DG_i(T)\right]+\sum_{j=1}^{q}v_jDH_j(T)\right\}\right)\geqq 0,$$

$$\forall S\in A^n \tag{9}$$

$$\sum_{j\in J_t}v_jH_j(T)\geqq 0,\ t\in M,\ T\in A^n,\ u\in U,\ v\in\mathbb{R}_+^q. \tag{10}$$

In the following we consider $\mathscr{F}(S,T;\cdot):L_1^n(X,A,\mu)\to\mathbb{R}$ a convex function and $\Lambda_t(\cdot,v^*):A^n\to\mathbb{R}$, $\Lambda_t(T,v^*)=\sum_{j\in J_t}v_j^*H_j(T)$, $t\in M$.

Theorem 2. *(Weak duality). Let S and (T,u,v) be arbitrary feasible solutions to (P) and (D), respectively, and assume that any one of the following sets of hypotheses is satisfied:*

(a) (i) $(2k\Omega_t(\cdot,T,u,v);\ 2(m-k)\Lambda_t(\cdot,v))$ is $(\mathscr{F},b,\phi_t,\rho_t,\theta)$-strictly pseudo quasi univex type-I at the point T, according to the partition $\{K,M\backslash K\}$, ϕ_t is increasing and $\phi_t(0)=0$ for each $t\in M$;

(ii) $\frac{1}{k}\sum_{t\in K}\rho_t+\sum_{t\in M\backslash K}\frac{\rho_t}{m-k}\geqq 0$;

(b) (i) $(2k\Omega_t(\cdot,T,u,v);\ 2(m-k)\Lambda_t(\cdot,v))$ is $(\mathscr{F},b,\phi_t,\rho_t,\theta)$- prestrictly quasi strictly pseudo univex type-I at the point T, according to the partition$\{K,M\backslash K\}$, ϕ_t is increasing and $\phi_t(0)=0$ for each $t\in M$;

(ii) $\frac{1}{k}\sum_{t\in K}\rho_t+\sum_{t\in M\backslash K}\frac{\rho_t}{m-k}\geqq 0$;

(c) (i) $(2k\Omega_t(\cdot,T,u,v);2(m-k)\Lambda_t(\cdot,v))$ is $(\mathscr{F},b,\phi_t,\rho_t,\theta)$-prestrictly quasi quasi univex-type-I at the point T, according to the partition$\{K,M\backslash K\}$,ϕ_t is increasing and $\phi_t(0)=0$ for each $t\in M$;

(ii) $\frac{1}{k}\sum_{t\in K}\rho_t+\sum_{t\in M\backslash K}\frac{\rho_t}{m-k}>0$;

(d) (i) $(3k_1\Omega_t(\cdot,T,u,v);\ 3(m-k)\Lambda_t(\cdot,v))$ is $(\mathscr{F},b,\phi_t,\rho_t,\theta)$-strictly pseudo quasi univex-type-I at the point T according to the partition$\{K_1,M\backslash K\}$, $(3k_2\Omega_t(\cdot,T,u,v)\ ;\ 3(m-k)\Lambda_t(\cdot,v)$ is $(\mathscr{F},b,\phi_t,\rho_t,\theta)$-prestrictly quasi quasi univex type-I at the point T according to the partition$\{K_2,M\backslash K\}$, ϕ_t is increasing and $\phi_t(0)=0$ for each $t\in M$, where $\{K_1,K_2\}$ is a partition of K, $K_1\neq\emptyset$, $k_1=|K_1|$ and $K_2\neq\emptyset$, $k_2=|K_2|$;

(ii) $\frac{1}{k_1}\sum_{t\in K_1}\rho_t+\frac{1}{k_2}\sum_{t\in K_2}\rho_t+\sum_{t\in M\backslash K}\frac{\rho_t}{m-k}\geqq 0$;

(e) (i) $(3k\Omega_t(\cdot,T,u,v);3(m_1-k_1)\Lambda_t(\cdot,v))$ is $(\mathscr{F},b,\phi_t,\rho_t,\theta)$-prestrictly quasi strictly pseudo univex-type-I at the point T, according to the partition $\{K,(M\backslash K)_1\}$, $(3k\Omega_t(\cdot,T,u,v);3(m_2-k_2)\Lambda_t(\cdot,v))$ is $(\mathscr{F},b,\phi_t,\rho_t,\theta)$-prestrictly quasi quasi univex-type-I at the point T, according to the partition$\{K,(M\backslash K)_2\}$, ϕ_t is increasing and $\phi_t(0)=0$,for each $t\in M$, where $\{(M\backslash K)_1,(M\backslash K)_2\}$ is a partition of $M\backslash K$, $(M\backslash K)_1\neq\emptyset$, $m_1=|(M\backslash K_1)|$ and $(M\backslash K)_2\neq\emptyset$, $m_2=|(M\backslash K)_2|$;

(ii) $\frac{1}{k}\sum_{t\in K}\rho_t + \sum_{t\in(M\backslash K)_1}\frac{\rho_t}{m_1-k_1} + \sum_{t\in(M\backslash K)_2}\frac{\rho_t}{m_2-k_2} \geqq 0;$

(f) *(i)* $(4k_1\Omega_t(\cdot,T,u,v);4(m_1-k_1)\Lambda_t(\cdot,v))$ *is* $(\mathscr{F},b,\phi_t,\rho_t,\theta)$*-strictly pseudo strictly pseudo univex-type-I at the point* T*, according to the partition* $\{K_1,(M\backslash K)_1\}$, $(4k_2\Omega_t(\cdot,u,v);4(m_2-k_2)\Lambda_t(\cdot,v))$ *is* $(\mathscr{F},b,\phi_t,\rho_t,\theta)$*-prestrictly quasi quasi univex-type-I at the point* T*, according to the partition* $\{K_2,(M\backslash K)_2\},\phi_t$ *is increasing and* $\phi_t(0)=0$ *for each* $t\in M$*, where* $\{K_1,K_2\}$ *is a partition of* K*,* $K_1\neq\emptyset$*,* $k_1=|K_1|$ *and* $K_2\neq\emptyset$*,* $k_2=|K_2|$*;* $\{(M\backslash K)_1,(M\backslash K)_2\}$ *is a partition of* $M\backslash K$*,* $(M\backslash K)_1\neq\emptyset$*,* $m_1=|(M\backslash K_1)|$ *and* $(M\backslash K)_2\neq\emptyset$*,* $m_2=|(M\backslash K)_2|$*;*

(ii) $\frac{1}{k_1}\sum_{t\in K_1}\rho_t+\frac{1}{k_2}\sum_{t\in K_2}\rho_t + \sum_{t\in(M\backslash K)_1}\frac{\rho_t}{m_1-k_1} + \sum_{t\in(M\backslash K)_2}\frac{\rho_t}{m_2-k_2} \geqq 0;$

(iii) $K_1\neq\emptyset$ *or* $(M\backslash K)_1\neq\emptyset$ *or*

$$\frac{1}{k_1}\sum_{t\in K_1}\rho_t+\frac{1}{k_2}\sum_{t\in K_2}\rho_t + \sum_{t\in(M\backslash K)_1}\frac{\rho_t}{m_1-k_1} + \sum_{t\in(M\backslash K)_2}\frac{\rho_t}{m_2-k_2} > 0.$$

Then $\varphi(S)\nleq\delta(T,u,v)$.

Proof. (a) Suppose to the contrary that $\varphi(S)\leq\delta(T,u,v)$. This implies that

$$G_i(T)F_i(S)-F_i(T)G_i(S)\leqq 0, \tag{11}$$

for each $i\in \underline{p}$, with strict inequality for at least one subscript $l\in\underline{p}$.

From the inequalities (10), (11), nonnegativity of u and primal feasibility of S we deduce that

$$2k\Omega_t(S,T,u,v)\leqq 2k\Omega_t(T,T,u,v) \text{ for each } t\in K,$$

with strict inequality holding for at least one $t\in K$ since $u>0$.

It follows from the properties of ϕ_t (ϕ_t is increasing and $\phi_t(0)=0$), that for each $t\in M$,

$$\phi_t\left(2k\Omega_t(S,T,u,v)-2k\Omega_t(T,T,u,v)\right)\leqq 0, \tag{12}$$

Since for each $t\in M\backslash K$,

$$0\leqq 2(m-k)\Lambda_t(T,v),$$

it follows from the properties of ϕ_t that

$$\phi_t\left(-2(m-k)\Lambda_t(T,v)\right)\leqq 0. \tag{13}$$

From (12) and (13) and assumption (i), we deduce that

$$\mathscr{F}\left(S,T;2kb(S,T)\left\{\sum_{i\in I_t}u_i\left[G_i(T)DF_i(T)-F_i(T)DG_i(T)\right]+\sum_{j\in J_t}v_jDH_j(T)\right\}\right)$$
$$<-\rho_t d^2(\theta(S,T)),\ t\in K. \tag{14}$$

and

$$\mathscr{F}(S,T;b(S,T)2\,(m-k)\sum_{j\in J_t}v_jDH_j(T)\leqq-\rho_t d^2(\theta(S,T)),t\in M\backslash K. \tag{15}$$

From (14), summing after $t\in K$ we obtain

$$\sum_{t\in K}\mathscr{F}\left(S,T;kb(S,T)\left\{\sum_{i\in I_t}2u_i\left[G_i(T)DF_i(T)-F_i(T)DG_i(T)\right]+\sum_{j\in J_t}2v_jDH_j(T)\right\}\right)$$
$$<-\sum_{t\in K}\rho_t d^2(\theta(S,T)).$$

But from the convexity of $\mathscr{F}\,(S,T;\cdot)$ we have

$$\mathscr{F}\left(S,T;b(S,T)\left\{\sum_{t\in K}\sum_{i\in I_t}2u_i\left[G_i(T)DF_i(T)-F_i(T)DG_i(T)\right]+\sum_{t\in K}\sum_{j\in J_t}2v_jDH_j(T)\right\}\right)$$
$$=\mathscr{F}\left(S,T;b(S,T)\frac{1}{k}\sum_{t\in K}2k\left\{\sum_{i\in I_t}u_i\left[G_i(T)DF_i(T)-F_i(T)DG_i(T)\right]+\sum_{j\in J_t}v_jDH_j(T)\right\}\right)$$
$$\leqq\frac{1}{k}\sum_{t\in K}\mathscr{F}\left(S,T;b(S,T)2k\left\{\sum_{i\in I_t}u_i\left[G_i(T)DF_i(T)-F_i(T)DG_i(T)\right]+\sum_{j\in J_t}v_jDH_j(T)\right\}\right)$$
$$<-\frac{1}{k}\sum_{t\in K}\rho_t d^2(\theta(S,T)),$$

i.e.

$$\mathscr{F}\left(S,T;b(S,T)\left\{\sum_{t\in K}\sum_{i\in I_t}2u_i\left[G_i(T)DF_i(T)-F_i(T)DG_i(T)\right]+\sum_{t\in K}\sum_{j\in J_t}2v_jDH_j(T)\right\}\right)$$
$$<-\frac{1}{k}\sum_{t\in K}\rho_t d^2(\theta(S,T)). \tag{16}$$

From (15), summing after $t \in M\backslash K$ we obtain

$$\sum_{t\in M\backslash K} \mathscr{F}(S,T;b(S,T)2\,(m-k)\sum_{j\in J_t} v_j DH_j(T) \leqq - \sum_{t\in M\backslash K} \rho_t d^2(\theta(S,T))$$

But from the convexity of $\mathscr{F}(S,T;\cdot)$ we have

$$\begin{aligned}
&\mathscr{F}(S,T;b(S,T)2\sum_{t\in M\backslash K}\sum_{j\in J_t} v_j DH_j(T))\\
&= \mathscr{F}(S,T;b(S,T)\sum_{t\in M\backslash K}\frac{1}{m-k}2\,(m-k)\sum_{j\in J_t} v_j DH_j(T))\\
&\leqq \frac{1}{m-k}\sum_{t\in M\backslash K}\mathscr{F}(S,T;b(S,T)2\,(m-k)\sum_{j\in J_t} v_j DH_j(T))\\
&\leqq -\sum_{t\in M\backslash K}\frac{\rho_t}{m-k} d^2(\theta(S,T)),
\end{aligned}$$

i.e.

$$\mathscr{F}(S,T;b(S,T)2\sum_{t\in M\backslash K}\sum_{j\in J_t} v_j DH_j(T)) \leqq -\sum_{t\in M\backslash K}\frac{\rho_t}{m-k} d^2(\theta(S,T)) \tag{17}$$

Now, using (9), the convexity of $\mathscr{F}(S,T;\cdot)$,(16) and (17), we obtain

$$\begin{aligned}
0 &\leqq \mathscr{F}\left(S,T;b(S,T)\left\{\sum_{t\in K}\sum_{i\in I_t} 2u_i\left[G_i(T)DF_i(T) - F_i(T)DG_i(T)\right]\right.\right.\\
&\qquad \left.\left. + \sum_{t\in K}\sum_{j\in J_t} 2v_j DH_j(T)\right\}\right)\\
&\qquad + \mathscr{F}(S,T;b(S,T)2\sum_{t\in M\backslash K}\sum_{j\in J_t} v_j DH_j(T))\\
&< -\frac{1}{k}\sum_{t\in K}\rho_t d^2(\theta(S,T)) - \sum_{t\in M\backslash K}\frac{\rho_t}{m-k} d^2(\theta(S,T))\\
&= -\left(\frac{1}{k}\sum_{t\in K}\rho_t + \sum_{t\in M\backslash K}\frac{\rho_t}{m-k}\right) d^2(\theta(S,T))
\end{aligned}$$

which contradicts (a) (iii). Therefore, we conclude that $\varphi(S) \nleq \delta(T,u,v)$.

The proofs of (b)–(f) can be obtained following similar arguments to that of part (a) and Zalmai [18]. □

Theorem 3. *(Strong duality). Let $S^* \in \mathscr{P}$ be a regular efficient solution of (P), let $\mathscr{F}(S,S^*;DF(S^*)) = \sum_{k=1}^{n} \left\langle D_kF(S^*), \chi_{S_k} - \chi_{S_k^*} \right\rangle$ for any differentiable function $F: A^n \to \mathbb{R}$ and $S \in A^n$, and assume that any one of the sets of hypotheses specified in Theorem 2 holds for all feasible solutions of (D). Then there exist $u^* \in U$ and $v^* \in \mathbb{R}_+^q$ such that (S^*,u^*,v^*) is an efficient solution of (D) and $\varphi(S^*) = \delta(S^*,u^*,v^*)$.*

Proof. By Theorem 1 there exist $u^* \in U$ and $v^* \in \mathbb{R}_+^q$ such that (S^*,u^*,v^*) is a feasible solution of (D) and $\varphi(S^*) = \delta(S^*,u^*,v^*)$. That (S^*,u^*,v^*) is efficient for (D) follows from the corresponding parts of Theorem 2 . □

Remark 1. Using the previous Theorems 2 and 3 and techniques from [9] and [17], we can obtain also a converse duality result.

4 Conclusions

In this paper, we have introduced a new class of generalized $(\mathscr{F},b,\phi,\rho,\theta)$-type-I univex functions. Based on these functions a semiparametric dual model for a multiobjective fractional subset programming problem was introduced. Weak, strong and converse duality theorems were derived.

References

1. Bătătorescu, A., Preda, V., Beldiman, M.: On higher order duality for multiobjective programming involving generalized $(\mathscr{F},\rho,\gamma,b)$-convexity. Math. Rep. **9(59)**, 161–174 (2007)
2. Bector, C.R., Bhatia, D., Pandey, S.: Duality for multiobjective fractional programming involving n-set functions. J. Math. Anal. Appl. **186**(3), 747–768 (1994)
3. Chinchuluun, A., Yuan, D., Pardalos, P.M.: Optimality conditions and duality for nondifferentiable multiobjective fractional programming with generalized convexity. Ann. Oper. Res. **154**(1), 133–147 (2007)
4. Corley, H.W.: Optimization theory for n-set functions. J. Math. Anal. Appl. **127**(1), 193–205 (1987)
5. Lee, T.Y.: Generalized convex set functions. J. Math. Anal. Appl. **141**(1), 278–290 (1989)
6. Lin, L.J.: On the optimality of differentiable nonconvex n-set functions. J. Math. Anal. Appl. **168**(2), 351–366 (1992)
7. Mishra, S.K.: Duality for multiple objective fractional subset programming with generalized $(\mathscr{F},\rho,\sigma,\theta)$-V-type-I functions. J. Global Optim. **36**(4), 499–516 (2006)
8. Morris, R.J.T.: Optimal constrained selection of a measurable subset. J. Math. Anal. Appl. **70**(2), 546–562 (1979)
9. Preda, V.: On minmax programming problems containing n-set functions. Optimization **22**(4), 527–537 (1991)
10. Preda, V.: On efficiency and duality for multiobjective programms. J. Math. Anal. Appl. **166**(2), 365–377 (1992)
11. Preda, V.: On duality of multiobjective fractional measurable subset selection problems, J. Math. Anal. Appl. **196**(2), 514–525 (1995)

12. Preda, V., Stancu-Minasian, I.M., Beldiman, M., Stancu, A.: Optimality and duality for multiobjective programming with generalized V-type I univexity and related n-set functions. Proc. Rom. Acad. Ser. A **6**(3), 183–191 (2005)
13. Preda, V., Stancu-Minasian, I.M., Beldiman, M., Stancu, A.: Generalized V-univexity type-I for multiobjective programming problems with n-set function. J. Global Optim. **44**(1), 131–148 (2009)
14. Stancu-Minasian, I.M., A sixth bibliography of fractional programming. Optimization **55**(4), 405–428 (2006)
15. Stancu-Minasian, I.M., Preda, V.: Optimality conditions and duality for programming problems involving set and n-set functions: a survey. J. Stat. Manag. Syst. **5**(1–3), 175–207 (2002)
16. Yuan, D.H., Liu, X.L., Chinchuluun, A., Pardalos, P.M., Nondiferentiable minimax fractional programming problems with (C; α ; ρ; d)-convexity. J. Optim. Theor. Appl. **129**(1), 185–199 (2006)
17. Zalmai, G.J.: Efficiency conditions and duality models for multiobjective fractional subset programming problems with generalized $(\mathscr{F},\alpha,\rho,\theta)$-V-convex functions. Comp. Math. Appl. **43**(12), 1489–1520 (2002)
18. Zalmai, G.J.: Generalized $(\mathscr{F},b,\phi,\rho,\theta)$-univex n-set functions and global semiparametric sufficient efficiency conditions in multiobjective fractional subset programming. Int. J. Math. Math. Sci. **6**, 949–973 (2005)

A Markov-Based Decision Model of Tax Evasion for Risk-Averse Firms in Greece

Nikolaos D. Goumagias and Dimitrios Hristu-Varsakelis

Abstract We develop a Markov-based optimization model that captures the process via which a risk-averse firm in Greece decides whether to engage in tax evasion. The firm seeks to maximize the expected utility of its wealth, the latter viewed as a function of the portion of profits which the firm attempts to conceal from the government. Our model takes into account the basic features of the Greek tax system, including random audits and tax penalties applied when the audit reveals any wrongdoing. The proposed model is used to (1) show that the parameters currently in place are conducive to tax evasion and (2) "chart" the problem's parameter space in order to identify "virtuous" combinations (from the point of view of the government), and obtain a relationship between audit probability, tax penalty and likelihood of the firm engaging in tax evasion.

Key words Markov Chains • Optimization • Taxation • Greece

1 Introduction

Greece is currently under severe economic stress, facing perhaps its most serious crisis in its modern history. The government's plan for coping with the country's high debt and budget deficits has included a "rescue package" backed by the ECB and the IMF, combined with a series of austerity measures. One of the most talked-about and widely agreed-upon measures—for which, however, there has been little in the way of implementation—has to do with combating tax evasion, which is openly acknowledged as a sizeable drain on the country's finances, as well as one of its most persistent problems. The purpose of this paper is to explore a Markov-based

N.D. Goumagias (✉) • D. Hristu-Varsakelis
University of Macedonia, Egnatia Av. 156 Thessaloniki, Greece
e-mail: ngoum@uom.gr; dcv@uom.gr

A. Migdalas et al. (eds.), *Optimization Theory, Decision Making, and Operations Research Applications*, Springer Proceedings in Mathematics & Statistics 31,
DOI 10.1007/978-1-4614-5134-1_15,

optimization model which may be used to (1) investigate the expected behavior of Greek firms[1] with respect to tax evasion and (2) test candidate tax policies before they are implemented.

In this work we are interested in obtaining a simple and adaptable optimization model that captures the salient features of the Greek tax system and can be used as a decision support tool by the government, as the latter seeks a framework that neither rewards "cheating" nor penalizes firms any more than is necessary to curb tax evasion. We adopt the perspective of a "representative" firm, which is assumed to act selfishly in order to optimize its wealth via tax evasion, by leaving some of its profits unreported. The firm can be viewed as choosing to allocate its profits between two assets (as in [1]). One is risk-free and involves the firm declaring its profit and paying the proper tax, meaning that it ends up with a somewhat lower wealth as a result. The other asset is risky: profit is concealed (zero tax payed); however, the firm's wealth may be reduced in the future (more than if it had behaved honestly) if the firm is audited. In these circumstances, a risk-averse self-interested firm is expected to maximize the expected utility of its wealth by choosing to conceal some fraction of its profits. We wish to find the optimal allocation which the firm will adopt, as a function of its risk aversion level and the parameters of the tax system (audit probability and tax penalty). Furthermore, we would like to identify alternative tax parameters which could lead to higher government revenues by eliminating or reducing the incentive for tax evasion. Towards that end, we "chart" the parameter space and compute a surface which quantifies the firm's incentive to cheat (i.e., its optimal percentage of profits to conceal). The proposed model allows us to estimate the relative efficiency of the tax penalty and audit probability as tax evasion deterrents under different levels of risk aversion.

The remainder of this paper is structured as follows. After a brief literature review, we give a brief description of the Greek tax system in Sect. 2 and formulate a corresponding Markov-based model from the point of view of a risk-averse firm that would like to increase its net wealth by evading taxes but is worried about tax penalties in the event of an audit. We describe the objective function the firm seeks to optimize and discuss our choice of model parameters. Section 3 presents the firm's optimal strategy together with numerical results on its expected behavior and offers a brief discussion of the relevant policy implications.

1.1 Related Work

The approach adopted here is to examine the problem of optimal tax evasion (and by extension, optimal taxation) from the perspective of a self-interested firm. To the best of our knowledge, most of the optimization models aimed at taxation have adopted a macroeconomic viewpoint. Early work on begins with [1] who proposed

[1]The word "firms" refers to incorporated entities in Greece, operating according to the general accounting principles commonly known in Europe as S.A.—Société Anonyme).

the portfolio allocation idea used in this work, but optimized a macroeconomic equilibrium-based model. Subsequent work [2], concentrated on the effects of increased probability of detection, or of tax rates [4]. For an optimal control viewpoint on taxation, see [13]. The criteria based on which tax evasion decisions are made are discussed in [5]. Some authors, e.g., [7, 8] addressed the morality of taxpayers and auditors as variables, or investigated the idea of bonuses to auditors that reveal tax evasion [11].

The model in [15] captured the trade-off between fines and audit probabilities while proposing different treatment for risk-neutral versus risk-averse firms. Other work relevant to our setting includes [14] and [3] who applied Bedford s law to tax evasion and other types of financial fraud. With respect to Greece in particular, the tax evasion literature (e.g., [12] and [16]) provides some good theoretical and empirical discussion but no rigorous analysis with respect to how tax policy should be shaped. A recent exception is [9].

2 Model Description

We proceed to give a brief description of the Greek tax system, leading to the formulation of an optimization problem which the firm is faced with each year.

2.1 *The Greek Tax System for Firms*

In Greece, firms report their profits at the end of each fiscal year, and pay a tax rate of 24%. Typically, the government does not have adequate information on the firm's true profits, which may be manipulated through a variety of methods. Two of the most often used include (1) manipulation of financial statements to under-report income and (2) invoices (often issued by another, usually short-lived firm) that document supposed expenses and are used to offset profits. To discourage "cheating," firms are subject to random audits, and there are monetary penalties which apply for unreported profits.

A tax audit can reveal a firm's true profit but is costly and resource-intensive. When an audit occurs, it can cover up to a 5-year period in the past, meaning that the government retains the right to audit a tax statement for up to 5 years from its submission. After that 5-year window, any unreported profit, unpaid tax, etc., is essentially capitalized by the firm. Although there are no official data published on the number of audits performed each year, information obtained from the press and tax professionals suggests that the probability of a firm being audited is approximately 5%, with the distibution being skewed towards firms which are approaching the 5-year mark (and therefore a past tax statement which is about to go beyond the reach of the audit process).

When an audit does reveal tax evasion, the penalties imposed depend on the amount of unreported income, and the time elapsed since the offense occurred. The total cost to the firm is the tax originally due on the unreported income, plus a 2% monthly penalty on that tax. Thus, "older" tax evasion decisions are potentially more costly than recent ones. The total penalty amount is subject to a 2/3 discount if the firm agrees to pay promptly once the evasion is discovered. Finally, the penalty amount cannot exceed twice the original tax owed.

The above is, of course, not an exhaustive description of the Greek tax code, but does include the features which are most germane. Some aspects, such as VAT payments (collected via an independent mechanism) are not considered here, but could be incorporated into the model at later stages.

2.2 *The Model*

We will describe the possible tax status of the firm in any given year via a Markov chain which evolves on a set $\mathscr{S}$ with $\mathscr{S} = \{A, N_1, \ldots, N_4\}$. States in $\mathscr{S}$ are labeled as follows:

- A: the firm is being audited,
- N_j: the firm's last audit was $j = 1, \ldots, 4$ years ago.

These labels will be used mainly to facilitate the discussion below. Otherwise, for notational convenience, we will refer to states in $\mathscr{S}$ by integer, in order of appearance (i.e., $A \to 1, \ldots, N_4 \to 5$).

Let F be the firm's annual profit. Each year, k, the firm decides the fraction $u_k \in [0,1]$ of its profit to conceal (thereby declaring to the government only $(1-u_k)F$), and then transitions to a new state in $\mathscr{S}$, with probabilities $a_{ij} = P(s_k = i | s_{k-1} = j)$, where the indices i and j indicate the ith and jth states in $\mathscr{S}$. Here, we have assumed that the transition probabilities are independent of the firm's actions and that the annual profit is constant. These assumptions can be removed, but we will not take up the issue here, mainly because of space considerations.

Based on the discussion of Sect. 2.1, we can express the Markov chain as $x_{k+1} = Ax_k$, where x_k is the state's probability distribution at time k. The stochastic matrix $A = [a_{ij}]$ is not written down here, but can be easily read off from the transition diagram shown in Fig. 1, where p denotes the overall audit probability (i.e., the fraction of tax returns that the government is able to audit each year). There is little-to-none official data on p; we estimate its value at $p = 0.05$ (and its distribution among states as per Fig. 1), based on reports in the Greek financial press and discussions with individuals familiar with the inner workings of the tax authority and audit mechanism in Greece. As the transition diagram indicates, the probability of an audit is heavily skewed towards firms in the last (N_4) state; there, the firm has been unaudited for 4 years and is about to file its fifth consecutive tax statement. Therefore, if it is not audited in the next time period, the oldest of these five statements will go beyond the reach of any future audits.

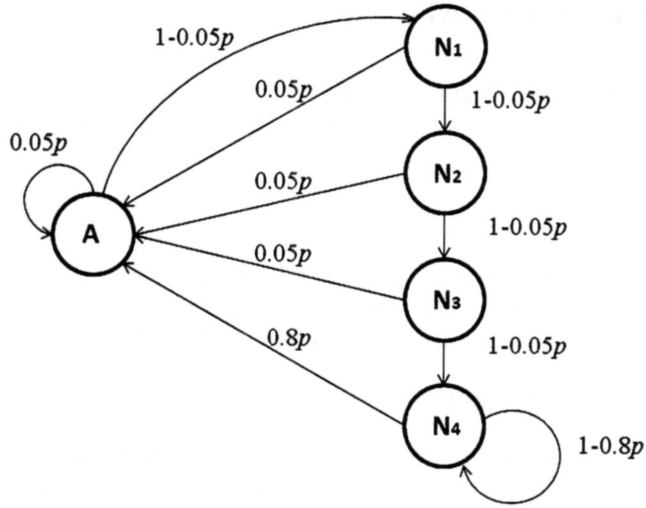

Fig. 1 Markov transition diagram for our simplified model of the Greek tax system as it pertains to firms. The overall probability of a tax audit each year is p, distributed so that 80% of audits involve firms with at least one tax statement whose 5-year statute of limitations is about to elapse

The firm allocates its yearly profits between two "assets," as discussed in Sect. 1. One is a risky asset, $R(j)$, which corresponds to concealing profit when being at the jth state. The payoff in that case is that the firm gets to keep more of its profit (since it pays no tax on the amount concealed), at a risk of being audited sometime in the next five time periods, in which case it will have to pay the tax back, plus a penalty. The alternative to R is the risk-free asset, S whose payoff is the firm's after-tax profit, i.e., $S = F(1-r)$, where $r = 0.24$ is the tax rate. The firm then has a portfolio whose worth at time k depends on the allocation of profits between the two assets:

$$W(u_k, j) = (1-u_k)S + u_k R(j) \tag{1}$$

We note that R (and thus W) depends on the state $j = 1,\ldots,5$ in which the firm is at when making its decision, because (the probability distribution of) the number of years that will pass until the firm is audited depends on its initial state j.

Based on our description of the Greek tax system, we can specify the risky asset's return rate as follows:

$$R(j) = \begin{cases} F(1-\gamma^n r(1+3/5n\beta)) & \text{the firm is audited in year } k+n,\ n \leq 5 \\ F & \text{no audit before year } k+6. \end{cases} \tag{2}$$

Notice that $R(j)$ (and thus $W(u_k, j)$) is a random variable whose value depends on when the next audit will occur. In (2), $\gamma \in [0,1]$ is a discount coefficient, used to capture the present value of any tax and penalties the firm might pay in the future. The term $-F\gamma^n r$ corresponds to the tax the firm will pay for every percentage of its profit it is found to have concealed, while $-F\gamma^n \beta 3/5$ is the penalty rate.

3 Optimal Firm Behavior

We can now proceed with computing the behavior expected of the firm. We will assume that the latter acts according to a constant relative risk aversion utility function:

$$U(x) = \frac{x^{1-\lambda}}{1-\lambda}, \tag{3}$$

where λ is the "average" firm's risk aversion coefficient. The firm's objective is then to select the fraction of profit, u_k, that it will attempt to hide from the government at time k, in order to maximize the expected utility of its portfolio:

$$\max_u \{\mathbb{E}(U(W(u,j)))\}, \quad j = 1,2,\ldots,5. \tag{4}$$

where for simplicity we have dropped the subscript k in u, and the expectation in (4) is taken with respect to the probability distribution on the number of years $\{1,2,\ldots,5,\infty\}$ from the time the firm makes a decision until the next audit occurs (either within 5 years, or never). This is essentially the distribution of first passage probabilities from each state j of our Markov chain, to the first (audit) state in precisely n steps.

$$f_{1j}^{(n)} = p(s_{k+n} = 1 | s_k = j, s_{k+i} \neq 1 \text{ for } i = 1,\ldots,n-1).$$

It is well known that for finite n, the $f_{ij}^{(n)}$ can be computed by solving the following upper-triangular system of algebraic equations:

$$a_{ij}^{(n)} = \sum_{r=1}^{n} f_{ij}^{(r)} a_{ii}^{(n-r)} \tag{5}$$

where $a_{ij}^{(k)}$ denotes the (i,j)th element the Markov transition matrix after it is raised to the kth power, A^k. Consequently (4) leads to the firm's optimal policy:

$$u^*(j) = \arg\max_u \left\{ \sum_{n=1}^{5} f_{1j}^{(n)} U\left((1-u)(1-r)F + u(1-\gamma^n r(1+3/5n\beta))F\right) \right.$$
$$\left. + \left(1 - \sum_{n=1}^{5} f_{1j}^{(n)}\right) U\left((1-u)(1-r)F + uF\right) \right\}. \tag{6}$$

It must be noted that in the last equation it is possible for the argument of $U(\cdot)$ to be negative when, for example, u and the penalty coefficient β are sufficiently high (e.g., the penalty is so high that it exceeds the firm's annual profit). This can be dealt with in various ways, but perhaps the simplest one is to notice that there is always a choice of u that results in positive wealth (namely $u = 0$). Thus, since U is continuous and the firm presumably prefers positive wealth to negative wealth we can simply restrict the maximization in (6) to the range of values for u that result in $W(u,j) > 0$.

3.1 Charting the (p,β) Parameter Space

We obtained first-order optimality conditions from (6), and solved them for a range of tax penalty and audit probability values to obtain the firm's optimal tax evasion level in each case. Before examining the firm's behavior, we explain some of our assumptions and justify the parameter ranges that we consider meaningful to explore.

3.2 Parameter Selection

The discount coefficient γ in (6) was based on an assumed interest rate of 3%, i.e., $\gamma = 1/(1+0.03) = 0.9709$. In order to isolate the effect of tax-penalties and audit probabilities on firm behavior, the tax rate r was be kept fixed at its current levels. However, our model can easily be used to examine the effects of changing the tax rate as well.

The firm's profit, F, was estimated based on published data from the Greek Secretariat of Information Systems [10] which indicates that in 2009, the average firm declared approximately €75,700. The true F is, of course, known only to the firm itself. Studies on tax evasion estimate Greece's "hidden economy" at 25–40%, of what is documented, depending on the assumptions used (see, for example [6, 12]), implying a true average profit in the range of 95,000–106,000.

Regarding the range of values for the fraction of profit, $u \in [0,1]$, which may be hidden, it may be practically impossible for a firm to claim zero profits by overstating expenses and/or hiding income. At least some income will be documented, and the firm may be under pressure to show profits for shareholders or capital markets. To account for this, u could be interpreted in a "marginal" sense, i.e., viewed as the fraction of profits that *can* be concealed; alternatively, one could apply (6) by replacing F with the portion of profits that are concealable and thus at stake in the portfolio. Assuming that $F = 100,000$ and that the firm has the option of hiding 30–40% of that amount, the risk aversion coefficient λ that results in the level of tax evasion estimated in the literature must be in the range of 0–12.

The figures estimated above are clearly approximations. Nevertheless what is presented here could serve as the basis for a decision support model at the hands of official entities which are in a position to have more precise knowledge of the required parameters and can thus "tune" the model appropriately. In the following, we chart the firm's tax evasion behavior for $\lambda = 0$ and $\lambda = 6$.

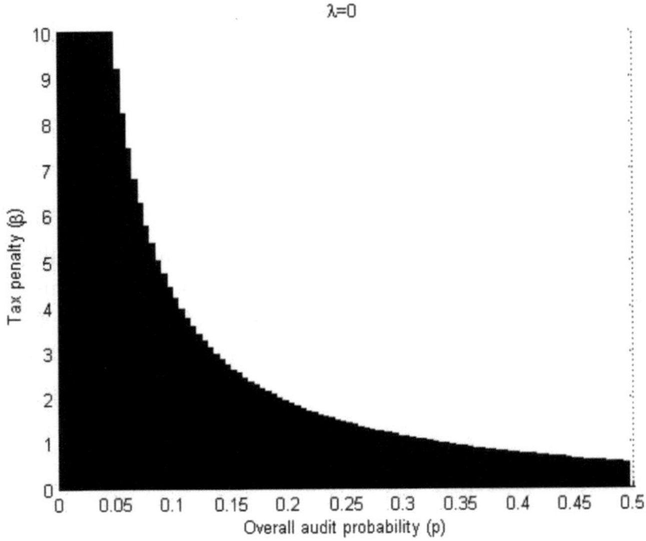

Fig. 2 Tax evasion for a risk-neutral firm. *Dark area*: the firm conceals as much profit as possible ($u^* = 1$). *White area*: it is optimal for the firm to disclose all profit ($u^* = 0$). The non-smoothness of the boundary is due to discretization

3.3 Risk-Neutral Firm ($\lambda = 0$)

We computed the firm's optimal decision u^* in the (β, p) space when $\lambda = 0$, i.e., U is linear and the firm is risk-neutral. Linearity makes the firm's behavior simple to describe, because the solution of (6) is either $u = 0$ or $u = 0$. Figure 2 illustrates the resulting mapping assuming the firm is in the first (audit) state (i.e., $u^*(1)$). The situation is qualitatively similar when the firm is in the other four states.

We observe is a kind of boundary in the p-vs-β space, below which the firm always attempts to hide as much profit as possible (($u^* = 1$, black region). On the other hand, it is best for the firm to disclose all profit ($u^* = 0$) at points above the boundary. Notice the very high tax penalty coefficients required to "induce" honest behavior. At the current $p = 0.05$ audit probability, the tax penalty needs to be approximately 10, which, after the 2/5 discount is applied corresponds to a net tax penalty rate of 600% per annum on unpaid taxes, a rate significantly higher than the baseline 24%. This suggests that tax penalties may be ineffective without the backup of an effective audit mechanism. Finally, a tax penalty rate lower than $\beta = 0.6$ is ineffective in eliminating tax evasion for a risk neutral firm, even for unrealistically high audit probabilities (up to almost 50%).

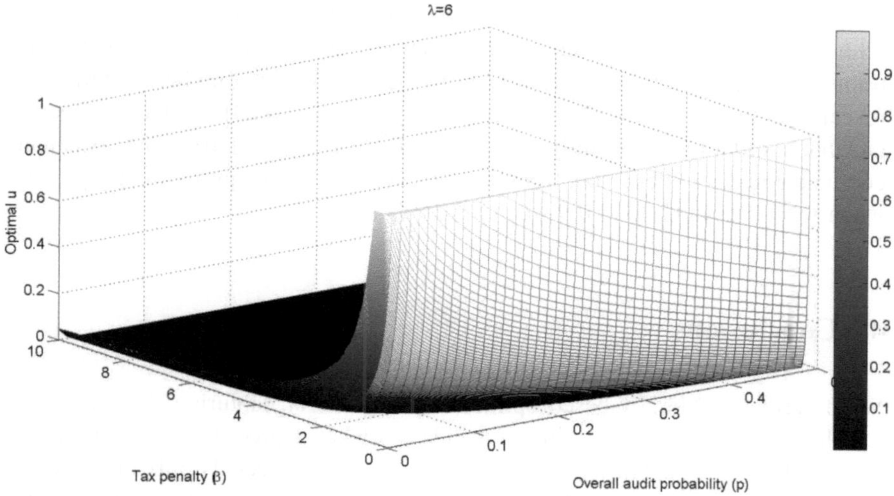

Fig. 3 Tax evasion mapping for a risk-averse firm ($\lambda = 6$) assuming the firm is deciding immediately after an audit. *Gray* levels represent tax evasion between 0% (*dark*) and 100% (*bright*)

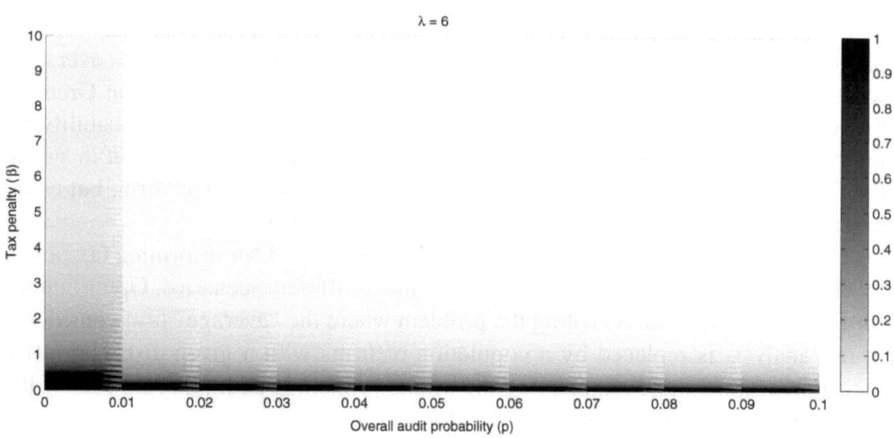

Fig. 4 Top view of figure on the *left*, focusing on audit probabilities between 0.005 and 0.1. *Gray* levels represent tax evasion between 0% (*white*) and 100% (*black*)

3.4 Risk-Averse Firm ($\lambda = 6$)

When the firm is risk averse, its decisions regarding tax evasion are no longer "all-or-nothing." Figures 3 and 4 illustrate the $u^*(1)$ vs. p vs. b surface in the region $\beta \in [0,5]$ and $p \in [0.005, 0.5]$. Other values of λ generate surfaces of similar geometry.

In Fig. 4 the surface is viewed along the u^* axis, with values of u^* encoded in the gray levels, and we have zoomed into the region $\beta \in [0,5]$ and $p \in [0.005, 0.1]$.

The range for p is of practical interest because in the case of audits, for example, a high p is not easy to implement (mode audits require hiring of new personnel, training, etc.). Inspection of the surface for the $\lambda = 6$ case at higher magnification reveals that the current set of parameters $(p = 0.05, \beta = 0.24)$ is rather ineffective, and that in order to significantly curb tax evasion (say at under 10%) the tax penalty coefficient needs to be raised, from the current $\beta = 0.24$ to approximately $\beta = 1.5$ when $p = 0.01$, and to $\beta = 0.9$ when $p = 0.05$.

4 Conclusions

We described a Markov-based optimization model for capturing the process by which a risk-averse firm in Greece decides to what degree it will engage in tax evasion. Each year, the firm acts so as to maximize the expected utility of its wealth, based on the first passage probabilities of transitioning to an "audit" state before the statute of limitations on its decision expires. Our model captures the basic features of the Greek tax system, in which a firm is either audited or accumulates up to five unaudited tax statements, and audits can cover up to 5 years in the past. Some of the model's parameters (tax penalties, tax rates, average firm profit) were set based on government reports, while others (audit probabilities, firm's risk aversion) were estimated implicitly from publicly available data and estimates on Greece's hidden economy. The proposed model was used to "chart" the audit probability vs. tax penalty space in order to compute the expected level of tax evasion in which the firm engages. Such charts were produced here for an "average" firm, but could easily be tailored to specific sectors or even individual firms.

The work presented here can be viewed as a basic tool for informing tax policy by elucidating the firm's expected behavior under different scenarios. Opportunities for future work include revisiting the problem where the "average" firm considered in our analysis is replaced by a population of firms with a given distribution for their risk aversion, and augmenting the model to include additional aspects of the tax system, such as VAT payments and closure.

References

1. Allingham, P., Sandmo, H.: Income tax evasion: a theoretical analysis. J. Public Econ. **1**(6), 988–1001 (1972)
2. Baldry, J.C.: Tax evasion and labour supply. Econ. Lett. **3**, 53–56 (1979)
3. Bhattacharya, S., Xu, D., Kumar, K.: An ANN-based auditor decision support system using Benford's law. Decis. Support Syst. **50**, 576–584 (2011)
4. Clotfelter, C.T.: Tax evasion and tax rates: an analysis of individual returns. Rev. Econ. Stat. **65**, 363–373 (1983)
5. Eide, E.: Tax evasion with rank dependent expected utility. Unpublished paper, University of Oslo, Department of Private Law (2002) eale2002.phs.uoa.gr

6. Feld, L.P., Schneider, F.: Survey on the shadow economy and undeclared earnings in oecd countries. German Econ. Rev. **11**, 109–149 (2010)
7. Frey, B., Feld, L.P.: Deterrence and morale in taxation: an empirical analysis. CESifo working paper, 760 (2002)
8. Gordon, J.P.F.: Individual morality and reputation costs as deterrence to tax evasion. Eur. Econ. Rev. **33**, 797–805 (1989)
9. Goumagias, N., Hristu-Varsakelis, D., Saraidaris, A.: A decision support model for tax revenue collection in Greece. Decis. Support Syst. **53**(1), 76–96 (2012)
10. Greek Secretariat for Information Systems: Statistical report of tax data for the 2009 fiscal year (in Greek) (2009)
11. Hindriks, J., Keen, M., Muthoo, A.: Corruption, extortion and evasion. J. Public Econ. **88**, 161–170 (1996)
12. Kanellopoulos, K., Kousoulakos, I., Rapanos, B.: The underground economy in Greece: What official data show. Greek Econ. Rev. **14**, 215–236 (1992)
13. Kydland, F.E., Prescott, E.C.: Dynamic optimal taxation, rational expectations and optimal control. J. Econ. Dyn. Contr. **2**, 79–91 (1980)
14. Nigrini, M.J.: A taxpayer compliance analysis of benford's law. J. Am. Taxat. **18**(1) (1996)
15. Polinsky, A.M., Shavell, S.: The optimal tradeoff between the probability and magnitude of fines. Am. Econ. Rev. **69**(5), 880–891 (1979)
16. Tatsos, N.: Economic fraud and tax evasion in greece (in Greek). Papazisis Publishings (2001)

Stochastic Decentralized Control of a Platoon of Vehicles Based on the Inclusion Principle

Srdjan S. Stanković, Milorad J. Stanojević, and Dragoslav D. Šiljak

Abstract In this paper the Stochastic Inclusion Principle is applied to decentralized Linear Quadratic Gaussian (LQG) suboptimal longitudinal control design of a platoon of automotive vehicles. Starting from a stochastic linearized platoon state model, input/state overlapping subsystems are identified and extracted after an adequate expansion. An algorithm for approximate LQG optimization of these subsystems is developed in accordance with their hierarchical lower-block-triangular (LBT) structure. Vehicle controllers obtained after contraction, which leaves the local Kalman filters uncontracted, provide high performance tracking and noise immunity. Simulation results illustrate characteristic properties of the proposed algorithm.

1 Introduction

The problem of design of automated highway systems (AHS) has attracted a considerable attention among researchers, e.g. [2, 12]. AHS control architecture proposed in [2, 12, 19] is based on the introduction of a notion of platoons, groups of vehicles following the leading vehicle with small intra-platoon separation. Control of platoons has been studied from different viewpoints [7, 8, 18]. Main theoretical contributions are related to the stability problem [7, 18]. It has been shown that

S.S. Stanković (✉)
Faculty of Electrical Engineering, University of Belgrade, Serbia
e-mail: stankovic@etf.rs

M.J. Stanojević
Faculty of Transport and Traffic Engineering, University of Belgrade, Serbia
e-mail: milorad@sf.bg.ac.rs

D.D. Šiljak
School of Engineering, Santa Clara University, Santa Clara, CA, USA
e-mail: dsiljak@scu.edu

A. Migdalas et al. (eds.), *Optimization Theory, Decision Making, and Operations Research Applications*, Springer Proceedings in Mathematics & Statistics 31,
DOI 10.1007/978-1-4614-5134-1_16, © Springer Science+Business Media New York 2013

an efficient decentralized control law can be formulated when each individual vehicle is supplied with data representing its acceleration, velocity, and position of the preceding vehicle, as well as the references for velocity, acceleration, and inter-vehicle separation [8]. However, tuning of the local proportional regulator parameters has been based, apparently, on the arguments related to relative stability, without taking into account optimality in any predefined sense, structural and signal uncertainties and possibilities to improve the performance by introducing dynamics into the regulator. In [15, 17] a systematic procedure for the design of decentralized overlapping platoon controller on the basis of LQ optimization [4] has been described. In [11] this procedure is applied within the context of output feedback design by introducing reduced order observers.

In this paper a generalization of the approach in [11, 15, 17] to the stochastic case is presented. Namely, Stochastic Inclusion Principle [14, 16] is applied to the design of decentralized Linear Quadratic Gaussian (LQG) suboptimal longitudinal control of a platoon of vehicles, taking into account uncertainty resulting from the influence of the environment and measuring devices. The first part of the paper contains the results related to platoon modelling, formulated in accordance with [2, 8, 12, 17, 19], taking into account stochastic disturbances and measurement noise. A linearized stochastic state model for a string of moving vehicles is derived on the basis of [5, 8, 15]. Each vehicle is described by a state model, with accelerations, velocities, and distances to preceding vehicles as state variables. In the second part an outline of the theory of the Stochastic Inclusion Principle is presented. It is shown that a suitable expansion of the obtained platoon model which possesses the overlapping structure enables formal extraction of "subsystems" for which local quadratic performance indices can be formulated. Having in mind both the subsystem model structure and the available data set [8], an optimization technique resembling to the methodology for deriving Linear Quadratic (LQ) suboptimal control for systems with the hierarchical lower-block-triangular (LBT) structure proposed in [4, 6, 9, 17] is developed and presented in the third part of the paper. Each subsystem controller contains a specific Kalman type estimator, together with the corresponding state feedback gain. Contraction to the original space provides a decentralized controller for the whole platoon, leaving all local state estimators uncontracted. Experimental results are given in order to illustrate main properties of the proposed methodology.

2 Model Formulation

It will be adopted in this paper that *i*th automotive vehicle in a close formation platoon consisting of n vehicles can be represented by the following dynamic model:

$$\begin{aligned} \dot{d}_i &= v_{i-1} - v_i, \quad \dot{v}_i = a_i, \\ a_i &= f_a^i(y_i - k_1^i v_i^2 - k_2^i - e_i), \quad \dot{y}_i = f_j^i(\alpha_i(u_i - y_i)), \end{aligned} \tag{1}$$

where $d_i = x_{i-1} - x_i$ is the distance between two consecutive vehicles, x_{i-1} and x_i represent their positions, v_i, a_i and $\dot{y}_i$ are velocity, acceleration and jerk, respectively, $f_a^i(.)$ and $f_j^i(.)$ are static nonlinearities of saturation type, α_i represents the inverse time-constant of the basic vehicle dynamics, k_1^i and k_2^i are the constants defining rolling resistance, u_i is the corresponding control input, while e_i represents the white random noise force input with variance r_i^e, resulting from wind gusts and road roughness. A slightly modified version of (1) is taken in [8, 15] as the basic model of individual vehicles in a platoon. There are several possibilities for constructing linearized models in the state-space form, depending on the choice of state variables, e.g. [5, 8, 12, 15, 17]. A convenient form follows directly from (1). Supposing, for the sake of simplicity, that $n = 3$ and that all the vehicles have the same models, we obtain

$$\begin{bmatrix}\dot{X}_1\\ \dot{X}_2\\ \dot{X}_3\end{bmatrix} = \begin{bmatrix}A_v & 0 & 0\\ A_d & A_v & 0\\ 0 & A_d & A_v\end{bmatrix}\begin{bmatrix}X_1\\ X_2\\ X_3\end{bmatrix} + \begin{bmatrix}B_v & 0 & 0\\ 0 & B_v & 0\\ 0 & 0 & B_v\end{bmatrix}\begin{bmatrix}u_1\\ u_2\\ u_3\end{bmatrix} + \begin{bmatrix}G_e & 0 & 0\\ 0 & G_e & 0\\ 0 & 0 & G_e\end{bmatrix}\begin{bmatrix}e_1\\ e_2\\ e_3\end{bmatrix}, \qquad (2)$$

where $X_i^{\mathrm{T}} = [d_i \; v_i \; a_i]$ ($x_0 = 0$ in d_1) and

$$A_v = \begin{bmatrix}0 & -1 & 0\\ 0 & 0 & 1\\ 0 & 0 & -\alpha\end{bmatrix}, \quad A_d = \begin{bmatrix}0 & 1 & 0\\ 0 & 0 & 0\\ 0 & 0 & 0\end{bmatrix},$$

$$B_v^{\mathrm{T}} = \begin{bmatrix}0 & 0 & \alpha\end{bmatrix}, \quad G_e^{\mathrm{T}} = \begin{bmatrix}0 & 1 & 0\end{bmatrix}.$$

Control design for (2) can, obviously, be based on various methodologies. However, any attempt to formulate a globally optimal control law for the entire platoon is faced with the problem that control of each vehicle depends, in general, on the states of all the remaining vehicles. Permissible control strategies should essentially be decentralized, having in mind the supposed information structure [8], i.e. the local control u_i is to be calculated on the basis of the noisy measurements of the local vehicle state variables $\{d_i, v_i, a_i\}$, together with the noisy information about the velocity and acceleration of the preceding vehicle $\{v_{i-1}, a_{i-1}\}$, which is assumed to be transmitted by appropriate communication channels. Each vehicle is also supplied with the information about the spacing, velocity, and acceleration reference command $\{d_r, v_r, a_r\}$. The theory of large-scale systems abounds with methodologies for both decentralized design of complex control structures and decentralized design of completely decentralized control structures, e.g. [13, 15, 17]. One of the elegant and powerful methodologies is based on the Stochastic Inclusion Principle [3, 14, 16].

3 Stochastic Inclusion Principle

3.1 General Aspects

Consider a pair $(\mathbf{S},\tilde{\mathbf{S}})$ of linear stochastic continuous-time dynamic systems represented by

$$\begin{aligned}\mathbf{S}:\ \mathrm{d}x &= Ax\mathrm{d}t + Bu\mathrm{d}t + \Gamma\mathrm{d}\xi, \quad \mathrm{d}z = Cx\mathrm{d}t + \mathrm{d}\eta \\ \tilde{\mathbf{S}}:\ \mathrm{d}\tilde{x} &= \tilde{A}\tilde{x}\mathrm{d}t + \tilde{B}\tilde{u}\mathrm{d}t + \tilde{\Gamma}\mathrm{d}\tilde{\xi}, \quad \mathrm{d}\tilde{z} = \tilde{C}\tilde{x}\mathrm{d}t + \mathrm{d}\tilde{\eta}\end{aligned} \tag{3}$$

where $x(t_0) = x_0$ and $\tilde{x}(t_0) = \tilde{x}_0$. The first equations in (3) are Ito stochastic differential equations describing the evolution of state vectors $x(t) \in R^n$ and $\tilde{x}(t) \in R^{\tilde{n}}$ of $\mathbf{S}$ and $\tilde{\mathbf{S}}$, respectively, driven by control inputs $u(t) \in R^m$ and $\tilde{u}(t) \in R^{\tilde{m}}$ (it is straightforward to connect model (2) with model (3)). Stochastic disturbances are modelled by Wiener processes $\xi(t) \in R^r$ and $\tilde{\xi}(t) \in R^{\tilde{r}}$ with incremental covariances $R_{\xi}\mathrm{d}t$ and $R_{\tilde{\xi}}\mathrm{d}t$, respectively. The second equations are the observation equations, where $\eta(t) \in R^q$ and $\tilde{\eta}(t) \in R^{\tilde{q}}$ are Wiener processes with incremental covariances $R_{\eta}\mathrm{d}t$ and $R_{\tilde{\eta}}\mathrm{d}t$, respectively. Vectors x_0 and $\tilde{x}_0$ are assumed to be Gaussian with means m_0 and $\tilde{m}_0$, and covariances R_0 and $\tilde{R}_0$, respectively. It is assumed that $\xi(t),\eta(t)$ and x_0, as well as $\tilde{\xi}(t),\tilde{\eta}(t)$ and $\tilde{x}_0$ are mutually independent. Matrices $A,B,\Gamma,C,\tilde{A},\tilde{B},\tilde{\Gamma}$ and $\tilde{C}$ are assumed to be constant. The basic assumption is that $n \leq \tilde{n}, p \leq \tilde{p}$ and $q \leq \tilde{q}$.

In general, for a stochastic process $\alpha(t)$ we shall denote the mean by $m_{\alpha}(t)$ and covariance by $R_{\alpha}(t_1,t_2)$. If $\alpha(t) = T\,\beta(t)$ $(\forall t \geq t_0)$, where $\alpha(t)$ and $\beta(t)$ are n_{α}- and n_{β}-dimensional stochastic processes, respectively, and T a full rank matrix, we shall say that $\alpha(t)$ is a strong (strict) expansion of $\beta(t)$, i.e. $\alpha(t) = E_s[\beta(t);T]$, if $n_{\alpha} > n_{\beta}$, and that $\alpha(t)$ is a strong (strict) contraction of $\beta(t)$, i.e. $\alpha(t) = C_s[\beta(t);T]$, if $n_{\alpha} < n_{\beta}$. If, for the same processes, $m_{\alpha}(t) = Tm_{\beta}(t)$ and $R_{\alpha}(t_1,t_2) = TR_{\beta}(t_1,t_2)T^{\mathrm{T}}$ $(\forall t,t_1,t_2 \geq t_0)$, we shall say that $\alpha(t)$ is a weak expansion of $\beta(t)$, i.e. $\alpha(t) = E_w[\beta(t);T]$ if $n_{\alpha} > n_{\beta}$, and a weak contraction, i.e. $\alpha(t) = C_w[\beta(t);T]$, if $n_{\alpha} < n_{\beta}$.

Definition 1. The system $\tilde{\mathbf{S}}$ includes the system $\mathbf{S}$ if there exists a quadruplet of full rank matrices $\{U_{n\times\tilde{n}}, V_{\tilde{n}\times n}, R_{\tilde{p}\times p}, S_{q\times\tilde{q}}\}$ satisfying $UV = I_n$, such that for any x_0 and $u(t)$ in $\mathbf{S}$ the conditions $\tilde{x}_0 = E_w[x_0;V]$ and $\tilde{u}(t) = E_s[u(t);R]$ imply $x(t) = C_w[\tilde{x}(t);U]$ and $z(t) = C_w[\tilde{z}(t);S]$ $\quad(\forall t \geq t_0)$.

The expansion $\tilde{\mathbf{S}}$ contains all necessary information about $\mathbf{S}$ expressed in terms of second-order statistics, having in mind the Gauss-Markov properties of $x(t)$, $\tilde{x}(t)$, $z(t)$ and $\tilde{z}(t)$. Weak contractions/expansions are related to the states and outputs, and strong contractions/expansions to control inputs.

Restriction and aggregation represent two important special cases of inclusion.

Definition 2. The system $\mathbf{S}$ is a restriction (type c, according to [16]) of the system $\tilde{\mathbf{S}}$ if there exists a full rank matrix $V_{\tilde{n}\times n}$ such that for any x_0 the conditions $\tilde{x}_0 = E_w[x_0;V]$ and $u(t) = C_s[\tilde{u}(t);Q]$ imply $\tilde{x}(t) = E_w[x(t);V]$ and $\tilde{z}(t) = E_w[z(t);T]$ $(\forall t \geq t_0)$.

Theorem 1. *The system* $\mathbf{S}$ *is a restriction (type c) of* $\tilde{\mathbf{S}}$ *if there exist full rank matrices* V,Q *and* T *such that*

$$\tilde{A}V = VA,\ \ V\Gamma R_\xi \Gamma^{\mathrm{T}} V^{\mathrm{T}} = \tilde{\Gamma} R_{\tilde{\xi}} \tilde{\Gamma}^{\mathrm{T}},$$
$$\tilde{B} = VBQ,\ \ \tilde{C}V = TC,\ \ TR_\eta T^{\mathrm{T}} = R_{\tilde{\eta}}. \tag{4}$$

Definition 3. The system $\mathbf{S}$ is an aggregation (type c) of $\tilde{\mathbf{S}}$ if there exist a triplet of full rank matrices (U,R,S) such that for any $\tilde{x}_0$ and $u(t)$ the conditions $x_0 = C_w[\tilde{x}_0;U]$ and $\tilde{u}(t) = E_s[u(t);R]$ imply $x(t) = C_w[\tilde{x}(t);U]$ and $z(t) = C_w[\tilde{z}(t);S]$ $(\forall t \geq t_0)$.

Theorem 2. *The system* $\mathbf{S}$ *is an aggregation (type c) of* $\tilde{\mathbf{S}}$ *if there exist full rank matrices* U,R *and* S *such that*

$$U\tilde{A} = AU,\ \ \Gamma R_\xi \Gamma^{\mathrm{T}} = U\tilde{\Gamma} R_{\tilde{\xi}} \tilde{\Gamma}^{\mathrm{T}} U^{\mathrm{T}},$$
$$U\tilde{B}R = B,\ \ S\tilde{C} = CU,\ \ SR_{\tilde{\eta}} S^{\mathrm{T}} = R_\eta. \tag{5}$$

3.2 Inclusion of Estimators

Consider time-invariant estimators $\mathbf{E}$ and $\tilde{\mathbf{E}}$ for $\mathbf{S}$ and $\tilde{\mathbf{S}}$, respectively,

$$\mathbf{E}:\ \ \mathrm{d}w = Fw\mathrm{d}t + Gu\mathrm{d}t + L\mathrm{d}z$$
$$\tilde{\mathbf{E}}:\ \ \mathrm{d}\tilde{w} = \tilde{F}\tilde{w}\mathrm{d}t + \tilde{G}\tilde{u}\mathrm{d}t + \tilde{L}\mathrm{d}\tilde{z}, \tag{6}$$

where $w(t) \in R^s$ and $\tilde{w}(t) \in R^{\tilde{s}}$ are the estimator outputs satisfying $s \leq \tilde{s}$. State models for $\mathbf{S^e} = (\mathbf{S},\mathbf{E})$ and $\tilde{\mathbf{S}}^{\mathbf{e}} = (\tilde{\mathbf{S}},\tilde{\mathbf{E}})$ are, respectively,

$$\mathbf{S^e}:\ \ \mathrm{d}X = A^e X\mathrm{d}t + B^e u\mathrm{d}t + \Gamma^e \mathrm{d}\Xi$$
$$\tilde{\mathbf{S}}^{\mathbf{e}}:\ \ \mathrm{d}\tilde{X} = \tilde{A}^e \tilde{X}\mathrm{d}t + \tilde{B}^e \tilde{u}\mathrm{d}t + \tilde{\Gamma}^e \mathrm{d}\tilde{\Xi}, \tag{7}$$

where $X = [x^{\mathrm{T}} w^{\mathrm{T}}]^{\mathrm{T}}$, $\tilde{X} = [\tilde{x}^{\mathrm{T}} \tilde{w}^{\mathrm{T}}]^{\mathrm{T}}$ $\Xi = [\xi^{\mathrm{T}} \eta^{\mathrm{T}}]^{\mathrm{T}}$, $\tilde{\Xi} = [\tilde{\xi}^{\mathrm{T}} \tilde{\eta}^{\mathrm{T}}]^{\mathrm{T}}$

$$A^e = \begin{bmatrix} A & 0 \\ LC & F \end{bmatrix},\ \ B^e = \begin{bmatrix} B \\ G \end{bmatrix},\ \ \Gamma^e = \begin{bmatrix} \Gamma & 0 \\ 0 & L \end{bmatrix};$$

matrices $\tilde{A}^e, \tilde{B}^e$ and $\tilde{\Gamma}^e$ are defined analogously. It will be assumed that $X_0 = X(t_0)$ and $\tilde{X}_0 = \tilde{X}(t_0)$ are Gaussian with means M_0 and $\tilde{M}_0$ and covariances R_0^X and $R_0^{\tilde{X}}$, respectively.

Definition 4. The pair $(\tilde{\mathbf{S}}, \tilde{\mathbf{E}})$ includes the pair $(\mathbf{S}, \mathbf{E})$ if $\tilde{\mathbf{S}}$ includes $\mathbf{S}$, and there exists a pair of full rank matrices $(D_{s\times\tilde{s}}, E_{\tilde{s}\times s})$ satisfying $DE = I_s$, such that for any given X_0 and $u(t)$ the conditions $\tilde{X}_0 = E_w[X_0; V^*]$ and $\tilde{u}(t) = E_s[u(t); R]$ imply $X(t) = C_w[\tilde{X}(t); U^*]$ $(\forall t \geq t_0)$, where $U^* = \text{diag}\{U, D\}$ and $V^* = \text{diag}\{V, E\}$.

Theorem 3. *The pair* $(\mathbf{S}, \mathbf{E})$ *is a restriction (type c) of the pair* $(\tilde{\mathbf{S}}, \tilde{\mathbf{E}})$ *if the conditions of Theorem 1 are satisfied, together with* $\tilde{F}E = EF$*, where* E *is a full rank matrix, and* $\tilde{G} = EGQ$*,* $\tilde{L}T = EL$*.*

Theorem 4. *The pair* $(\mathbf{S}, \mathbf{E})$ *is an aggregation (type c) of the pair* $(\tilde{\mathbf{S}}, \tilde{\mathbf{E}})$ *if the conditions of Theorem 2 are satisfied, together with* $D\tilde{F} = FD$*, where* D *is a full rank matrix, and* $D\tilde{G}R = G$*,* $D\tilde{L} = LS$*.*

3.3 Contractibility of Dynamic Controllers

Let $\mathbf{S^f} = (\mathbf{S}, \mathbf{E}, \mathbf{F})$ and $\tilde{\mathbf{S}}^\mathbf{f} = (\tilde{\mathbf{S}}, \tilde{\mathbf{E}}, \tilde{\mathbf{F}})$, where $\mathbf{F}$ and $\tilde{\mathbf{F}}$ are feedback mappings added to the pairs $(\mathbf{S}, \mathbf{E})$ and $(\tilde{\mathbf{S}}, \tilde{\mathbf{E}})$

$$\mathbf{F}: \quad u = Kw + v; \quad \tilde{\mathbf{F}}: \tilde{u} = \tilde{K}\tilde{w} + \tilde{v} \tag{8}$$

where K and $\tilde{K}$ are constant matrices, and v and $\tilde{v}$ reference signals. Obviously, we have

$$\begin{aligned} \mathbf{S^f}: \ \mathrm{d}X &= A^* X \mathrm{d}t + B^* v \mathrm{d}t + \Gamma^* \mathrm{d}\Xi \\ \tilde{\mathbf{S}}^\mathbf{f}: \ \mathrm{d}\tilde{X} &= \tilde{A}^* \tilde{X} \mathrm{d}t + \tilde{B}^* \tilde{v} \mathrm{d}t + \tilde{\Gamma}^* \mathrm{d}\tilde{\Xi} \end{aligned} \tag{9}$$

where

$$A^* = \begin{bmatrix} A & BK \\ LC & F + GK \end{bmatrix}, \ B^* = \begin{bmatrix} B \\ G \end{bmatrix}, \ \Gamma^* = \begin{bmatrix} \Gamma & 0 \\ 0 & L \end{bmatrix};$$

matrices $\tilde{A}^*, \tilde{B}^*$ and $\tilde{\Gamma}^*$ are defined analogously.

Definition 5. We say that the dynamic controller $(\tilde{\mathbf{E}}, \tilde{\mathbf{F}})$ for $\tilde{\mathbf{S}}$ is contractible to the dynamic controller $(\mathbf{E}, \mathbf{F})$ for $\mathbf{S}$ if $\tilde{\mathbf{S}}^\mathbf{f}$ includes $\mathbf{S^f}$ in the sense of Definition 1.

Theorem 5. *The controller* $(\tilde{\mathbf{E}}, \tilde{\mathbf{F}})$ *is contractible to the controller* $(\mathbf{E}, \mathbf{F})$ *when* $(\mathbf{S}, \mathbf{E})$ *is a restriction (type c) of* $(\tilde{\mathbf{S}}, \tilde{\mathbf{E}})$ *and the condition* $K = Q\tilde{K}E$ *is satisfied.*

Theorem 6. *The controller* $(\tilde{\mathbf{E}}, \tilde{\mathbf{F}})$ *is contractible to the controller* $(\mathbf{E}, \mathbf{F})$ *when* $(\mathbf{S}, \mathbf{E})$ *is an aggregation (type c) of* $(\tilde{\mathbf{S}}, \tilde{\mathbf{E}})$ *and the condition* $\tilde{K} = RKD$ *is satisfied.*

The above results show that K can be obtained for any given $\tilde{K}$ in the case of restriction, while L can be obtained from any given $\tilde{L}$ in the case of aggregation. When $F = A - LC, G = B, D = U$ and $E = V$ the corresponding explicit contraction mappings are $L = U\tilde{L}T$ and $K = Q\tilde{K}V$ [10].

3.4 Inclusion of Performance Indices

Consider the following pair of steady-state performance indices for $\mathbf{S}$ and $\tilde{\mathbf{S}}$, respectively,

$$J(u) = \lim_{T\to\infty} \frac{1}{T} E\{\int_0^{\mathrm{T}} (x^{\mathrm{T}} W_x x + u^{\mathrm{T}} W_u u) \mathrm{d}t\}$$

$$\tilde{J}(\tilde{u}) = \lim_{T\to\infty} \frac{1}{T} E\{\int_0^{\mathrm{T}} (\tilde{x}^{\mathrm{T}} \tilde{W}_x \tilde{x} + \tilde{u}^{\mathrm{T}} \tilde{W}_u \tilde{u}) \mathrm{d}t\} \tag{10}$$

where the matrices W_x, W_u, $\tilde{W}_x$, and $\tilde{W}_u$ are symmetric and positive semidefinite.

Definition 6. The pair $(\tilde{\mathbf{S}}, \tilde{J})$ includes the pair $(\mathbf{S}, J)$ in sense of the optimal feedback control law if the controller $(\tilde{\mathbf{E}}^*, \tilde{\mathbf{F}}^*)$ minimizing $\tilde{J}$ includes the controller $(\mathbf{E}^*, \mathbf{F}^*)$ minimizing J and

$$J(\mathbf{E}^*, \mathbf{F}^*) = \tilde{J}(\tilde{\mathbf{E}}^*, \tilde{\mathbf{F}}^*). \tag{11}$$

Theorem 7. *If* $\mathbf{S}$ *is a restriction (type c) of* $\tilde{\mathbf{S}}$*, then the pair* $(\tilde{\mathbf{S}}, \tilde{J})$ *includes the pair* $(\mathbf{S}, J)$ *in the sense of the optimal feedback control law if*

$$V^{\mathrm{T}} M_x V = 0, \quad W_u^{-1} = Q \tilde{W}_u^{-1} Q^{\mathrm{T}}, \tag{12}$$

where M_x *is obtained from* $\tilde{W} = U^{\mathrm{T}} W_x U + M_x$*.*

If $\mathbf{S}$ is an aggregation (type c) of $\tilde{\mathbf{S}}$, the pair $(\tilde{\mathbf{S}}, \tilde{J})$ includes the pair $(\mathbf{S}, J)$ if

$$\tilde{W}_x = U^{\mathrm{T}} W_x U; \quad W_u^{-1} = Q \tilde{W}_u^{-1} Q^{\mathrm{T}}. \tag{13}$$

If $\mathbf{S}$ is a restriction of $\tilde{\mathbf{S}}$, it follows that the optimal feedback gain matrix is contractible to the original space by $K = Q\tilde{K}V$.

3.5 Overlapping Decentralized Control

The essence of the application of the above exposed inclusion principle to the decentralized control design of systems with the overlapping structure $\mathbf{S}$, lies in the application of such an expansion which results into $\tilde{\mathbf{S}}$ in which subsystems of $\mathbf{S}$ appear as disjoint, e.g. [10, 13, 14, 16, 17]. For example, if $\mathbf{S}$ is defined by (3), where $A = [A_{ij}]$, $B = [B_{ij}]$, $C = [C_{ij}]$, $\Gamma = \text{diag}\{\Gamma_1, \Gamma_2, \Gamma_3\}$, $R_\xi = \text{diag}\{R_{\xi,1}, R_{\xi,2}, R_{\xi,3}\}$ and $R_\eta = \text{diag}\{R_{\eta,1}, R_{\eta,2}, R_{\eta,3}\}$, $(i,j = 1,2,3)$, then we can consider, under certain conditions concerning submatrices A_{13}, A_{31}, B_{13}, B_{31}, C_{13} and C_{31} in A, B and C, that it is composed of two overlapping subsystems $\tilde{\mathbf{S}}_1$ and $\tilde{\mathbf{S}}_2$ defined by system matrices $\tilde{A}^1 = [A_{ij}]$, $\tilde{B}^1 = [B_{ij}]$, $\tilde{C}^1 = [C_{ij}]$, $\tilde{\Gamma}^1 = \text{diag}\{\Gamma_1, \Gamma_2\}$, $\tilde{R}^1_\xi = \text{diag}\{R_{\xi,1}, R_{\xi,2}\}$, $\tilde{R}^1_\eta = \text{diag}\{R_{\eta,1}, R_{\eta,2}\}$, $(i,j = 1,2)$ and $\tilde{A}^2 = [A_{jk}]$, $\tilde{B}^2 = [B_{jk}]$, $\tilde{C}^2 = [C_{jk}]$, $\tilde{\Gamma}^2 = \text{diag}\{\Gamma_2, \Gamma_3\}$, $\tilde{R}^2_\xi = \text{diag}\{R_{\xi,2}, R_{\xi,3}\}$, $\tilde{R}^2_\eta = \text{diag}\{R_{\eta,2}, R_{\eta,3}\}$, $(j,k = 2,3)$, respectively. After performing an appropriate expansion and extracting the corresponding subsystems from $\mathbf{S}$, we shall look for decentralized dynamic controllers $(\tilde{\mathbf{E}}_1, \tilde{\mathbf{F}}_1)$ and $(\tilde{\mathbf{E}}_2, \tilde{\mathbf{F}}_2)$, characterized, in the case of local LQG optimal control, by the gain pairs $(\tilde{L}_1, \tilde{K}_1)$ and $(\tilde{L}_2, \tilde{K}_2)$, which, after being contracted back to the original space, result into a suboptimal controller $(\mathbf{E}, \mathbf{F})$ for $\mathbf{S}$.

In the above context the main point is to find such pairs of matrices (U,V), (Q,R) and (S,T) which enable an expansion with satisfactory decoupling effects, as well as a direct contraction to the original space. We shall consider different restriction and aggregation relations between $\mathbf{S}$ and $\tilde{\mathbf{S}}$ obtained by using these matrices in two characteristic forms, e.g.:

$$V_1 = \begin{bmatrix} I & 0 & 0 \\ 0 & I & 0 \\ 0 & I & 0 \\ 0 & 0 & I \end{bmatrix}, \quad U_1 = \begin{bmatrix} I & 0 & 0 & 0 \\ 0 & \beta I & (1-\beta)I & 0 \\ 0 & 0 & 0 & I \end{bmatrix} \tag{14}$$

and $V_2 = U_1^{\mathrm{T}}$, $U_2 = V_1^{\mathrm{T}}$, where β is a scalar satisfying $0 < \beta < 1$; matrices R and T are analogous to V, while Q and S are analogous to U. Starting from matrices $A, B, C, \Gamma, R_\xi, R_\eta$, matrices $\tilde{A}, \tilde{B}, \tilde{C}, \tilde{\Gamma}, R_{\tilde{\xi}}, R_{\tilde{\eta}}$ can be obtained by choosing e.g. matrix M_A in $\tilde{A} = VAU + M_A$, matrix M_B in $\tilde{B} = VBQ + M_B$, matrix M_C in $\tilde{C} = TCU + M_C$, etc. For example, conditions for both restriction and aggregation are satisfied for the following matrix $\tilde{A}$, obtained by using (U_1, V_1):

$$\tilde{A} = \begin{bmatrix} A_{11} & \beta A_{12} & (1-\beta)A_{12} & A_{13} \\ A_{21} & A_{22} & 0 & A_{23} \\ A_{21} & 0 & A_{22} & A_{23} \\ A_{31} & \beta A_{32} & (1-\beta)A_{32} & A_{33} \end{bmatrix}.$$

After expansion, the dynamic controller for the resulting $\tilde{\mathbf{S}}$ is designed by optimizing in the LQG sense separately $\tilde{\mathbf{S}}_1$ and $\tilde{\mathbf{S}}_2$, obtained by cutting $\tilde{A}$ adequately (as well as the remaining matrices in $\tilde{\mathbf{S}}$). The resulting estimator and feedback gain matrices $\tilde{L}_1, \tilde{L}_2, \tilde{K}_1$ and $\tilde{K}_2$ give $\tilde{L}_D = \text{diag}\{\tilde{L}_1, \tilde{L}_2\}$ and $\tilde{K}_D = \text{diag}\{\tilde{K}_1, \tilde{K}_2\}$, defining the overall controller $(\tilde{\mathbf{E}}, \tilde{\mathbf{F}})$ for $\tilde{\mathbf{S}}$. The global performance index $\tilde{J}$ for $\tilde{\mathbf{S}}$ is constructed by using weighting matrices $\tilde{W}_x = \text{diag}\{\tilde{W}^1_x, \tilde{W}^2_x\}$ and $\tilde{W}_u = \text{diag}\{\tilde{W}^1_u, \tilde{W}^2_u\}$, where the local weighting matrices $\tilde{W}^1_x, \tilde{W}^2_x, \tilde{W}^1_u, \tilde{W}^2_u$ are chosen in accordance with (10), in order to satisfy inclusion of the performance indices $\tilde{J}$ and J. Contraction to the original space is done by $L = U\tilde{L}T$ and $K = Q\tilde{K}V$ after an eventual modification of either $\tilde{L}_D$ or $\tilde{K}_D$ aimed at satisfying contractibility conditions ($U\tilde{L} = U\tilde{L}TS$ or $Q\tilde{K} = Q\tilde{K}VU$), having in mind that in the case of restriction we can never have a block-diagonal $\tilde{L}$, and in the case of aggregation a block-diagonal $\tilde{K}$. The resulting controller $(\mathbf{E}, \mathbf{F})$ is suboptimal with the suboptimality degree μ, i.e. $\mu^{-1}J^* = J$, where J^* is the minimal value of J corresponding to the globally optimal controller in the original space.

4 Decentralized LQG Suboptimal Platoon Control

Following the above exposed line of thought, a decentralized LQG suboptimal control strategy will be developed by considering a platoon of vehicles as a concatenation of overlapping "subsystems." The ith subsystem is defined by the following state model (see [15, 17] for the deterministic case)

$$\dot{\xi}_i = \begin{bmatrix} A_L & 0 \\ \bar{A}_d & A_v \end{bmatrix} \xi_i + \begin{bmatrix} B_L & 0 \\ 0 & B_v \end{bmatrix} \begin{bmatrix} u_{i-1} \\ u_i \end{bmatrix} + \begin{bmatrix} G_L & 0 \\ 0 & G_e \end{bmatrix} \begin{bmatrix} e_{i-1} \\ e_i \end{bmatrix} \tag{15}$$

where

$$A_L = \begin{bmatrix} 0 & -1 \\ 0 & -\alpha \end{bmatrix}, \quad \bar{A}_d^{\mathrm{T}} = \begin{bmatrix} 1 & 0 & 0 \\ 0 & 0 & 0 \end{bmatrix}$$

$$B_L^{\mathrm{T}} = \begin{bmatrix} 0 & \alpha \end{bmatrix}, \quad G_L^{\mathrm{T}} = \begin{bmatrix} 1 & 0 \end{bmatrix}$$

and $\xi_i^{\mathrm{T}} = [v_{i-1}\ a_{i-1}\ d_i\ v_i\ a_i]$. According to (2), the overlapping part in the state matrix is, obviously, A_L with both the preceding and the following subsystems. Having in mind the formalism of the Inclusion Principle, the above subsystems can be extracted from the basic model by expanding the state using a matrix V, which has, for two subsystems, the form (14), with appropriate dimensions (generalization to n vehicles is straightforward). The above "subsystems" can hardly be given any precise physical interpretation; notice, however, that, formally, the noisy state vectors of the subsystems are supposed to be available in each vehicle [8]. The subsystems are not only state overlapping, but also input overlapping (they have one input in common), so that the input expansion is needed, as well; the corresponding transformation matrix R has the form analogous to V. As u_i is essentially the physical control signal in the ith vehicle, then u_{i-1} in the corresponding subsystem could be considered to represent, together with the corresponding part of the subsystem dynamics, the preceding part of the platoon, as viewed by ith vehicle (for the second vehicle in the platoon this is exactly the leading vehicle dynamics). Therefore, u_i depends on the entire subsystem state, and u_{i-1} only on the part of the subsystem state vector overlapping with the preceding subsystem. After expansion, the subsystems in the platoon model appear as disjoint. Application of the LQG methodology based on the definition of local performance indices leads to local state feedback control (depending on the appropriate sets of measurements). Contraction to the original space provides a physically implementable control law.

As the leading vehicle dynamics represents formally a part of each subsystem, we shall describe the proposed control strategy consecutively, starting from the leading vehicle.

4.1 Leading Vehicle Control

The leading vehicle is supplied with the reference command and uses its own state vector for control design. Formally, if the leading vehicle model is represented by

$$\dot{X}_L = A_L X_L + B_L u_1 + G_L e_1 \tag{16}$$

where $X_L^{\mathrm{T}} = [v_1\ a_1]$, then the optimal feedback control law using noisy measurements $Y_L^{\mathrm{T}} = [v_1 + n_1^v \quad a_1 + n_1^a]$ (where n_1^v and n_1^a are mutually independent white noise terms with variances r_1^v and r_1^a, respectively) should be found from the condition for the minimum of the performance index

$$J_L = E\{\int_{t_0}^{\infty} [(X_L - X_{1r})^{\mathrm{T}} Q_L (X_L - X_{1r}) + R_L u_1^2] \mathrm{d}t\} \tag{17}$$

where $X_{1r}^{\mathrm{T}} = [v_r\ a_r]$ is a time-varying reference supplied to the first vehicle, known entirely in advance, and $Q_L \geq 0$ and $R_L > 0$ are the corresponding weights. This is, in fact, an LQG optimal tracking problem, which can be solved in the following way [1]:

$$\begin{aligned}
u_1 &= -K_1 \hat{X}_L - M_1 X_{1r} \\
K_1 &= R_L^{-1} B_L^{\mathrm{T}} P_L \\
M_1 &= R_L^{-1} B_L^{\mathrm{T}} (A_L - B_L K_1)^{-T} Q_L \\
P_L A_L &+ A_L^{\mathrm{T}} P_L - P_L B_L R_L^{-1} B_L^{\mathrm{T}} P_L + Q_L = 0
\end{aligned} \tag{18}$$

where $\hat{X}_L$ is obtained by the locally optimal Kalman filter obtained from (16). This control law is suboptimal, since the feedforward block is reduced to a constant matrix; this is, however, a very reasonable solution, having in mind characteristic forms of the reference command signals. A priori choice of the criterion weights can provide different tracking properties. Notice that the static steady-state error reduces to zero, having in mind that A_L is singular [1].

4.2 General Subsystem Control

Control of the second vehicle assumes that the leading vehicle control is appropriately designed. Consequently, control design for the general subsystem model (15) can be decomposed into two parts: first, u_{i-1} is found and the corresponding regulator is implemented and, second, u_i is found for the resulting system by using the complete feedback starting from the noisy state measurements. According to (18), we have

$$u_{i-1} = -K_1 [\hat{v}_{i-1}\hat{a}_{i-1}]^{\mathrm{T}} - M_1 X_{1r}. \tag{19}$$

After implementing (19), one comes to the following subsystem model

$$\dot{\xi}_i = \begin{bmatrix} A_L - B_L K_1 & 0 \\ \bar{A}_d & A_v \end{bmatrix} \xi_i + \begin{bmatrix} 0 \\ B_v \end{bmatrix} u_i + \begin{bmatrix} G_L & 0 \\ 0 & G_e \end{bmatrix} \begin{bmatrix} e_{i-1} \\ e_i \end{bmatrix} + \begin{bmatrix} K_1 \\ 0 \end{bmatrix} \varepsilon_{i-1} + \begin{bmatrix} -M_1 \\ 0 \end{bmatrix} X_{1r} \tag{20}$$

where ε_{i-1} is the estimation error for $[\hat{v}_{i-1}\hat{a}_{i-1}]^{\mathrm{T}}$ obtained by the Kalman filter belonging to leading vehicle control law. Now, u_i is found from (20) by minimizing

$$J_i = E\{\int_{t_0}^{\infty} [(\xi_i - X_{2r})^{\mathrm{T}} Q_i (\xi_i - X_{2r}) + R_i u_i^2] \mathrm{d}t\} \tag{21}$$

where $Q_i \geq 0$ and $R_i > 0$, while $X_{2r}^{\mathrm{T}} = [d_r \; v_r \; a_r]$ is the complete set of reference commands. The state weighting matrix is assumed to have the following specific form, coming out basically from the regulator structure adopted in [8]:

$$Q_i = \begin{bmatrix} p_1 & 0 & 0 & -p_1 & 0 \\ 0 & p_2 & 0 & 0 & -p_2 \\ 0 & 0 & q_{33} & 0 & 0 \\ -p_1 & 0 & 0 & q_{44}+p_1 & 0 \\ 0 & -p_2 & 0 & 0 & q_{55}+p_2 \end{bmatrix} \tag{22}$$

In (22), q_{33} influences the spacing reference tracking, p_1 and p_2 influence tracking of the velocity and acceleration of the preceding vehicle, respectively, while q_{44} and q_{55} influence velocity and acceleration reference tracking. The problem posed belongs to the class of LQG optimal tracking problems with a priori known disturbances [1]. An approximately optimal solution, in the sense that all the gains are assumed to be constant, is given by

$$\begin{aligned} u_i &= -K_2 \hat{X}_i - M_2 X_{2r} - M_3 X_{1r} \\ K_2 &= R_i^{-1} B_i^{\mathrm{T}} P_2 \\ M_2 &= R_i^{-1} B_i^{\mathrm{T}} (A_i - B_i K_2)^{-T} Q_i \\ M_3 &= R_i^{-1} B_i^{\mathrm{T}} (A_i - B_i K_2)^{-T} P_2 B_M \\ P_2 A_i &+ A_i^{\mathrm{T}} P_2 - P_2 B_i R_i^{-1} B_i^{\mathrm{T}} P_2 + Q_i = 0 \end{aligned} \tag{23}$$

where

$$A_i = \begin{bmatrix} A_L - B_L K_1 & 0 \\ \bar{A}_d & A_v \end{bmatrix}$$

$$B_i^{\mathrm{T}} = \begin{bmatrix} 0 \ B_v \end{bmatrix}$$

$$B_M^{\mathrm{T}} = \begin{bmatrix} -M_1 \ 0 \end{bmatrix}.$$

$\hat{X}_i$ represents the estimate of the subsystem state obtained by using the Kalman filter derived from (20), taking into account specific properties of the input disturbance. Notice that one disturbance term in (20) comes out from the first optimization step, i.e. from the optimal tracking problem solved by u_{i-1}. Consequently, M_2 represents the feedforward gain for the complete reference X_{2r}, while M_3 compensates the effects of the disturbance.

The state feedback gain $K_i^{\mathrm{T}} = [K_1^{\mathrm{T}} \ K_2^{\mathrm{T}}]$ has the LBT structure, in accordance with the information supposed to be locally available. The overall feedforward gain matrix M_i, which can be obtained simply from M_1, M_2 and M_3, multiplies essentially X_{2r}, since X_{1r} is a subset of X_{2r}. It is important to notice that the steady-state error is again zero for constant references.

4.3 Platoon Control

Local regulators formulated for the subsystems are to be contracted to the original space before implementation. The state feedback gains are obtained by using the transformation matrix Q analogous to U with $\beta = 0.5$, i.e. after contraction, one gets $K_M = Q\tilde{K}_i V$. The feedforward gains multiplying the reference signals are not contracted in accordance with the Inclusion Principle, since they are out of the feedback loop. The estimator gains are not contracted, as well, having in mind that all the local subsystem estimators remain uncontracted in the original system state space; formally, $D = E = I$ in terms of the inclusion of the estimators. The main additional requirement is here to keep the steady-state error at zero. It can be easily shown that the structure of M_2 and M_3 in (23) is such that the only nonzero elements are M_2^{51} and M_3^{51}; the only nonzero element in M_1 in (18) is M_1^{31}. It is possible to show that the required modification aimed at reducing the steady-state error to zero is to increment M_3^{51} in (23) by $\Delta M_3^{51} = -(A_K^{32} M_1^{31} + A_K^{35} M_3^{51})/A_K^{35}$, where $A_K = (A_i - B_i K_M)^{-1}$ and $K_M^{\mathrm{T}} = 0.5(K_2^{\mathrm{T}} + [0\ 0\ 0\ K_1^{\mathrm{T}}])$. The corresponding overall feedforward gain M_i (multiplying X_{2r}) contains only three nonzero elements: $M_i^{32} = M_1^{31}, M_i^{51} = M_2^{51}$ and $M_i^{52} = M_3^{51}$. The overall platoon tracks the command reference in a suboptimal way in the LQG sense, preserving the predefined information structure and ensuring the correct steady-state regime. Note that reduced order observers have been applied in [11], within the deterministic context.

5 Experimental Results

Numerous simulations have been undertaken; the platoon has been assumed to obey the nonlinear model (1) and control has been generated according to the described algorithm. Attention has been focused on the choice of the weights in (17) and (22) and noise influence. Figures 1 and 2 give time histories for a platoon of eight vehicles, containing velocities and inter-vehicle spacings; the first velocity and spacing plots correspond to a direct application of LQ feedback (not containing the estimators [17]), while the second plots are obtained by using the whole proposed LQG suboptimal algorithm, including the local Kalman filters. Figure 1 corresponds to the parameter $\alpha = 2$, and Fig. 2 to $\alpha = 10$. The remaining design parameters have been $Q_L = \text{diag}\{200, 10\}$, $R_L = 10$, $p_1 = 100$, $p_2 = 50$, $q_{33} = 500$, $q_{44} = 300$, $q_{55} = 10$, $R_i = 10$, so that we obtained the following feedback and feedforward gains: $K_1 = [17.3205\ 4.3217]$, $K_2 = [-9.3869\ \ -1.6108\ \ -22.3607\ 21.8018\ 4.9121]$, $M_1^{31} = 34.6410$, $M_2^{51} = 24.8298$, $M_3^{51} = -44.7214$ (for $\alpha = 2$) and $K_1 = [4.4721\ 0.7103]$, $K_2 = [-4.0615\ \ -1.2581\ \ -7.0711\ 6.7287\ 1.2909]$, $M_1^{31} = 44.7214$, $M_2^{51} = 26.6721$, $M_3^{51} = -70.7107$ (for $\alpha = 10$). Tracking capabilities and noise immunity of the proposed algorithm are obvious, compared to the results obtainable by using the LQ methodology directly, illustrated by the responses (a) and (c) in both Figs. 1 and 2 [15, 17]. Responses (b) and (d) are obtained by the proposed algorithm; they are faster for $\alpha = 10$. Also, comparison with the results presented in [8] shows a substantial advantage of the proposed approach. It is to be noted that it is important to make decision about the relative importance of tracking the preceding vehicle velocity and the reference command, as well as about the weight of tracking the desired inter-vehicle spacing. The choice of the control weights influences the jerk level, which is important having especially in mind the introduced nonlinearities. In Fig. 3 a part of a real spacing measurement signal is represented, together with one typical autocorrelation function, providing parameters necessary for tuning the applied Kalman filters (the authors are grateful to the staff of the PATH Program, University of Berkeley, for providing real experimental data).

6 Conclusion

In this paper the Stochastic Inclusion Principle has been applied to LQG suboptimal control of a platoon of automotive vehicles. Identification of input/state overlapping stochastic subsystems and their extraction by an appropriate expansion have led to approximate LQG optimization, adapted to the LBT structure of the subsytem model. Simulation results show a high efficiency of the proposed algorithm, from the point of view of both tracking precision and noise immunity. One of the main problems for further investigations is the tracking precision in the case of long platoons.

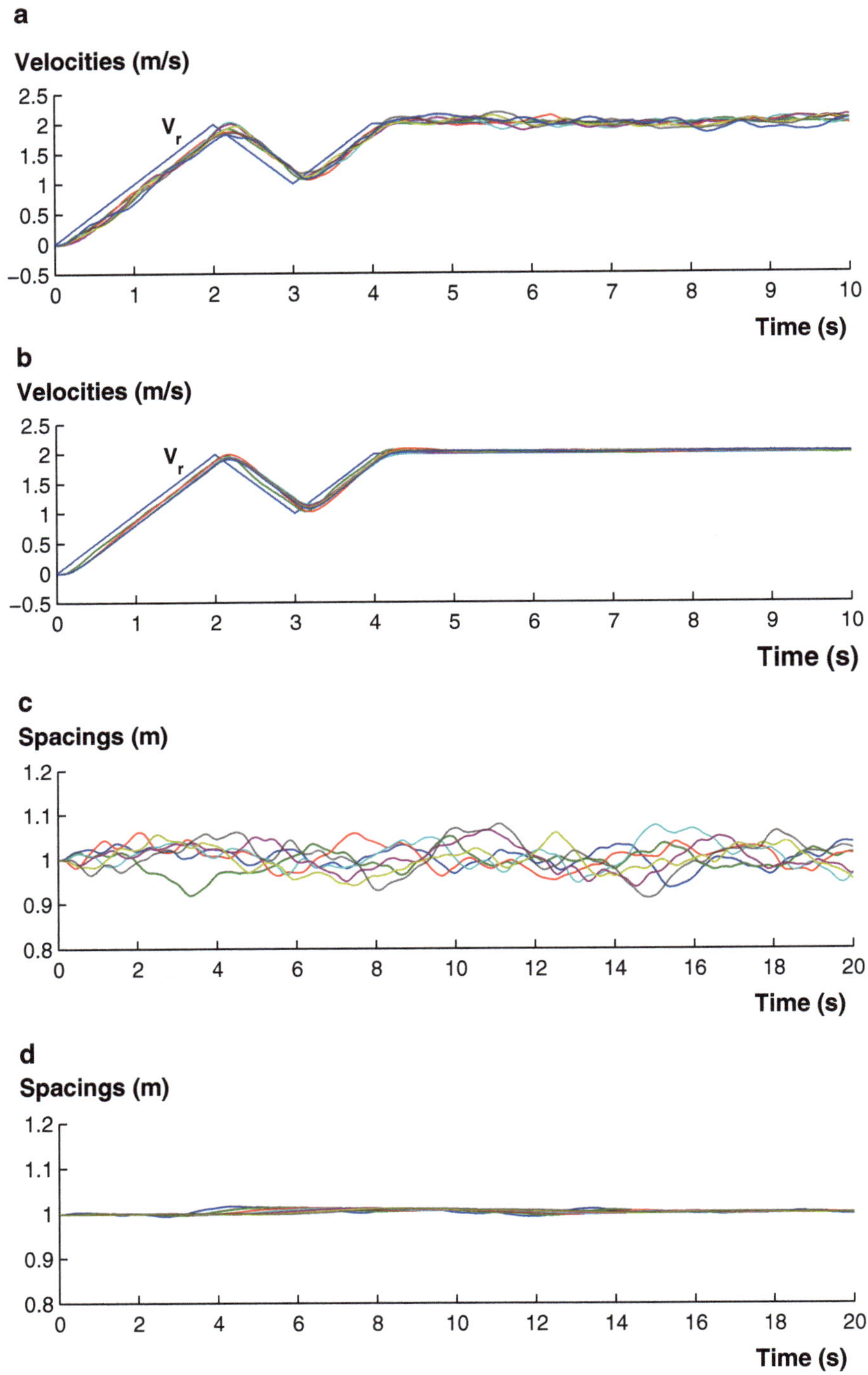

Fig. 1 $\alpha = 2$; Velocities: (**a**) LQ, (**b**) Proposed algorithm; Spacings: (**c**) LQ, (**d**) Proposed algorithm

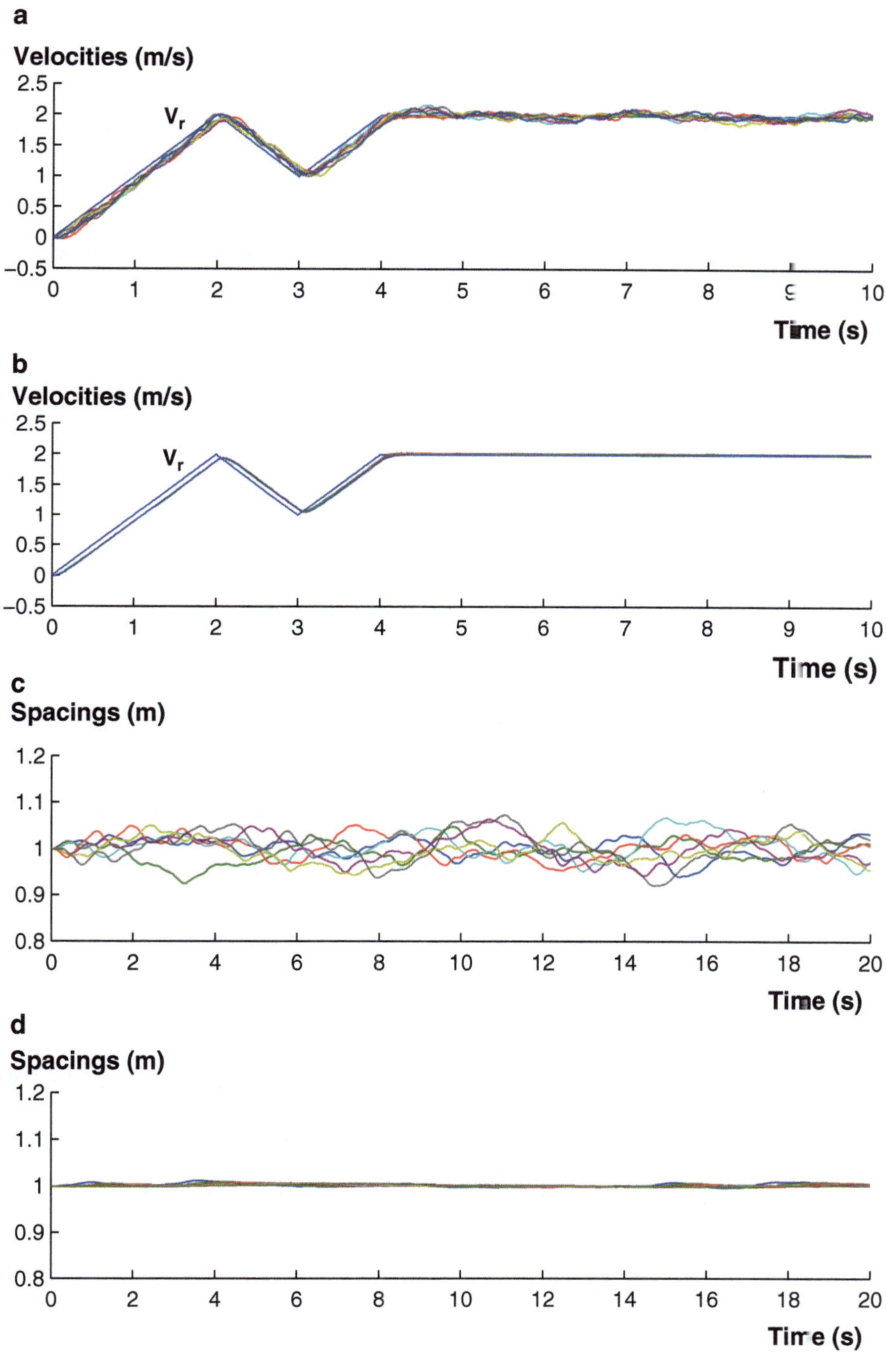

Fig. 2 $\alpha = 10$; Velocities: (**a**) LQ, (**b**) Proposed algorithm; Spacings: (**c**) LQ, (**d**) Proposed algorithm

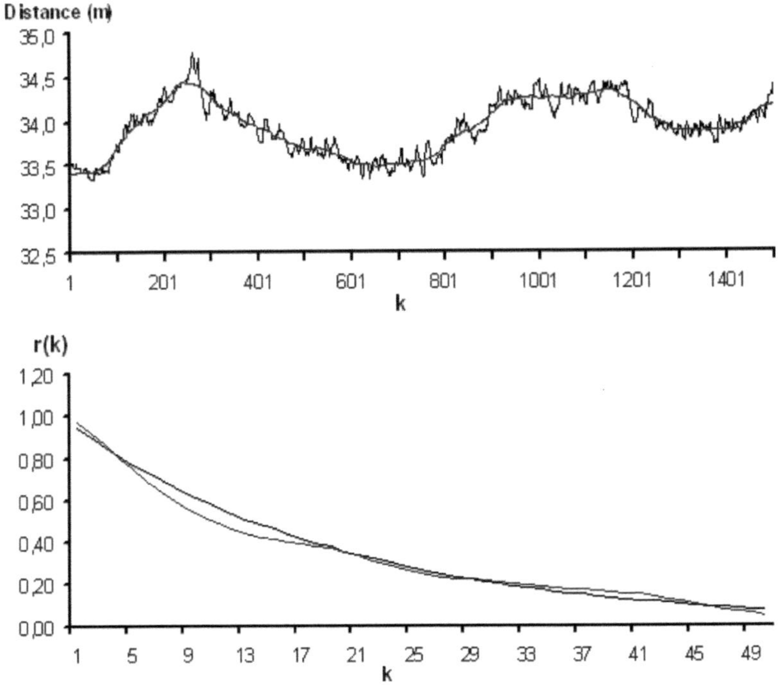

Fig. 3 Real measurement signal and its autocorrelation function

Acknowledgments We are grateful to the staff of the PATH Program from the University of Berkeley for real measurement data.

This work is supported by Project Grant III44004 (2011–2014) financed by the Ministry of Education and Science, Republic of Serbia.

References

1. Anderson, B.D.O., Moore, J.B.: Optimal Control: Linear Quadratic Methods. Prentice Hall, NJ (1990)
2. Godbole, D.N., Eskafi, F.H., Varaiya, P.P, In: Proceedings of 13th IFAC Congress, San Francisco, CA, **L**, 121–126 (1996)
3. Iftar, A., Özgüner, Ü.: Contractible controller design and optimal control with state and input inclusion. Automatica **26**, 593–597 (1990)
4. Kwakernaak, H., Sivan, R.: Linear Optimal Control Systems. Wiley, New York (1972)
5. Levine, W.S., Athans, M.: On the optimal error regulation of a string of moving vehicles. IEEE Trans. Autom. Contr. **11**, 355–361 (1966)
6. Özgüner, Ü., Perkins, W.R.: Optimal control of multilevel large-scale systems. Int. J. Contr. **28**, 967–980 (1978)
7. Sheikholeslam, S., Desoer, C.A.: Control of interconnected nonlinear dynamic systems, the platoon problem. IEEE Trans. Autom. Contr. **37**, 806–810 (1992)

8. Shladover, S.E.: Longitudinal control of automotive vehicles in close formation platoons. J. Dyn. Syst. Meas. Contr. **113**, 231–241 (1991)
9. Stanković, S.S., Šiljak, D.D.: Sequential LQG optimization of hiererchically structured systems. Automatica **25**, 545–559 (1989)
10. Stanković, S.S., Šiljak, D.D.: Contractibility of overlapping decentralized control. Syst. Contr. Lett. **44**, 189–199 (2001)
11. Stanković S.S., Stanojević, M.J.: Decentralized control of a platoon of vehicles based on a reduced measurement set. In: Proceedings of the 1st International Symposium and 10th Balkan Conference on Operations Research, (BALCOR 2011), Thessaloniki, 22–25 September, Greece, vol. 2, pp. 417–424 (2011)
12. Shladover, S.E., et al.: Automatic vehicle control developments in the PATH program. IEEE Trans. Veh. Tech. **40**, 114–130 (1991)
13. Šiljak, D.D.: Decentralized Control of Complex Systems. Academic, London (1991)
14. Stanković, S.S., Chen, X.B., Šiljak, D.D.: Stochastic inclusion principle applied to decentralized overlapping suboptimal LQG control. In: Proceedings of the 13th IFAC Congress, San Francisco, vol. L, pp. 12–18 (1996)
15. Stanković, S. S., M. J. Stanojević, Šiljak, D.D.: Decentralized suboptimal LQ control of a platoon of vehicles. In: Proceedings of the 8th IFAC/IFIP/IFORS Symp. Trans. Syst., Chania, Greece, vol. 1, pp. 81–86 (1997)
16. Stanković, S.S., Chen, X.B., Mataušek, M.R., Šiljak, D.D.: Stochastic inclusion principle applied to decentralized automatic generation control. Int. J. Contr. **72**, 276–288 (1999)
17. Stanković, S.S., Stanojević M.J., Šiljak, D.D.: Decentralized overlapping control of a platoon of vehicles. IEEE Trans. Contr. Syst. Tech. **8**(5), 816–832 (2000)
18. Swaroop, D., Hedrick, J.K.: String stability of interconnected systems. IEEE Trans. Autom. Contr. **41**, 349–357 (1996)
19. Varaiya, P.: Smart cars on smart roads: problems of control. IEEE Trans. Autom. Contr. **38**, 195–207 (1993)

Homogeneous and Non-homogeneous Algorithms

Ioannis Paparrizos

Abstract Motivated by recent best case analyses for some sorting algorithms and based on the type of complexity we partition the algorithms into two classes: *homogeneous* and *non-homogeneous* algorithms.[1] Although both classes contain algorithms with worst and best cases, homogeneous algorithms behave uniformly on all instances. This partition clarifies in a completely mathematical way the previously mentioned terms and reveals that in classifying an algorithm as homogeneous or not best case analysis is equally important with worst case analysis.

Key words Algorithm analysis • Algorithm complexity • Algorithm classification

1 Introduction

In the 1970s and 1980s a lot of discussion was going on regarding the right use of the asymptotic symbols O, Θ and Ω used to analyze algorithms and compare their theoretical efficiency. Some researchers use these symbols to denote the rate of growth of functions and others to denote sets of functions; see relevant comments in [3, 10, 13]. Following the approach of using the asymptotic symbols as sets of functions we partition the class of algorithms into two non-empty subclasses: *homogeneous* and *non-homogeneous* algorithms. Both classes are wide. They contain iterative and recursive algorithms. Although both classes contain

[1]This paper was also presented at local proceedings of PCI'09 [Paparrizos, Homogeneous and Non-Homogeneous Algorithms (2009)].

I. Paparrizos (✉)
Computer Science Department, Columbia University, New York, NY, USA
e-mail: jopa@cs.columbia.edu

A. Migdalas et al. (eds.), *Optimization Theory, Decision Making, and Operations Research Applications*, Springer Proceedings in Mathematics & Statistics 31,
DOI 10.1007/978-1-4614-5134-1_17, © Springer Science+Business Media New York 2013

algorithms with worst and best cases, homogeneous algorithms behave uniformly on all instances of the problem being solved. The partition clarifies in a completely mathematical way the terms of algorithm, worst and best case complexity, the only difference between them being the sets of instances they referred to.

This classification of algorithms was triggered by recent theoretical result concerning best case analysis of some heapsort algorithms [2, 4–8, 20] and [21]. Also, computational results indicate that best case analysis might have practical value too, see, for example, [7] and [21]. Our results indicate that in order to classify an algorithm as homogeneous or not the complexity of the exact, up to a set of functions defined by the asymptotic symbol Θ, best case and worst case must be computed. When the classification is accomplished the analysis of the complexity of the algorithm is complete, indicating, from a theoretical point of view, that best case analysis is equally important with the worst-case analysis.

The term inhomogeneity has been used by Nadel [17] who characterizes the imprecision of an analysis of an algorithm in terms of the difference $\Delta_C = c_w - c_b$ between the worst and best case complexity, where C is a proper measure of complexity. In particular, for the sorting problem, C is the number of comparisons. Using various combinations of disorder parameters, Nadel [17] partitions the set of instances in big, medium, small, tiny and singleton subclasses and computes the inhomogeneity in each subclass. Other relevant results for other problems are presented in [11, 14–16, 18]. Our approach is different in the sense that the set of algorithms is partitioned and not the set of instances of the problem.

In the next section we formally describe the two classes of algorithms. Some details regarding the algorithm classification are presented in Sect. 3. Recursive and divide and conquer homogeneous and non-homogeneous algorithms are discussed and some side results are also presented in the last section.

2 Description of the Two Classes

We derive our results using the Random Access Machine (RAM) model in which every elementary operation such as addition, subtraction, multiplication, and division of two numbers, comparison of two numbers, reading and writing a number in the memory, calling a function, etc., is executed in constant time. It is well known that all constant functions belong to the set $\Theta(1)$. Recall that $\Theta(g(n))$ denotes a set of functions defined as follows:

Definition 1. Given a function $g(n)$ we denote by $\Theta(g(n))$ the set of functions $t(n)$ for which there exists constants $a > 0$ and $b > 0$ and a positive integer n_0 such that

$$bg(n) \leq t(n) \leq ag(n) \tag{1}$$

for every $n \geq n_0$.

All functions used in this paper denote time and therefore they are positive. The argument n denotes the dimension of the problem and, hence, it is a positive integer.

The sets of functions $O(g(n))$ and $\Omega(g(n))$ are similarly defined. Simply, in the definition of $O(g(n))$ the left inequality of (1) is missing, while in the definition of $\Omega(g(n))$ the right one. Observe that $\Theta(g(n))$ is strictly contained in the sets $O(g(n))$ and $\Omega(g(n))$. As a result the assumption that the basic operations are executed in $\Theta(1)$ time (instead of $O(1)$ or $\Omega(1)$ time) provides a more precise algorithm analysis.

It is well known that the symbol Θ considered as a binary relation between functions, is reflexive, symmetric and transitive and therefore it partitions the set of functions into disjoined classes. In other words, if $f(n)$ and $g(n)$ are two different functions, then either $\Theta(f(n)) = \Theta(g(n))$ or $\Theta(f(n)) \cap \Theta(g(n)) = \oslash$. In particular the following two results are well known.

Theorem 1. *If $f(n) \in \Theta(g(n))$, then $\Theta(f(n)) = \Theta(g(n))$.*

Theorem 2. *The sets $\Theta(1)$ and $\Theta(n)$ are disjoint.*

Given a computational problem we denote the set of instances of dimension n by $I(n)$. Consider now an algorithm A solving the problem under consideration. The time taken by algorithm A to solve instance i of dimension n is denoted by $t_A(i,n)$. In algorithm analysis we try to describe in a nice way the set of time functions

$$S = \{t_A(i,n) : i \in I(n)\}$$

One way to do this is via the sets of functions defined by the asymptotic symbols O, Θ, Ω. We are completely satisfied if we can determine a function $g(n)$ such that

$$S \subseteq \Theta(g(n)). \tag{2}$$

Once again, observe that we use the set $\Theta(g(n))$ which is strictly contained in the sets $O(g(n))$ and $\Omega(g(n))$, and therefore the description of set S is more precise. This preference though leads us naturally to the following definition.

Definition 2. An algorithm is *homogeneous* if there exists a function $g(n)$ such that relation (2) holds. Otherwise, the algorithm is *non-homogeneous*.

Theorem 3. *The class of algorithms is partitioned into two non-empty and disjoined subclasses, the subclasses of homogeneous and non-homogeneous algorithms.*

Proof. Let U be the class of all algorithms, H the class of homogeneous and NH the class of non-homogeneous algorithms. It is obvious from Definition 2 that

$$H \cap NH = \oslash \text{ and } H \cup NH = U.$$

It remains to show that $H \neq \oslash$ and $NH \neq \oslash$. This proof is done by providing a simple algorithm for each class.

Algorithm 1: MIN

```
a ← T(1)
for j = 2 → n do
    if T(j) < a then
        a ← T(j)
    end if
end for
```

Firstly, consider the problem of finding the smallest among n given numbers stored as elements of an array T.

The algorithm *min* (Algorithm 1) solves this problem and is homogeneous. Indeed assuming that an element of an array is reached in constant time $\Theta(1)$ in the computational model of constant times, it is easy to conclude that

$$t_{\min}(i,n) \in \Theta(n)$$

for every instance $i \in I(n)$. Hence, algorithm min is homogeneous and $H \neq \oslash$.

Secondly, consider the following problem. Given an array T of n elements sorted in increasing order, i.e.

$$T(j) \leq T(j+1) \text{ for } i = 1,2,\ldots,n-1$$

and a number x, sort all elements of T and the number x in increasing order. This problem is solved by the algorithm *insert* (Algorithm 2).

Denote by i_b the instance $T = [1,2,3,\ldots,n-1,n]$ and $x = n+1$. When algorithm *insert* is applied on instance i_b, the while loop is executed once and hence,

$$t_{\text{insert}}(i_b,n) \in \Theta(1). \tag{3}$$

Denote now by i_w the instance $T = [1,2,3,\ldots,n-1,n]$ and $x = 0$. When algorithm *insert* is applied on instance i_w, the while loop is executed $\Theta(n)$ times and therefore

$$t_{\text{insert}}(i_w,n) \in \Theta(n). \tag{4}$$

This is so because the first two assignments of the pseudo code insert are executed in $\Theta(1)$ time and each execution of the while loop takes $\Theta(1)$ time. We show now that there is no function $g(n)$ such that relation (2) holds. This in turn shows that algorithm insert is non-homogeneous. Suppose, on the contrary, that such a function $g(n)$ does exist. By relation (2) we conclude that

$$t_{\text{insert}}(i_b,n) \in \Theta(g(n)) \text{ and } t_{\text{insert}}(i_w,n) \in \Theta(g(n)). \tag{5}$$

By Theorem 1 and relations (3) and (4) we conclude that

$$\Theta(t_{\text{insert}}(i_b,n)) = \Theta(1) \text{ and } \Theta(t_{\text{insert}}(i_w,n)) = \Theta(n). \tag{6}$$

Algorithm 2: INSERT

$j \leftarrow n$
$T(n+1) \leftarrow x$
while $j \geq 1$ and $T(j) > T(j+1)$ **do**
  $temp \leftarrow T(j)$
  $T(j) \leftarrow T(j+1)$
  $T(j+1) \leftarrow temp$
  $j \leftarrow j-1$
end while

Combining Theorem 1 and relations (5) we conclude that

$$\Theta(t_{\text{insert}}(i_b,n)) = \Theta(t_{\text{insert}}(i_w,n)) = \Theta(g(n)) \tag{7}$$

Finally, from relations (6) and (7) we conclude that $\Theta(1) = \Theta(n)$, which contradicts Theorem 2. This completes the proof of the Theorem.

In the proof of Theorem 3 we used two simple algorithms to show that the classes of homogeneous and non-homogeneous algorithms are non-empty. In fact both classes are wide and include recursive and iterative algorithms. The class of non-homogeneous algorithms includes plenty of iterative algorithms. The great majority of recursive and divide and conquer algorithms are homogeneous. Among the exceptions is the well-known recursive sorting algorithm quick sort [12] and Euclid's algorithm for computing the greatest common divisor of two numbers.

3 Algorithm Classification

The instances i_b and i_w used in Theorem 3 are the well-known best and worst cases, respectively. We call i_b *minimum time instance* and i_w *maximum time instance*. More precisely, we give the following definition.

Definition 3. An instance i is a *minimum (maximum)* time instance for an algorithm A, if the total number of elementary operations executed when algorithm A is applied on it is the *minimum (maximum)* possible.

The analysis so far and particularly algorithm *min* used in the proof of Theorem 3 might mislead someone to conclude that homogeneous algorithms do not contain minimum and maximum time instances. This is not correct. A striking example of an iterative homogeneous algorithm containing minimum and maximum time instances is the well-known Floyd's classical algorithm [9] for building an initial heap. A heap is a data structure introduced in [22] to develop an efficient general iterative sorting algorithm known today as heapsort. A recursive homogeneous algorithm containing worst and best cases is the well-known algorithm in [1], which computes order statistics in linear time.

Some algorithms are obviously homogeneous. If this is not clear for a new algorithm with unknown complexity, using Definition 3 we can set

$$S_b = \{\ i_b : i_b \in I(n) \text{ is a minimum time instance}\},$$
$$S_w = \{\ i_w : i_w \in I(n) \text{ is a maximum time instance}\}.$$

In the worst (best) case analysis of an algorithm we try to determine a set $\Theta(g(n))$ ($\Theta(f(n))$) containing the set S_w (S_b) and say that the worst (best) case complexity of the algorithm is $\Theta(g(n))$ ($\Theta(f(n))$). Observe the similarities among the worst and best case complexities of a non-homogeneous algorithm and the complexity of a homogeneous algorithm. In particular, the only difference is the set of instances on which they are referred to. Therefore, all these complexities should be described by sets of the form $\Theta(g(n))$.

It is now of interest to determine the complexity of a non-homogeneous algorithm, i.e, to find a set of functions including set S. Since a set of the form $\Theta(g(n))$ does not exist, we generalize Definition 1 as follows.

Definition 4. Given two (proper) functions $f(n)$ and $g(n)$ we denote by $\Theta(f(n),g(n))$ the set of functions $t(n)$ for which there exist constants $a > 0$ and $b > 0$ and a positive integer n_0 such that

$$bf(n) \leq t(n) \leq ag(n)$$

for $n \geq n_0$.

It is easy to see that $\Theta(f(n),g(n)) = \Omega(f(n)) \cap O(g(n))$. It is also easy to see that the sets $\Theta(0,\infty) = \Omega(0) = O(\infty)$ include always set S. However, in order to be as precise as possible, we are always looking for a minimal set containing set S. In the case of non-homogeneous algorithms we are seeking the minimal set $\Theta(f(n),g(n))$ containing set S. Obviously, the set $\Theta(f(n),g(n))$ is minimal if there exist worst and best case instances i_w and i_b such that $t(i_w,n) \in \Theta(g(n))$ and $t(i_b,n) \in \Theta(f(n))$, respectively. Recall that the set $\Theta(1,n)$ describing the complexity of algorithm *insert* in Theorem 3 is minimal. Observe also that the classification of an algorithm as homogeneous or not is not possible unless the set $\Theta(f(n),g(n))$ describing its complexity is minimal. As the set $\Theta(f(n),g(n))$ is described by best and worst case complexities, both complexities are equally important from the theoretical point of view.

4 Additional Results and Discussion

We mentioned earlier that homogeneous algorithms contain worst and best cases. Hence, the average complexity of a homogeneous algorithm is easily defined. Clearly, the mean time of the algorithm on a random instance is,

$$t(n) = \frac{\sum_{i \in I(n)} t(i,n)}{|I(n)|}$$

where $|I(n)|$ denotes the number of elements of set $I(n)$. If the complexity of the homogeneous algorithm is $\Theta(g(n))$, it is natural to expect that $t(n) \in \Theta(g(n))$. Indeed, this is the case.

Theorem 4. *The average complexity of a homogeneous algorithm of complexity $\Theta(g(n))$, is also $\Theta(g(n))$.*

Proof. Let t(n) be the expected time to solve a random instance. Then

$$t(n) = \frac{\sum_{i \in I(n)} t(i,n)}{|I(n)|} \in \frac{\sum_{i \in I(n)} \Theta(g(n))}{|I(n)|} = \frac{|I(n)|\Theta(g(n))}{|I(n)|} = \Theta(g(n))$$

and the proof is complete. □

Observe that this result is independent of the distribution of the instances.

So far we focused our attention on iterative algorithms. Recursive algorithm can be homogeneous and non-homogeneous too. But how recursive homogeneous and non-homogeneous algorithms look like? A recursive or divide and conquer algorithm makes a fixed number of calls to itself. Therefore, if each call is made on a problem with fixed dimension, the algorithm is homogeneous provided the work required to solve all subproblems dominates the remaining work. On the contrary, if the dimensions of the subproblems on which calls are made are not fixed and depend on the instance, the algorithm might very well be non-homogeneous. Recall that this is the case for the algorithm quicksort [12]. A recursive or divide and conquer algorithm can be non-homogeneous if the number of calls to subproblems is not fixed and depends on the instance. This is the case for Euclid's algorithm computing the greatest common divisor.

Acknowledgments We thank an anonymous referee for useful suggestions and for bringing to our attention the reference [17].

References

1. Blum, M., Floyd, R., Pratt, V., Rivest, R., Tarjan, R.: Time bounds for selection. J. Comp. Syst. Sci. **7**(4), 448–461 (1973)
2. Bollobas, B., Fenner, T.I., Frieze, A.M.: On best case of heapsort. J. Algorithms **20**, 205–217 (1996)
3. Brassard, G.: Crusade for a better notation. ACM Sigact News **17**(1), 60–64 (1985)
4. Ding, Y., Weiss, M.A.: Best case lower bounds for Heapsort. Computing **49**, 1–9 (1992)
5. Dutton, R.: Weak-heapsort. BIT **33**, 372–381 (1993)
6. Edelkamp, S., Wegener, I.: On the performance of weak heasort, STACS. Lecture Notes in Computer Science, pp. 254–266. Springer, Berlin (2000)
7. Edelkamp, S., Stiegeler, P.: Implementing heapsort with $nlogn - 0.9n$ and quicksort with $nlogn + 0.2n$ comparisons. ACM J. Exp. Algorithmics (JEA) **7**(1), 1–20 (2002)
8. Fleischer, R.: A tied lower bound for the worst case of bottom-up heapsort. Algorithmica **11**, 104–115 (1994)
9. Floyd, R. Algorithm 245: treesort 3. Comm. ACM **7**, 701 (1964)

10. Gurevich, Y.: What does $O(n)$ mean? ACM Sigact News **17**(4), 61–63 (1986)
11. Haralick, R.M., Elliot, G.L.: Increase tree search efficiency for constraint satisfaction problems. Artif. Intell. **14**, 263–313 (1980)
12. Hoare, A.: Quicksort. Comp. J. **5**, 10–15 (1962) s
13. Knuth, D.: Big omicron and big theta and big omega. ACM Sigact News **8**(2), 18–23 (1976)
14. Nadel, B.A.: The consistent labeling problem and its algorithms: Towards exact-case complexities and theory-based heuristics. Ph.D. dissertation, Department of Computer Science, Rutgers University, New Brunswick, NJ, May (1986)
15. Nadel, B.A.: The complexity of constraint satisfaction in Prolog. In: Proceedings of the 8th National Conference Artificial Intell. (AAAI'90)pp. 33–39, Boston, MA, August 1990. An expanded version is available as Technical Report CSC-89-004, Department of Computer Science, Wayne State University, Detroit, MI (1989)
16. Nadel, B.A.: Representation selection for constrain satisfaction: a case study using n-queens. IEEE Expert **5**(3), 16–23 (1990)
17. Nadel, B.A.: Precision complexity analysis: a case study using insertion sort. Inf. Sci. **73**, 139–189 (1993)
18. Nudel, B.A.: Solving the general consistent labeling (or constraint satisfaction) problem: two algorithms and their expected complexities. In: Proceedings of the 3rd National Conference Artificial Intell. (AAAI'83), pp. 292–296, Washington, DC, Aug (1983)
19. Paparrizos, I.: Homogeneous and non-homogeneous algorithms. In: Proceedings of the 13th Panhellenic Conference on Informatics (PCI'09), September (2009)
20. Schaffer, R., Sedgwick, R.: The analysis of heapsort. J. Algorithms **15**, 76–100 (1993)
21. Wang, X.D., Wu, Y.J.: An improved heapsort algorithm with $nlogn - 0.788928n$ comparisons in the worst case. J. Comp. Sci. Tech. **22**(6), 898–903 (2007)
22. Williams, J.W.J.: Algorithm 232: heapsort. Comm. ACM **6**, 347–348 (1964)

Service Quality Evaluation in the Tourism Industry: A SWOT Analysis Approach

Marianna Tsitsiloni, Evangelos Grigoroudis, and Constantin Zopounidis

Abstract The quality evaluation of the provided tourism services constitutes the most important issue for the viability of this particular sector and the improvement of the total tourism product. This paper presents the results of a tourist satisfaction survey that took place in the island of Mykonos during the period of May–September 2009. The final sample consists of 1,026 questionnaires that were distributed to Greek and foreign tourists during their departure from the island (harbor and airport). The main objective of this paper is to evaluate tourists' satisfaction and identify the strong and the weak points of the tourism services offered. These results may help the development of a strategic plan for the quality improvement of the overall tourism product. Beyond descriptive statistical techniques, the analysis of the collected data is based on the multicriteria method MUSA. The method is able to combine satisfaction importance and performance results and provides a SWOT (Strengths-Weaknesses-Opportunities-Threats) analysis for the whole set of the tourist satisfaction criteria. The presented analytical results reveal that the main strong points of the offered tourist product are the fame and the natural beauties of the island, as well as the high level of expenses. On the other hand, the most important weak points concern the small duration of stay, as well as the low level of satisfaction in specific service quality criteria (local transports, information, and environment).

Key words Tourist satisfaction • MUSA method • SWOT analysis • Service quality

M. Tsitsiloni (✉) • E. Grigoroudis • C. Zopounidis
Department of Production Engineering and Management, Technical University of Crete, University Campus, Kounoupidiana, GR73100 Chania, Greece
e-mail: mtsitsiloni@yahoo.gr; vangelis@ergasya.tuc.gr; kostas@dpem.tuc.gr

A. Migdalas et al. (eds.), *Optimization Theory, Decision Making, and Operations Research Applications*, Springer Proceedings in Mathematics & Statistics 31,
DOI 10.1007/978-1-4614-5134-1_18,

1 Introduction

The tourism industry constitutes one of the most important sectors in many local economies in Greece, mainly not only because of its constant increasing contribution to the income of these regions but also due to the opportunities offered for further growth [14]. The importance of tourism sector is presented in the work of Naisbitt [17], who emphasizes that the world economy in the current century will be dominated by three sectors: information technology, telecommunications, and tourism.

Modern business organizations consider service quality as the most reliable source of market information. Service quality is considered as the main determinant of customer satisfaction, which in turn influences purchase intentions [4, 29]. The importance of service quality evaluation through customer satisfaction measurement is reinforced by the necessity of adopting a "continuous improvement" philosophy and understanding customer perceptions (e.g., needs, expectations).

Generally, the main reasons for measuring customer satisfaction are summarized in the following [8]:

- Customer satisfaction constitutes the most reliable market information. This way, a business organization is able to evaluate its current position against competition, and design its future plans accordingly.
- A large number of customers avoid expressing their complaints or their dissatisfaction from the product or service provided, either due to a particular attitude or because they are not sure that the company will perform any corrective action.
- Customer satisfaction measurement is able to identify potential market opportunities.
- The main principles of continuous improvement require the development of a specific customer satisfaction measurement process. This way, any improvement action is based on standards that take into account customer expectations and needs.
- Customer satisfaction measurement may help business organizations to understand customer behavior, and in particular to identify and analyze customer expectations, needs, and desires.
- The application of a customer satisfaction measurement program may reveal potential differences in the service quality perceptions between the customer and the management of the business organization.

The necessity of customer satisfaction measurement in the tourism industry is justified by the importance of the tourism sector for local economies and the intense competition among alternative tourism destinations that is evident in recent years. Furthermore, tourism sector is heavily influenced by significant external factors from the global economic environment, and it is, therefore, necessary to improve the quality of the services offered in order to gain competitive advantages and increase tourist loyalty levels.

However, tourist satisfaction from a destination area is a general and ambiguous notion, since tourism goods and services should be treated as a subset of goods and services in general. For this reason, as noted by Yuksel [37], a large number of researchers have studied components of experiences, which contribute to tourist satisfaction within different tourism and hospitality contexts (e.g., guest satisfaction with hotels and restaurant services, satisfaction with destination services, satisfaction with recreational services, satisfaction with tours or cruise travel). As suggested by Pizam et al. [24], tourist satisfaction is the result of the interaction between the tourist's experience at the destination area and the expectations she/he had about that destination. This confirmation/disconfirmation approach is rather common in tourist satisfaction research [2].

The HOLSAT model is a characteristic approach used to evaluate satisfaction from a particular destination [31]. The model is based on the disconfirmatory paradigm outlined before and adopts the philosophy of the SERVQUAL model [20–22]. The main results of the HOLSAT model focus on the difference between "expectation" and "experience" scores for each attribute, which gives a quantitative measure of the level of satisfaction shown by the vacationers [32]. Other research efforts in tourism management combine the disconfirmation paradigm with additional quality improvement tools, like QFD, Kano's model, etc. [23].

Despite the context and the multivariate nature of tourist satisfaction measurement, Multiple Criteria Decision Analysis (MCDA) has not been widely applied in evaluating service quality in the tourism industry. Rozman et al. [26] applied the DEX method, which combines traditional MCDA approaches and elements of expert systems and machine learning, in order to assess tourist farm service quality. An AHP model, combined with fuzzy TOPSIS, was applied by Hsu et al. [13] in a preference analysis for tourist choice of destination in Taiwan. The MUSA method has also been applied by Arabatzis and Grigoroudis [1] in order to examine the level of satisfaction of the visitors of the National Park of Dadia-Lefkimi-Soufion area.

The main objective of this paper is to present an application of an MCDA approach in tourist satisfaction measurement from a destination area. The presented tourist satisfaction survey took place in the island of Mykonos (Greece). Moreover, the presented study aims to demonstrate how a SWOT analysis approach may be applied in the context of tourism management.

The island of Mykonos is located in the Central Aegean Sea and is part of the group of islands known as the Cyclades. It is a well-known international tourist resort, which has experienced rapid tourist development during the last 30 years. Beyond the touristic characteristics of a traditional Greek destination (e.g., beaches, climate, archaeological sites), Mykonos is the most cosmopolitan Greek island. It attracts numerous celebrities and its name is connected with nightlife. These unique characteristics affect visitors' expectations and should be considered in the tourist satisfaction analysis.

The paper is organized into four more sections. Section 2 briefly presents the adopted methodology, including the development of the MUSA method and proposed gap analysis approach. The main results of the tourist satisfaction survey are presented in Sect. 3, giving emphasis on the determination of the strong and weak point of the services offered. Finally, Sect. 4 summarizes some concluding remarks.

2 Methodology

2.1 MUSA Method

The MUSA (MUlticriteria Satisfaction Analysis) method is a multicriteria preference disaggregation approach, which provides quantitative measures of customer satisfaction considering the qualitative form of customers' judgments [6, 28]. The main objective of the MUSA method is the aggregation of individual judgments into a collective value function, assuming that client's global satisfaction depends on a set of n criteria or variables representing service characteristic dimensions. This set of criteria is denoted as $\mathbf{X} = \{X_1, X_2, \ldots, X_n\}$, where a particular criterion i is represented as a monotonic variable X_i.

The MUSA method infers an additive collective value function Y^* and a set of partial satisfaction functions X_i^*, given customer's global satisfaction Y and partial satisfaction X_i according to criterion i (ordinal scaling). The main objective of the method is to achieve the maximum consistency between the value function Y^* and the customers' judgments Y. Based on the modeling of preference disaggregation approach, the ordinal regression equation becomes as follows:

$$\widetilde{Y}^* = \sum_{i=1}^{n} b_i X_i^* - \sigma^+ + \sigma^- \quad \text{with} \quad \sum_{i=1}^{n} b_i = 1$$

where $\widetilde{Y}^*$ is the estimation of Y^*, b_i is the weight of the ith criterion, n is the number of criteria, and σ^+, σ^- are the overestimation and the underestimation errors, respectively.

The most important results provided by the MUSA method are the estimated global and partial value functions, the criteria weights, and the average satisfaction, demanding, and improvement indices. In particular, regarding performance and importance results, the following should be noted:

- *Criteria weights*: They represent the relative importance of the assessed satisfaction dimensions. Their properties are also determined in the context of multicriteria analysis (e.g., the weights are value trade-offs among the criteria). The decision whether a satisfaction dimension is considered important by the customers is also based on the number of assessed criteria.
- *Average satisfaction indices*: They indicate the level of customers satisfaction in a range of 0–100%. They can be considered as the basic performance norms, since the average satisfaction indices are basically the mean value of the global and partial value functions.

- *Average demanding indices*: These indices are calculated according to the shape of global and partial value functions, which indicate customers' demanding level. They represent the average deviation of the estimated value functions from a "normal" (linear) function. The average global and partial demanding indices, D and D_i, respectively, are normalized in the interval $[-1, 1]$ and the following cases hold:
  - Neutral customers (D or $D_i \in [-0.33, +0.33]$): Their value function has an approximately linear shape. This means that the more satisfied they express to be, the higher the percentage of their fulfilled expectations is.
  - Demanding customers (D or $D_i \in [+0.33, +1.00]$): Their value function is convex. This means that they are not really satisfied, unless they receive the best quality level.
  - Non-demanding customers (D or $D_i \in [-1.00, -0.33]$): Their value function is concave. This means that they express satisfaction although only a small portion of their expectations is fulfilled.

  These indices are used in customer behavior analysis, but they may also indicate the extent of company's improvement efforts: the higher the value of the demanding index, the more the satisfaction level should be improved in order to fulfill customers' expectations.
- *Average improvement indices*: They represent the improvement efforts and they depend on the importance of satisfaction criteria and their contribution to dissatisfaction as well. These indices are normalized in the interval $[0, 1]$, and they indicate the improvement margins on a specific criterion.

Detailed presentation of the mathematical development of the MUSA method may be found in [6, 8, 28], while several applications to business organizations can be found in the literature [7, 9, 10, 16, 25, 28].

2.2 SWOT Analysis

SWOT analysis is widely used in management science to identify strengths, weaknesses, opportunities, and threats when studying a particular product/service or an entire company/organization. In service quality literature, SWOT analysis appears either as gap analysis or as performance-importance comparison [8]. In both cases, the main objective is to identify the quality gap of the service offered, i.e., identify the gap between what customers want and what customers get.

The MUSA method, using the previously described results, provides additional diagrams that may help in determining improvement actions. In particular, action diagrams are developed by combining weights and average satisfaction indices. These diagrams indicate the strong and the weak points of customer satisfaction, and define the required improvement efforts. Each of these diagrams is divided into

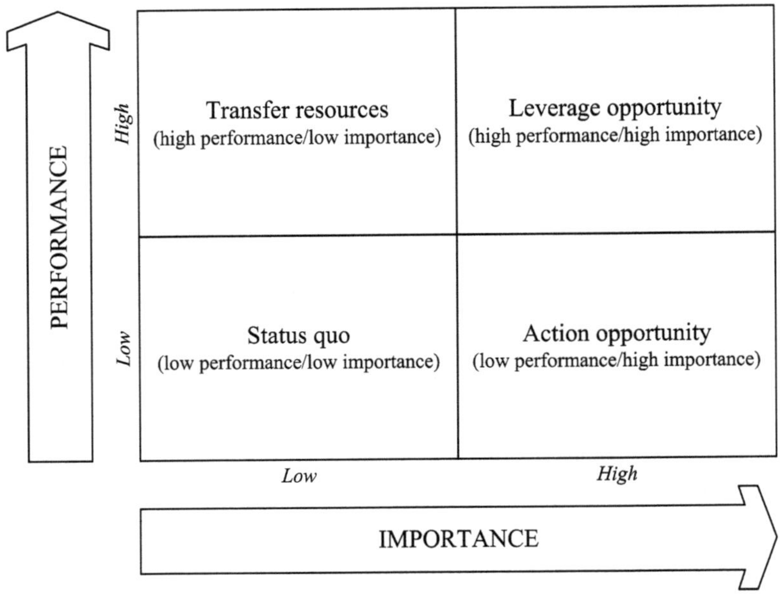

Fig. 1 Action diagram [8]

four quadrants, according to performance (high/low) and importance (high/low), which are used to classify actions (Fig. 1):

- *Status quo* (low performance and low importance): Generally, no action is required.
- *Leverage opportunity* (high performance/high importance): These areas can be used as advantage against competition.
- *Transfer resources* (high performance/low importance): Company's resources may be better used elsewhere.
- *Action opportunity* (low performance/high importance): These are the criteria that need attention.

These diagrams are similar to SWOT maps, since status quo quadrant refers to threats, leverage opportunity quadrant refers to strengths, transfer resources corresponds to opportunities, and action opportunity quadrant corresponds to weaknesses. In addition, they appear in the service quality literature as importance-performance analysis [5, 18] or gap analysis [12, 34, 36]. Similar gap analysis tools have been widely used in tourism research, mainly for the evaluation of hotel and restaurant services and facilities [14, 19], and the measurement of visitors satisfaction [11, 27, 30, 35].

MUSA provides also another type of diagrams, the improvement diagrams, which take into account customers' demanding level and are used in order to rank improvement priorities (Fig. 2). Similar to the previous ones, each of these diagrams is divided into four quadrants according to the demanding level (high/low), and

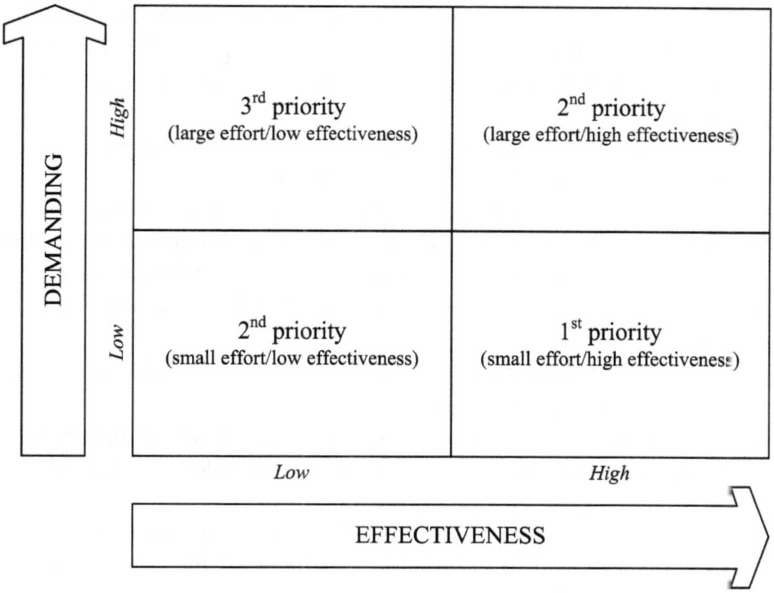

Fig. 2 Improvement diagram [8]

the effectiveness (high/low) of the considered satisfaction dimensions. Given a particular improvement diagram, the first priority should be given to satisfaction criteria having large improvement margins while requiring small effort and the last priority should be given to satisfaction dimensions with low dissatisfaction levels that need substantial effort to improve. All the remaining satisfaction dimensions having either a low demanding index or a high improvement index should be considered as second priorities.

3 Survey and Results

3.1 Satisfaction Criteria and Questionnaire

The most important phase in the implementation of the MUSA model is the assessment of the set of satisfaction criteria and the definition of the value hierarchy. Based on the previous applications of the MUSA method and customer satisfaction surveys in the tourism sector [1, 14, 24, 33, 37], the following set of satisfaction criteria was developed for this survey:

1. *Accommodation*: This criterion includes all the characteristics of accommodation such as the offered service, the facilities, the staff, the prices, etc.

2. *Food/Cuisine*: This criterion refers to the local cuisine and the food offered inside or outside the accommodation facilities. It includes the food quality, the variety of dishes, the environment (decoration, aesthetics), the provided services, the prices, etc.
3. *Natural environment*: This criterion concerns the natural environment, the climatic conditions, as well as the local architecture.
4. *Urban environment*: This criterion relates to the urban environment and the infrastructures of the island. It includes the cleanliness in public spaces, the noise pollution, the roads and the traffic conditions, the available parking, etc.
5. *Hospitality*: This criterion relates to the hospitality, behavior, and friendliness of the locals.
6. *Information*: The information available to tourists though desks, kiosks, signs, and maps is included in this criterion.
7. *Entertainment/Recreation*: This criterion refers to the entertainment/recreation choices offered to tourists during their stay. It includes the available choices, the offered services, the venues, the prices, etc.
8. *Transportation (from and to island)*: This particular criterion concerns the transportation from and to the island. It includes all the characteristics of the provided services in the island's port and airport.
9. *Local transportation means*: This last criterion concerns the local transportation means, i.e. the bus and taxi services, rented cars, etc. It includes all the characteristics of the provided services (availability, service from personnel, prices, etc.).

The final questionnaire was developed based on the aforementioned satisfaction criteria, for which tourists were asked to express their satisfaction using a 5-point Likert-type ordinal scale (dissatisfied, somehow dissatisfied, neither satisfied nor dissatisfied, somehow satisfied, satisfied). The first part of the questionnaire included questions about the tourist's personal characteristics (sex, age, income, purpose of trip). The second part was devoted to travel information (number and period of previous visits, alternative destination examined, reasons for choosing the island, sources of information), while the third part included questions about accommodation, length of stay, and expenses. The fourth part of the questionnaire concerned the satisfaction criteria, while the fifth and last part of the questionnaire contained loyalty-related questions.

3.2 Sample and Tourists' Profile

The final sample consisted of 1,026 questionnaires that were distributed to Greek and foreign tourists during their departure from the island (harbor and airport). The questionnaires were collected through personal interviews during summer 2009.

In order to formulate a customer profile, tourist's characteristics were studied. The sample was almost equally distributed between males and females (male 46.5%,

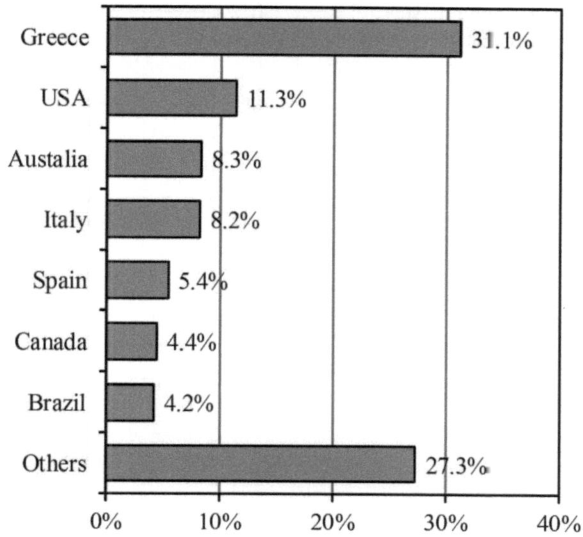

Fig. 3 Nationality groups

female 53.5%). In addition, the majority of the visitors were less than 40 years old (74.8%), while the group of older visitors was very small (less than 4% were older than 60 years).

Thirty one percent of the sample were Greek tourists, while the remaining 69% was constituted of foreign visitors mainly from the USA, Australia, Italy, Spain, Canada, and Brazil. Figure 3 shows that the distribution of the sample in the different nationalities was relatively high. Generally, beyond the Greek tourists, there was no other nationality group larger than approximately 10% of the sample, while almost 50 different nationality groups were identified in the final sample. Furthermore, it seems that the length of stay was relatively low, since the majority of tourists spent 1–3 nights in the island (almost 55%). As shown in Fig. 4, only 8% of the sample stayed more than 1 week in the island.

Additional analyses regarding other tourists' personal characteristics were also performed in order to develop a complete profile for the visitors (see more details in [33]). However, it seems that the most characteristic tourists' segments with distinguished preferences and behavior are formulated based on the nationality (Greek and foreign visitors).

3.3 *Satisfaction Analysis*

The results of the MUSA method reveal that the tourists give particular importance in the criterion of entertainment/recreation (weight 18.53%), while the importance of urban environment and transportation criteria is relatively lower (less than 10%). Moreover, it seems that generally, the visitors are relatively satisfied from

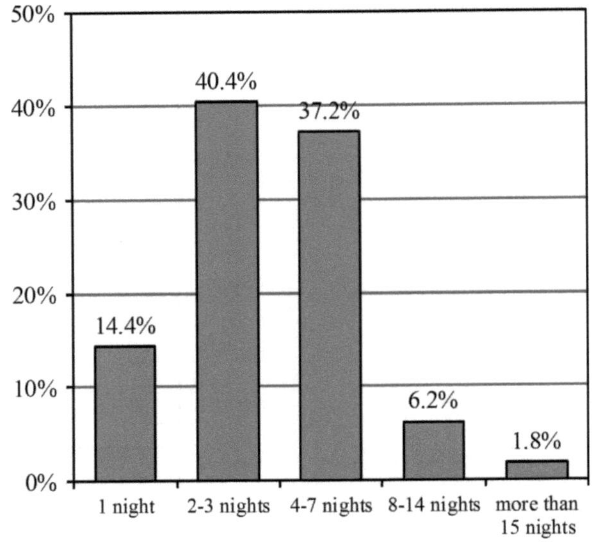

Fig. 4 Length of stay (number of nights spent)

Table 1 Average satisfaction indices and criteria weights

Satisfaction criteria	Weight (%)	Average satisfaction index (%)
Accommodation	10.49	82.85
Food/cuisine	10.90	80.87
Natural environment	11.11	89.11
Urban environment	9.13	72.61
Hospitality	10.12	80.19
Information	11.11	77.87
Entertainment/recreation	18.53	90.49
Transportation (airport/harbor)	9.90	80.48
Local transportation means	8.70	68.55
Overall satisfaction		86.40

their vacations in the island, since the estimated overall average satisfaction index is 86.40%. Although this overall satisfaction level is relatively high, significant improvement margins still exist.

Regarding the detailed satisfaction criteria, as Table 1 indicates, there are important differences regarding tourist satisfaction level. In particular, the results of Table 1 reveal the following:

1. Tourists seem to be quite satisfied by the criteria of entertainment/recreation and natural environment (the average satisfaction indices are approximately 90%), which are also the most important satisfaction dimensions.
2. In contrast, the level of tourist satisfaction is quite low regarding the criteria of environment, information, and local transportation (average satisfaction indices 70–78%).

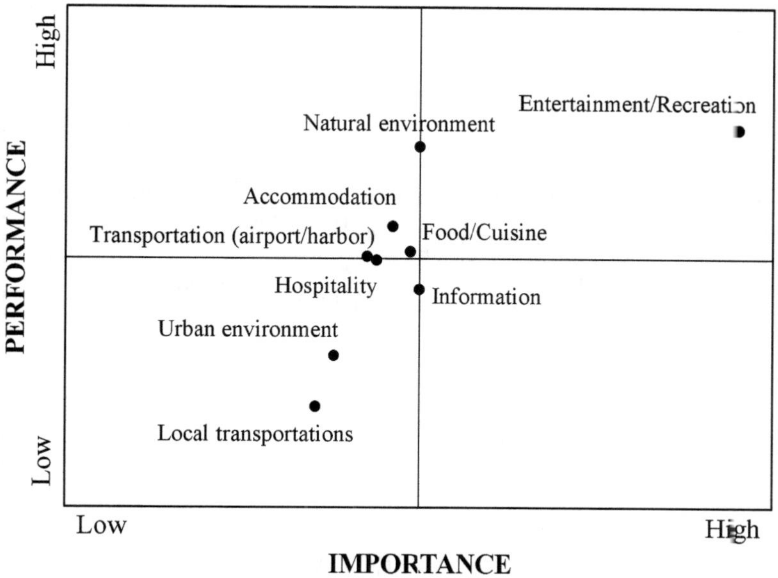

Fig. 5 Action diagram for the tourist satisfaction criteria

3. The rest of criteria present a medium level of satisfaction (80–83%), which is relatively lower than the average total satisfaction index.

Figure 5 displays the action diagram for the whole sample, which is used in order to develop a SWOT analysis map. This diagram indicates that the criterion of entertainment/recreation is the strongest point of the offered tourist product. Visitors are particularly satisfied by this characteristic, which they also consider very important. In addition, there are no satisfaction criteria in the action opportunity area (high importance and low satisfaction/performance), thus, no critical characteristics exist requiring for direct improvement actions. The criteria of urban environment (cleanliness in public spaces, roads, noise pollution, parking, etc.) and the local transportation means are the main threats of the tourist product, since they present a relatively low satisfaction level. They are currently considered as a threat and not as a weak point because of their lower importance. For the rest of the satisfaction criteria, the categorization is not easy, since they present a relatively medium satisfaction and importance level. However, it seems that the natural environment is a potentially strong point, while the information criterion is a potentially critical characteristic. In general, it seems that there is no "gap" regarding tourist satisfaction (i.e., what tourists want and what tourists get), since visitors seem to be more satisfied by those characteristics that they consider as important. These findings are consistent with the results from previous studies (see, for example, [3]).

Similarly, Fig. 6 displays the improvement diagram for the whole sample. This diagram takes into account the demanding level of tourists, as well as the

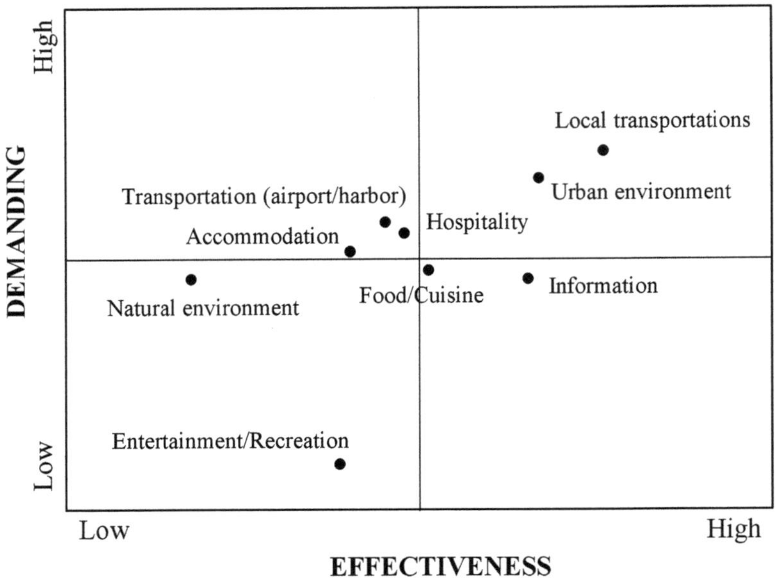

Fig. 6 Improvement diagram for the tourist satisfaction criteria

effectiveness of potential improvement actions. The most important results of Fig. 6 reveal the following:

1. Improvement actions should be focused firstly to information and food/cuisine, which are the most important satisfaction criteria with the lowest performance (see also Fig. 5).
2. The second priority should concern the improvement of the urban environment and the local transportation means, which have a relatively lower performance. Alternatively, further improvement action may concern the natural environment and the entertainment/recreation criteria.

3.4 *Statistical Analyses*

This section presents the results of additional statistical analyses in selected variables of the questionnaire. The results are based on a series of correlation analyses (i.e., chi-square tests), which have been used for identifying particular tourist clusters with distinctive preferences and expectations in relation to the total set.

Based on the results of Table 2, it seems that previous visit is related to nationality. In particular, the Greek tourists are the most loyal visitors, while there are many visitors from North and Central Europe, who had already visited the island. The 80–95% of the remaining nationalities was visiting the island for the first time.

Table 2 Results of chi-square tests for tourist characteristics

Variables		Chi-square	df	*p*-level
Previous visit	Nationality	298.353	7	0.000
	Income	12.458	3	0.006
	Age	7.526	3	0.057
Alternative destinations	Nationality	41.497	7	0.000
	Age	10.912	3	0.012
	Income	5.173	3	0.160
	Previous visit	14.378	1	0.000
Length of stay	Nationality	252.472	28	0.000
	Age	24.185	12	0.019
	Income	20.428	12	0.059
	Previous visit	57.296	4	0.000
Expenses	Nationality	78.479	21	0.000
	Age	15.751	9	0.072
	Income	59.879	9	0.000
	Previous visit	26.485	3	0.000
Overall satisfaction	Age	17.473	12	0.133
	Income	24.458	12	0.018
	Previous visit	16.712	4	0.002
	Length of stay	33.228	16	0.009
	Expenses	19.218	12	0.083

Some nationalities appear more loyal regarding the examination of alternative destinations (Table 2). In particular, Greeks, and Asians, in general, do not examine alternative destinations when deciding their holidays. Similarly, visitors over 60 years appear more loyal, since they do not examine other alternatives when choosing their holiday destination. Generally, there is no strong relation between income and examination of alternative destinations. Moreover, as expected, the tourists who had already visited the island were more loyal, regarding this particular characteristic.

Similarly, the length of stay is related to nationality (Table 2). Additional analyses indicated that tourists who stayed more days on the island were mostly Italians, Greeks and visitors from North and Central Europe. In contrast, Asians and Australians-New Zealanders were the tourists with the smallest length of stay. Age and length of stay do not seem to be strongly related, although the older tourists segment (more than 40 years old) affects the overall average length of stay of the whole sample, which appears rather low. Furthermore, there does not seem to be any strong relation between income and length of stay. On the contrary, a previous visit to the island significantly affects the duration of stay, since repeated visitors stay more days.

As Table 2 indicates, there is no significant relation between the level of expenses and the age of tourists. Moreover, nationality appears to affect the amount of expenses that tourists spent (except for tickets and accommodations), since the highest expenses are made by European tourists (Italy, North, and Central Europe). In contrast, the amount of expenditures is related to the annual family income of visitors, while repeated visitors seem to spend more, as expected.

Finally, the chi-square tests between overall satisfaction and several tourist characteristics are presented in Table 2. The results of these tests indicate that there is no strong relation between overall satisfaction and age or expenses. On the other hand, repeated tourists appear more satisfied, while a negative relation appears between overall satisfaction and income or length of stay. These results constitute a significant threat for the tourism services of the island.

Additional analyses, based on cross-tables, study the relation between nationality and reasons for choosing the island. The main reasons for Greeks and Italians include "service quality" and "entertainment-recreation," while "climate-natural beauty," "historical-archaeological monuments," and "relaxation" do not seem to play an important role. On the other hand, Europeans seem to choose the island for its "climate-natural beauties," "service quality," and "historical-architectural monuments," while "value for money" does not seem important. The reputation of the island seems to play an important role for groups originating from the outermost countries (North America, Australia, New Zealand, Asia, and Latin America). In general, it seems that there is no relation between age of income and reasons for choosing the island, although monuments and price/value appear important for the older and the younger tourists, respectively. Finally, repeated tourists give importance to service quality and entertainment, while first-time visitors give relatively greater emphasis on the historical-archaeological monuments and the fame of the island.

Regarding the sources of information, there is a clear grouping among Greek and foreign tourists. The Greek visitors prefer to collect information either from past personal experience, or from other media (magazines, newspapers, TV). Internet and tourist offices (tour operators) do not constitute a preferred source of information for this group. The opposite is observed in the case of foreign tourists. In addition, younger tourists prefer Internet from friends/relatives, while older visitors prefer personal experiences and tourist offices as sources of information.

Table 3 presents the chi-square tests regarding several loyalty measures included in the questionnaire. These tests indicate that the intention to repeat the visit is negatively related to age and previous visit, and positively related to expenses.

Similarly, the intention to recommend the island to friends/relatives is strongly related to age, income, and expenses (Table 3). In contrast, repeated visitors or tourists who stay longer do not seem to be more loyal according to this variable.

Table 3 shows that the confirmation of expectations is not related to the expenses or the length of stay, while repeated tourists, in general, think that their holidays were better or somehow better than expected. In addition a weak relation may be observed between the confirmation of expectations and age or income.

Finally, it should be emphasized that overall satisfaction is strongly related to all three loyalty measures (revisit intention, recommendation, and confirmation of expectations), fact that is consistent with the relative literature (see, for example, [8, 34]).

Consequently, it seems that nationality is the major discriminant variable that assesses the distinguished tourist segments. This is confirmed by several other studies, which emphasize that tourist perceptions of a destination or hospitality

Table 3 Results of chi-square tests for loyalty measures

Variables		Chi-square	df	*p*-level
Revisit intention	Age	27.594	12	0.006
	Income	12.736	12	0.389
	Previous visit	39.220	4	0.000
	Length of stay	21.976	16	0.144
	Expenses	28.834	12	0.004
	Overall satisfaction	283.812	16	0.000
Recommendation	Age	30.407	12	0.002
	Income	39.847	12	0.000
	Previous visit	6.504	4	0.165
	Length of stay	17.373	16	0.362
	Expenses	27.042	12	0.008
	Overall satisfaction	627.969	16	0.000
Confirmation of expectations	Age	23.809	12	0.022
	Income	25.522	12	0.013
	Previous visit	34.365	4	0.000
	Length of stay	17.371	16	0.362
	Expenses	19.139	12	0.085
	Overall satisfaction	508.674	16	0.000

businesses may vary according to the countries of origin (see, for example, [15]). These results are justified by the different languages, food consumption, and other national cultural differences (including values, ideas, attitudes, or symbols), and they can be used in the decision-making process of destination management regarding positioning and market segmentation strategies.

4 Conclusions

This paper presents an application of the multicriteria method MUSA for the service quality evaluation in the tourism industry. The results are based on a tourist satisfaction survey that took place in the island of Mykonos aiming at evaluating the tourists' satisfaction and identifying the strong and the weak points of the offered tourism services.

Combining satisfaction importance and performance results and taking into account additional results (tourist profiling, potential tourist segments, etc.) the study showed that it is possible to perform a SWOT analysis for the totally offered tourism product. In this context, the SWOT analysis revealed the following:

- The strong points (competitive advantages) of the total tourism product are the fame and the natural environment (natural beauties, climate, local architecture) of the island. The visitors seem to be loyal (it is more likely to revisit the island and/or suggest it to friends/relatives), and their expenses during their vacations are relatively high.

- The short period of stay on the island is the most important weak point of the tourism product. Another weak point concerns the relatively low satisfaction that is observed in specific characteristics such as the urban environment and the local transportation means.
- The most important threats include the intense competition from other Greek islands, as well as the high level of expectations created by the fame of the island. Another potential threat is the low satisfaction of repeated visitors.
- The opportunities concern the historical-archaeological monuments and the quality of provided services. These characteristics are not considered important by tourists, although they can be the competitive advantages of the island, due to their high performance.

The presented study may also reveal the advantages of the MCDA approaches in tourist satisfaction evaluation problems. In particular, the results provided by the MUSA method are able to give a complete set of tourist/customer behavior information. These results may help destination management organizations to analyze the problem of tourist satisfaction evaluation and determine potential improvement actions. Moreover, it should be emphasized that the MUSA method fully respects the qualitative form of input information (i.e., tourists' judgments on the defined satisfaction criteria). This way, the ordinal variables are not arbitrary quantified (this quantification is rather an output of the method).

Consequently, the MUSA method provides an important alternative for studying service quality gaps and performing SWOT analysis. Following service quality literature, SWOT analysis in the MUSA method is performed using a series of action (performance-importance) diagrams, which are able to analyze tourist perceptions and determine the strong and weak points of a destination.

References

1. Arabatzis, G., Grigoroudis, E.: Visitors' satisfaction, perceptions and gap analysis: the case of Dadia-Lefkimi-Souflion National park. Forest Pol. Econ. **12**(3), 163–172 (2010)
2. Bowen, D., Clarke, J.: Reflections on tourist satisfaction research: past, present and future. J. Vacation Market. **8**(4), 297–308 (2002)
3. Buhalis, D.: Tourism in Greece: strategic analysis and challenges. Curr. Issues Tourism **4**(5), 440–480 (2001)
4. De Ruyter, J.C., Bloemer, J.M.A., Peters, P.: Merging service quality and service satisfaction: An empirical test of an integrative framework. J. Econ. Psychol. **18**(4), 387–406 (1997)
5. Dutka, A.: AMA Handbook of Customer Satisfaction: A Complete Guide to Research, Planning and Implementation. NTC Business Books, Illinois (1995)
6. Grigoroudis, E., Siskos, Y.: Preference disaggregation for measuring and analysing customer satisfaction: the MUSA method. Eur. J. Oper. Res. **143**(1), 148–170 (2002)
7. Grigoroudis, E., Siskos, Y.: A survey of customer satisfaction barometers: results from the transportation-communications sector. Eur. J. Oper. Res. **152**(2), 334–353 (2004)
8. Grigoroudis, E., Siskos, Y.: Customer Satisfaction Evaluation: Methods for Measuring and Implementing Service Quality. Springer, New York (2010)

9. Grigoroudis, E., Siskos, Y., Saurais, O.: TELOS: a customer satisfaction evaluation software. Comp. Oper. Res. **27**(7–8), 799–817 (2000)
10. Grigoroudis, E., Politis, Y., Siskos, Y. Satisfaction benchmarking and customer classification: an application to the branches of a banking organization. Int. Trans. Opera. Res. **9**(5), 599–618 (2002)
11. Hanim, N., Salleh, M., Othman, R.: Importance-satisfaction analysis for Tioman Island Marine Park, MPRA paper, 22679, Munich University Library (2010)
12. Hill, N.: Handbook of Customer Satisfaction Measurement. Gower Publishing, Hampshire (1996)
13. Hsu, T.-K., Tsai, Y.-F., Wu, H.-H.: The preference analysis for tourist choice of destination: A case study of Taiwan. Tourism Manag. **30**(2), 288–297 (2009)
14. Karakitsiou, A., Mavrommati, A., Migdalas, A., Tsiakali, K.: Customer satisfaction evaluation in the tourism industry: the case of Chania. Found. Comput. Decis. Sci. **32**(2), 111–124 (2007)
15. Kozak, M.: Comparative assessment of tourist satisfaction with destinations across two nationalities. Tourism Manag. **22**(4), 391–401 (2001)
16. Mihelis, G., Grigoroudis, E., Siskos, Y., Politis, Y., Malandrakis, Y.: Customer satisfaction measurement in the private bank sector. Eur. J. Oper. Res. **130**(2), 347–360 (2001)
17. Naisbitt, J.: Global Paradox. Nicholas Brealey Publishing, London (1995)
18. Naumann, E., Giel, K.: Customer Satisfaction Measurement and Management. Thomson Executive Press, Cincinnati (1995)
19. Oh, H.: Revisiting importance-performance analysis. Tourism Manag. **22**(6), 617–627 (2001)
20. Parasuraman, A., Zeithaml, V.A., Berry, L.L.: A conceptual model of service quality and its implications for future research. J. Market. **49**(4), 41–50 (1985)
21. Parasuraman, A., Zeithaml, V.A., Berry, L.L.: SERVQUAL: a multiple item scale for measuring consumer perceptions of service quality. J. Retailing **64**(1), 14–40 (1988)
22. Parasuraman, A., Zeithaml, V.A., Berry, L.L.: Refinement and reassessment of the SERVQUAL scale. J. Retailing **67**(4), 420–450 (1991)
23. Pawitra, T.A., Tan, K.C.: Tourist satisfaction in Singapore: a perspective from Indonesian tourists. Manag. Serv. Qual. **13**(5), 399–411 (2003)
24. Pizam, A., Neumann, Y., Reichel, A.: Dimensions of tourist satisfaction with a destination area. Ann. Tourism Res. **5**(3), 314–322 (1978)
25. Politis, Y., Siskos, Y.: Multicriteria methodology for the evaluation of a Greek engineering department. Eur. J. Oper. Res. **156**(1), 223–240 (2004)
26. Rozman, C., Potočnik, M., Pažek, K., Borec, A., Majkoviž, D., Bohanec, M.: A multi-criteria assessment of tourist farm service quality. Tourism Manag. **30**(5), 629–637 (2009)
27. Ryan, C., Cessford, G.: Developing a visitor satisfaction monitoring methodology: quality gaps, crowding and some results. Curr. Issues Tourism **6**(6), 457–507 (2003)
28. Siskos, Y., Grigoroudis, E., Zopounidis, C., Saurais, O.: Measuring customer satisfaction using a collective preference disaggregation model. J. Global Optim. **12**(2), 175–195 (1998)
29. Spreng, R.A., McKoy, R.D.: An empirical examination of a model of perceived service quality and satisfaction. J. Retailing **72**(2), 201–214 (1996)
30. Tonge, J., Moore, S.A.: Importance-satisfaction analysis for marine park hinter-lands: a western Australian case study. Tourism Manag. **28**(1), 768–776 (2007)
31. Tribe, J., Snaith, T.: From SERVQUAL to HOLSAT: holiday satisfaction in Varacero, Cuba. Tourism Manag. **19**(1), 25–34 (1998)
32. Truong, T.-H., Foster, D.: Using HOLSAT to evaluate tourist satisfaction at destinations: The case of Australian holidaymakers in Vietnam. Tourism Manag. **27**(5), 842–855 (2006)
33. Tsitsiloni, M.: Evaluating service quality and developing improvement plans in the tourism sector. Diploma Thesis, Technical University of Crete, Greece (2010)
34. Vavra, T.G.: Improving your Measurement of Customer Satisfaction. ASQC Quality Press, Milwaukee (1997)

35. Wade, D.J., Eagles, P.: The use of importance-performance analysis and market segmentation for tourism management in parks and protected areas: an application to Tasmania's National Parks. J. Ecotourism **2**(3), 196–212 (2003)
36. Woodruff, R.B., Gardial, S.F.: Know your Customer: New Approaches to Under-standing Customer Value and Satisfaction. Blackwell, Oxford (1996)
37. Yuksel, A.: Managing customer satisfaction and retention: a case of tourist destinations, Turkey. J. Vacation Market. **7**(2), 153–168 (2001)

Correcting Certain Estimation Methods for the Generalized Pareto Distribution

Jelena Jocković

Abstract Generalized Pareto distributions (GPD) are widely used for modeling excesses over high thresholds. When its shape parameter is positive, the GPD has a finite upper bound that is a function of the distribution parameters. A well-known problem when estimating GPD parameters is inconsistency with the sample data, which is that one or more sample observations exceed the estimated upper bound. This paper proposes a new, general technique to overcome the inconsistency problem and improve performance of the existing GPD estimation methods. The technique is successfully applied to method-of-moments and method-of-probability-weighted-moments estimates, and, due to its flexibility, can be also applied to other estimation methods and distributions.

Key words Generalized Pareto distribution • Feasible estimates • Method of moments • Method of probability weighted moments

1 Introduction

The generalized Pareto distribution with shape parameter γ and scale parameter σ (denoted GPD(γ,σ)) is the distribution of a random variable X defined by $X = \sigma(1-e^{-\gamma Y})/\gamma$, where Y is a random variable with the standard exponential distribution. GPD(γ,σ) has the following cumulative distribution function

J. Jocković (✉)
Department of Physics and Mathematics, Faculty of Pharmacy, University of Belgrade, Vojvode Stepe 450, 11000 Belgrade, Serbia
e-mail: haustor@pharmacy.bg.ac.rs

A. Migdalas et al. (eds.), *Optimization Theory, Decision Making, and Operations Research Applications*, Springer Proceedings in Mathematics & Statistics 31,
DOI 10.1007/978-1-4614-5134-1_19,

$$F_{\gamma,\sigma}(x) = \begin{cases} 1 - \left(1 - \frac{\gamma}{\sigma}x\right)^{\frac{1}{\gamma}}, & \gamma \neq 0, \quad \sigma > 0 \\ 1 - \exp\left(-\frac{x}{\sigma}\right), & \gamma = 0, \quad \sigma > 0 . \end{cases} \tag{1}$$

The range is

$$\begin{cases} 0 \leq x < +\infty, & \gamma \leq 0 \\ 0 \leq x \leq \frac{\sigma}{\gamma}, & \gamma > 0 . \end{cases} \tag{2}$$

Many well-known probability distributions belong to GPD family. For example, GPD$(0,\sigma)$ is reduced to the exponential distribution with mean equal to σ, GPD$(1,\sigma)$ is the uniform $U[0,\sigma]$ distribution and for $\gamma < 0$, GPD reduces to the Pareto distribution.

Generalized Pareto distributions are the only continuous distribution functions that are stable with respect to excess over threshold operations (POT-stable). Precisely, if a random variable X has a GPD(γ,σ) distribution, then the conditional distribution of $X-u$ given $X>u$ is GPD$(\gamma,\sigma-\gamma u)$ [3,5,19]. POT-stability has a key role in the POT-approach to modeling extremes, which is based on fitting the GPD to the distribution of the excesses over a sufficiently high threshold. POT-framework was introduced through papers Balkema and de Haan [2] and Pickands [15]. It has numerous applications in hydrology [4, 17], insurance and finance [9, 17], ecology [4], and other fields.

During the last 30 years, several research papers have been dedicated to the problem of estimating GPD parameters and quantiles (see [3, 4, 8, 10–12, 14, 16]). A general review of this subject, which still attracts considerable attention, is given in [5]. Recent contributions to the field are given in [13, 19].

The present article proposes a correction technique for two-parameter GPD estimation methods, in cases when these methods are infeasible, i.e. when the estimated range fails to contain all observations. It is applied to method of moments (MOM) and method of probability weighted moments (PWM), which are known to suffer from the inconsistency problem. Performance of these corrected methods is evaluated under the simulation study, and a real data example is provided.

The paper is organized as follows: definitions and main properties of MOM and PWM estimation methods are given in Sect. 2, proposed corrections are derived in Sect. 3, simulation results are given in Sect. 4, an application to real data in Sect. 5, and conclusions in Sect. 6.

2 Estimating GPD Parameters and Quantiles

Method of moments and method of probability weighted moments are among the simplest and the most traditional methods for estimating GPD parameters and quantiles, as well as univariate distribution parameters in general.

2.1 Method of Moments

Moments of the two-parameter GPD(γ,σ) random variable X are defined as

$$E\left[\left(1-\gamma\frac{X}{\sigma}\right)^r\right]=\frac{1}{1+r\gamma}, \quad 1+r\gamma>0\,, \tag{3}$$

which leads to

$$E\left(X^r\right)=r!\frac{\sigma^r}{(-\gamma)^{r+1}}\frac{\Gamma\left(-\frac{1}{\gamma}-r\right)}{\Gamma\left(1-\frac{1}{\gamma}\right)}, \quad \gamma>-\frac{1}{r}\,, \tag{4}$$

with $\Gamma(\cdot)$ being the Gamma function. From (4) follows that the mean value and variance of the GPD(γ,σ) are

$$E(X)=\frac{\sigma}{1+\gamma}, \quad \gamma>-1\,, \tag{5}$$

and

$$\mathrm{VaR}(X)=\frac{\sigma^2}{(1+\gamma)^2(1+2\gamma)}, \quad \gamma>-\frac{1}{2}\,. \tag{6}$$

Let $x_1,x_2,\ldots,x_n$ be a random sample from GPD(γ,σ) and let $\bar{x}$ and s^2 be the sample mean and sample variance, respectively. MOM estimates for the parameters (γ,σ), defined by Hosking and Wallis [11], are obtained by replacing (5) and (6) with their sample equivalents as follows:

$$\hat{\gamma}_{\mathrm{MOM}}=\frac{1}{2}\left(\frac{\bar{x}^2}{s^2}-1\right) \quad \text{and} \quad \hat{\sigma}_{\mathrm{MOM}}=\frac{1}{2}\bar{x}\left(\frac{\bar{x}^2}{s^2}+1\right)\,. \tag{7}$$

MOM estimates $(\hat{\gamma}_{\mathrm{MOM}},\hat{\sigma}_{\mathrm{MOM}})$ exist if $\gamma>-0.5$. According to [11], they are asymptotically normal for $\gamma>-0.25$.

2.2 Method of Probability Weighted Moments

Probability weighted moments of a random variable X with cumulative distribution function F are given by

$$M_{p,r,s}=E[X^p(F(X))^r(1-F(X))^s]\,, \quad p,r,s\in\mathbb{R}\,. \tag{8}$$

In case of the two-parameter GPD(γ,σ) random variable X, there are simpler relations, given by

$$M_{1,0,s} = E[X(1-F(X))^s] = \frac{\sigma}{(s+1)(s+1+\gamma)}, \quad \gamma > -1, \quad s = 0,1,2,\ldots \quad (9)$$

Let $x_{1:n} \leq x_{2:n} \leq \cdots \leq x_{n:n}$ be a sorted random sample from GPD(γ,σ) and let

$$a_k = \frac{1}{n}\sum_{j=1}^{n}(1-p_{j:n})^k x_{j:n} \quad \text{with} \quad p_{j:n} = \frac{j-0.35}{n}. \quad (10)$$

PWM estimates for γ and σ, introduced by Greenwood et al. [10], obtained by substituting experessions (9) with their sample equivalents (10), are

$$\hat{\gamma}_{\text{PWM}} = \frac{a_0}{a_0-2a_1} - 2 \quad \text{and} \quad \hat{\sigma}_{\text{PWM}} = \frac{2a_0a_1}{a_0-2a_1}. \quad (11)$$

According to [10, 11], estimates $(\hat{\gamma}_{\text{PWM}},\hat{\sigma}_{\text{PWM}})$ are consistent and asymptotically normal for $\gamma > -0.5$.

2.3 *Estimating GPD Quantiles*

A problem that is closely related to fitting the GPD to data is estimating quantiles of the distribution. Quantiles of the GPD(γ,σ), given in terms of the parameters, are

$$x(F) = \begin{cases} \frac{\sigma}{\gamma}(1-(1-F)^{\gamma}), & \gamma \neq 0 \\ -\sigma\log(1-F), & \gamma = 0. \end{cases} \quad (12)$$

A quantile estimator, $\hat{x}(F)$ is obtained by substituting estimators for shape and scale parameters, $\hat{\gamma}$ and $\hat{\sigma}$, in (12).

3 Inconsistency with the Data and Correction Techniques

When $\gamma > 0$, the right endpoint of the support of GPD is σ/γ, i.e. depending on its shape and scale parameters. GPD estimation methods sometimes produce estimates which are inconsistent with the observed data, i.e. one or more sample observations exceed the estimated right endpoint. This problem occurs very often for MOM and PWM estimation methods, as pointed out by Dupuis [6]. The problem is also addressed in [3], and a detailed simulation study is given in [1]. Despite the existence of new and efficient estimation techniques, improving these two methods is valuable for their simplicity and computational speed.

A special type of inconsistency occurs when the estimated shape parameter is negative and there is a reason to believe (visual methods, some prior information that has become available, the nature of the problem itself ...) that it is really positive.

A way to overcome the infeasibility of MOM and PWM methods by introducing a simple auxiliary constraint is proposed in [8]. Shape parameters obtained in this way are always feasible and more efficient (in terms of bias and root mean squared error (RMSE)) than the original methods. The correction technique proposed in the present paper starts from a similar idea. However, it is more general (can be also applied to other estimation methods and other distributions) and can reduce the bias and RMSE of the shape and the scale parameter at the same time, which was not the case with the technique proposed in [8].

3.1 Proposed Corrections

Let $x_{1:n} \leq x_{2:n} \leq \cdots \leq x_{n:n}$ be a sorted random sample from the GPD, whose parameters need to be estimated. Let $\hat{\gamma}$ and $\hat{\sigma}$ be the shape and scale parameter estimates obtained with a given estimation method, and suppose that inconsistency exists, i.e. $\hat{\gamma} > 0$ and $x_{n:n} > \hat{\sigma}/\hat{\gamma}$. The goal is to find estimates for the scale and shape parameter, $\tilde{\sigma}$ and $\tilde{\gamma}$, such that $\tilde{\sigma}/\tilde{\gamma} \geq x_{n:n}$.
Estimates $\tilde{\gamma}$ and $\tilde{\sigma}$ are defined as

$$\tilde{\sigma} = \hat{\sigma} + \alpha p\,, \quad \tilde{\gamma} = \hat{\gamma} - (1-\alpha)p\,, \quad \alpha \in \mathbb{R}\,. \tag{13}$$

If the condition

$$\hat{\gamma} - (1-\alpha)p > 0 \tag{14}$$

holds, then from the inequality

$$\frac{\hat{\sigma} + \alpha p}{\hat{\gamma} - (1-\alpha)p} \geq x_{n:n}\,, \tag{15}$$

follows

$$p \geq \frac{x_{n:n}\hat{\gamma} - \hat{\sigma}}{\alpha + (1-\alpha)x_{n:n}}\,, \quad \text{if} \quad \alpha + (1-\alpha)x_{n:n} > 0\,, \tag{16a}$$

$$p \leq \frac{x_{n:n}\hat{\gamma} - \hat{\sigma}}{\alpha + (1-\alpha)x_{n:n}}\,, \quad \text{if} \quad \alpha + (1-\alpha)x_{n:n} < 0\,. \tag{16b}$$

Taking

$$p = \frac{x_{n:n}\hat{\gamma} - \hat{\sigma}}{\alpha + (1-\alpha)x_{n:n}} \tag{17}$$

solves the inconsistency problem in both cases.

Inequality (14), with p given by (17), holds for any $\hat{\gamma} > 0$, if

$$\alpha\hat{\gamma} + (1-\alpha)\hat{\sigma} > 0\,, \tag{18}$$

which is the condition for the proposed correction to be valid.

Condition (18) holds for

$$\alpha \in \left(-\frac{\hat{\sigma}}{|\,\hat{\sigma}-\hat{\gamma}\,|}, \frac{\hat{\sigma}}{|\,\hat{\sigma}-\hat{\gamma}\,|}\right), \quad \hat{\gamma} \neq \hat{\sigma}\,. \tag{19}$$

In case $\hat{\gamma} < 0$ ("falsely" negative shape parameter) corrected estimates have the same form.

Intuitively, the purpose of the parameter α is to control which of the estimates ($\hat{\gamma}$ or $\hat{\sigma}$) will be changed "more". "Natural" choices for α belong to the interval $[0,1]$. However, it was demonstrated (see Sect. 4) that some other choices from the interval (19) produce estimates with smaller bias and RMSE.

It is also possible to introduce another parameter, β, indicating the level of the correction, i.e. how far the largest observation will be from the newly estimated right endpoint. In that case, corrected estimates satisfy the equality

$$\frac{\tilde{\sigma}}{\tilde{\gamma}} = \frac{\hat{\sigma}+\alpha p}{\hat{\gamma}-(1-\alpha)p} = \beta x_{n:n}\,, \quad \text{with} \quad \beta \geq 1\,. \tag{20}$$

The overall estimates of the scale and shape parameter are now given by

$$\tilde{\sigma} = \begin{cases} \hat{\sigma}+\alpha p\,, & \text{if } x_{n:n}\hat{\gamma}-\hat{\sigma} > 0 \text{ and } \hat{\gamma} > 0\,, \text{ or } \hat{\gamma} < 0 \\ \hat{\sigma}, & \text{otherwise}\,, \end{cases} \tag{21a}$$

$$\tilde{\gamma} = \begin{cases} \hat{\gamma}-(1-\alpha)p, & \text{if } x_{n:n}\hat{\gamma}-\hat{\sigma} > 0 \text{ and } \hat{\gamma} > 0, \text{ or } \hat{\gamma} < 0 \\ \hat{\gamma}, & \text{otherwise}\,, \end{cases} \tag{21b}$$

with

$$p = \frac{\beta\hat{\gamma}x_{n:n}-\hat{\sigma}}{\alpha+(1-\alpha)\beta x_{n:n}}\,, \quad \text{and} \quad \beta \geq 1\,. \tag{22}$$

The condition $\hat{\gamma} < 0$ is optional. If it is removed, negative estimates of the shape parameter will not be corrected.

Some properties of the corrected estimates are:

1. If $\alpha = 0$, the scale parameter is estimated with the original (uncorrected) method and the inconsistency is removed. For $\beta = 1$ this is the hybrid estimator proposed in [8].
2. If $\alpha = 1$, the shape parameter is estimated with the original method and inconsistency is removed.
3. If $x_{n:n}\hat{\gamma}-\hat{\sigma} < 0$ (there is no inconsistency), the correction is not applied. Both parameters are estimated with the original method.

4 Simulation Results

In order to find adequate α values from the interval (19) for correcting MOM and PWM estimation methods, parameter α was defined as

$$\alpha = \begin{cases} i\hat{\sigma}\,, & \hat{\gamma}=\hat{\sigma} \\ \max\left\{i\hat{\sigma}\,, j\frac{\hat{\sigma}}{|\hat{\sigma}-\hat{\gamma}|}\right\}, & \hat{\gamma}\neq\hat{\sigma}\,, \end{cases} \tag{23}$$

with $i,j \in \{-1,-0.9,-0.8,\ldots,0.8,0.9,1\}$, and $\hat{\gamma},\hat{\sigma}$ being the estimates obtained with the original MOM or PWM methods. The term $i\hat{\sigma}$ is added to penalize against very large absolute values of α in cases when $\hat{\sigma}$ is close to $\hat{\gamma}$. Parameter β is defined to be in $[1,2]$.

After performing a large number of Monte Carlo experiments with simulated GPD data, it was noticed that acceptable α values are obtained for $i,j \leq 0$. They decrease both bias and RMSE of parameter estimates. Furthermore, they perform well even for extremely small, or extremely large values of σ. Values of β in the range $[1,1.5]$ seem to work good.

This part of the study is not included in the text, for space-saving purposes. In the second stage, a detailed simulation study was performed for particular α and β choices. Parameter α is defined as

$$\alpha = \begin{cases} -0.5\hat{\sigma}\,, & \hat{\gamma}=\hat{\sigma} \\ \max\left\{-0.5\hat{\sigma}\,, -\frac{0.9\hat{\sigma}}{|\hat{\sigma}-\hat{\gamma}|}\right\}, & \hat{\gamma}\neq\hat{\sigma}\,, \end{cases} \tag{24}$$

which was obtained by randomly choosing (i,j) pair from the acceptable range.

4.1 Simulation Study for Particular α and β Choices

For checking the performance of the proposed corrections for MOM and PWM estimation methods, 1,000 samples from GPD(γ,σ) are generated, for each combination (n,γ,σ,β) of sample sizes $n\in\{15,50,100\}$, distribution parameters $\gamma \in \{0.2,0.4,0.6,0.8,1,2\}$, $\sigma \in \{0.1,0.2,0.4,0.6,0.8,1,2,5,10,20\}$, and $\beta \in \{1,1.01,1.5\}$. Parameter α is defined as in (24). The original MOM and PWM estimates are compared with corrected methods for the following properties: shape parameter bias and RMSE, scale parameter bias and RMSE, bias and RMSE of 99% and 99.9% quantiles (scaled by the true values of quantiles being estimated), and number of datasets where the inconsistency occured.

Simulation results can be summarized as follows:

1. Number of datasets where inconsistency occurred (out of 1,000), obtained in this study (see Table 1), is in agreement with the results reported in [1,3,8].

Table 1 Number of datasets where inconsistency occurred for MOM and PWM methods

γ	n	MOM 15	PWM	MOM 50	PWM	MOM 100	PWM
0.2		87 (134)	106 (209)	34 (49)	94 (109)	45 (7)	86 (37)
0.4		129 (44)	136 (102)	163 (1)	201 (16)	166 (0)	206 (0)
0.6		228 (11)	240 (32)	306 (0)	325 (0)	302 (0)	322 (0)
0.8		283 (2)	249 (16)	352 (0)	352 (0)	382 (0)	381 (0)
1		339 (0)	291 (3)	395 (0)	377 (0)	444 (0)	437 (0)
2		438 (0)	301 (0)	473 (0)	400 (0)	507 (0)	465 (0)

Number of datasets where negative shape parameter estimates are obtained is given in the parentheses

2. All the estimates obtained with corrected MOM and PWM methods are feasible with the data, i.e. the inconsistency is completely removed.
3. For both corrected methods, the RMSE of shape and scale parameters decreased, comparing to original methods. For most of the samples (the only exception are samples from GPD$(2,\sigma)$), the bias also decreased.
4. For both corrected methods, the RMSE of 99% and 99.9% quantiles decreases, comparing to original methods. In some experiments, the bias also decreased, in others, slightly increased. Therefore, corrected methods compete well with the original methods in case of estimating high quantiles.
5. The most significant decrease in the bias and the RMSE for all parameter estimates is obtained for sample of size $n = 15$.

Results obtained for $\sigma = 1$ (chosen because that case is commonly presented in the literature) and $\beta = 1.01$ are summarized in Tables 1–4. The other simulation results lead to the same conclusions stated above and are available on request. All computations in this work are performed using MATLAB®, release 2007a.

Performance of the corrected estimator proposed in the present paper is also compared to the performance of the hybrid estimator proposed in [8]. The relative performances of the two methods, precisely, the ratios of absolute biases and RMSE's of shape, scale and quantiles estimates for newly proposed and hybrid method, are given in Table 5. In cases when the ratio is less than 1 newly corrected method outperforms the hybrid method, and in cases when the ratio is greater than 1 hybrid method works better.

5 An Example-Fish River Data

In the paper [1] authors provided a hydrological example, showing how both MOM and PWM methods produce infeasible estimates. The data consists of 42 observations (flows of the Fish River, Canada, registered during the period 1981–1999), below a suitably chosen threshold.

Table 2 Bias (RMSE) obtained when estimating shape and scale parameters

	n	γ	1	2	3	4
Shape par.						
	15	0.2	0.1(0.35)	0.09(0.32)	0.08(0.38)	0.06(0.34)
		0.4	0.1(0.43)	0.07(0.34)	0.08(0.43)	0.05(0.36)
		0.6	0.1(0.53)	0.05(0.40)	0.07(0.49)	0.02(0.41)
		0.8	0.16(0.65)	0.06(0.45)	0.1(0.56)	0.03(0.45)
		1	0.12(0.72)	−0.01(0.47)	0.04(0.58)	−0.03(0.47)
		2	0.46(1.97)	−0.31(0.77)	0.09(1.09)	−0.19(0.79)
	50	0.2	0.03(0.16)	0.02(0.15)	0.02(0.19)	0.01(0.17)
		0.4	0.02(0.17)	0.01(0.16)	0.02(0.20)	0(0.17)
		0.6	0.02(0.23)	−0.01(0.18)	0.01(0.24)	−0.02(0.20)
		0.8	0.03(0.28)	−0.02(0.21)	0.02(0.28)	−0.03(0.22)
		1	0.05(0.34)	−0.02(0.25)	0.03(0.32)	−0.02(0.25)
		2	0.1(0.72)	−0.23(0.43)	0.02(0.57)	−0.15(0.41)
	100	0.2	0.01(0.10)	0.01(0.10)	0.01(0.12)	0.01(0.11)
		0.4	0.01(0.12)	0(0.11)	0.01(0.14)	0(0.12)
		0.6	0(0.14)	−0.02(0.12)	0(0.16)	−0.03(0.13)
		0.8	0.01(0.19)	−0.02(0.15)	0.01(0.19)	−0.03(0.15)
		1	0.02(0.23)	−0.03(0.17)	0.01(0.23)	−0.04(0.18)
		2	0.06(0.47)	−0.16(0.31)	0.02(0.39)	−0.11(0.29)
Scale par.						
	15	0.2	0.1(0.44)	0.1(0.43)	0.09(0.46)	0.08(0.44)
		0.4	0.09(0.46)	0.08(0.42)	0.08(0.45)	0.06(0.42)
		0.6	0.09(0.49)	0.06(0.42)	0.07(0.45)	0.05(0.41)
		0.8	0.13(0.55)	0.08(0.44)	0.1(0.48)	0.06(0.42)
		1	0.09(0.52)	0.03(0.39)	0.05(0.44)	0.01(0.38)
		2	0.21(0.86)	−0.12(0.36)	0.06(0.50)	−0.05(0.37)
	50	0.2	0.03(0.21)	0.03(0.20)	0.02(0.22)	0.02(0.22)
		0.04	0.02(0.20)	0.02(0.19)	0.02(0.21)	0.01(0.20)
		0.6	0.02(0.22)	0.01(0.21)	0.02(0.23)	0(0.21)
		0.8	0.03(0.23)	0.01(0.21)	0.02(0.23)	0(0.21)
		1	0.04(0.25)	0.01(0.21)	0.03(0.24)	0.01(0.21)
		2	0.04(0.31)	−0.09(0.19)	0.02(0.25)	−0.05(0.19)
	100	0.2	0.01(0.14)	0.01(0.14)	0.01(0.15)	0.01(0.15)
		0.4	0.01(0.15)	0.01(0.14)	0.01(0.15)	0(0.15)
		0.6	0(0.14)	−0.01(0.14)	0(0.15)	−0.01(0.14)
		0.8	0.01(0.16)	0(0.14)	0.01(0.16)	0(0.14)
		1	0.01(0.17)	0(0.15)	0.01(0.16)	−0.01(0.15)
		2	0.02(0.20)	−0.06(0.14)	0.01(0.17)	−0.04(0.13)

Negative estimates of the shape parameter are not corrected. Methods: 1-MOM, 2-corrected MOM, 3-PWM, 4-corrected PWM

Sorted data, in days, are:

{7, 7, 9, 9, 11, 12, 15, 17, 18, 20, 20, 22, 22, 24, 28, 29, 30, 31, 31, 32, 34, 34, 35, 41, 41, 47, 49, 53, 57, 59, 59, 60, 62, 68, 72, 74, 76, 78, 79, 92, 101, 111}.

Table 3 Bias (RMSE) obtained for 99% and 99.9%-quantile estimation

	n	γ	1	2	3	4
99% q.						
	15	0.2 (3.01)	−0.03(0.29)	−0.03(0.29)	0.02(0.36)	0.03(0.36)
		0.4 (2.10)	−0.02(0.24)	−0.01(0.23)	0.02(0.30)	0.03(0.29)
		0.6 (1.56)	0.01(0.21)	0.03(0.20)	0.04(0.26)	0.06(0.25)
		0.8 (1.22)	0.02(0.19)	0.04(0.17)	0.04(0.22)	0.06(0.21)
		1 (0.99)	0.04(0.18)	0.06(0.16)	0.06(0.20)	0.08(0.19)
		2 (0.50)	0.03(0.13)	0.07(0.12)	0.05(0.13)	0.07(0.12)
	50	0.2 (3.01)	−0.01(0.16)	−0.01(0.16)	0.01(0.20)	0.02(0.19)
		0.4 (2.10)	−0.01(0.13)	0(0.12)	0(0.16)	0.02(0.14)
		0.6 (1.56)	0.01(0.11)	0.03(0.10)	0.02(0.14)	0.04(0.12)
		0.8 (1.22)	0.01(0.10)	0.03(0.08)	0.01(0.12)	0.04(0.10)
		1 (0.99)	0.01(0.09)	0.03(0.07)	0.01(0.09)	0.04(0.08)
		2 (0.50)	0.01(0.06)	0.03(0.05)	0.02(0.05)	0.03(0.05)
	100	0.2 (3.01)	−0.01(0.11)	0(0.11)	0(0.14)	0.01(0.13)
		0.4 (2.10)	0(0.09)	0.01(0.08)	0.01(0.11)	0.02(0.10)
		0.6 (1.56)	0.01(0.08)	0.02(0.07)	0.01(0.09)	0.03(0.08)
		0.8 (1.22)	0(0.07)	0.02(0.06)	0.01(0.08)	0.03(0.06)
		1 (0.99)	0.01(0.07)	0.03(0.05)	0.01(0.07)	0.03(0.06)
		2 (0.50)	0(0.04)	0.02(0.03)	0.01(0.03)	0.02(0.03)
99.9% q.						
	15	0.2 (3.74)	0.02(0.45)	0.03(0.45)	0.17(0.76)	0.18(0.76)
		0.4 (2.34)	0.04(0.36)	0.05(0.35)	0.13(0.56)	0.14(0.55)
		0.6 (1.64)	0.07(0.34)	0.09(0.33)	0.14(0.50)	0.16(0.49)
		0.8 (1.25)	0.06(0.27)	0.08(0.26)	0.14(0.37)	0.12(0.36)
		1 (1.00)	0.07(0.25)	0.10(0.24)	0.11(0.32)	0.13(0.32)
		2 (0.50)	0.04(0.15)	0.07(0.14)	0.06(0.15)	0.07(0.15)
	50	0.2 (3.74)	0.01(0.25)	0.01(0.25)	0.06(0.35)	0.07(0.34)
		0.4 (2.34)	0.01(0.19)	0.02(0.18)	0.04(0.26)	0.05(0.25)
		0.6 (1.64)	0.02(0.16)	0.04(0.15)	0.04(0.20)	0.06(0.19)
		0.8 (1.25)	0.02(0.13)	0.05(0.12)	0.03(0.15)	0.05(0.14)
		1 (1.00)	0.02(0.11)	0.05(0.09)	0.03(0.12)	0.05(0.11)
		2 (0.50)	0.01(0.07)	0.03(0.06)	0.02(0.05)	0.03(0.05)
	100	0.2 (3.74)	0(0.16)	0(0.16)	0.02(0.22)	0.03(0.22)
		0.4 (2.34)	0.01(0.13)	0.02(0.12)	0.02(0.17)	0.03(0.16)
		0.6 (1.64)	0.02(0.11)	0.03(0.10)	0.03(0.13)	0.05(0.12)
		0.8 (1.25)	0.01(0.09)	0.03(0.07)	0.01(0.10)	0.04(0.09)
		1 (1.00)	0.01(0.08)	0.03(0.06)	0.02(0.08)	0.04(0.07)
		2 (0.50)	0(0.04)	0.02(0.03)	0.01(0.03)	0.02(0.03)

The true values of the quantiles being estimated given in the parentheses. Negative estimates of the shape parameters are not corrected. Methods: 1-MOM, 2-corrected MOM, 3-PWM, 4-corrected PWM

In the present study, GPD parameters and quantiles for this dataset are estimated with the following methods: MOM, PWM, and corrected MOM and PWM. For comparison, estimates obtained with ML and EPM, methods that normally do not suffer from inconsistency problem, are also provided.

Table 4 Bias (RMSE) obtained when estimating shape and scale parameters in cases when negative estimates of the shape parameter are obtained

	n	γ	1	2	3	4
Shape par.						
	15	0.2	0.1(0.34)	0.14(0.27)	0.08(0.37)	0.16(0.30)
		0.4	0.09(0.40)	0.08(0.32)	0.07(0.41)	0.08(0.32)
		0.6	0.11(0.52)	0.05(0.38)	0.08(0.48)	0.04(0.37)
		0.8	0.13(0.64)	0.04(0.45)	0.08(0.56)	0.02(0.44)
		1	0.12(0.69)	−0.01(0.46)	0.04(0.57)	−0.03(0.46)
	50	0.2	0.02(0.15)	0.03(0.14)	0.02(0.18)	0.04(0.14)
		0.4	0.03(0.18)	0.01(0.15)	0.02(0.20)	0.01(0.17)
	100	0.2	0.02(0.11)	0.02(0.10)	0.01(0.13)	0.02(0.11)
Scale par.						
	15	0.2	0.11(0.44)	0.11(0.42)	0.09(0.46)	0.11(0.43)
		0.4	0.09(0.45)	0.08(0.41)	0.08(0.45)	0.07(0.41)
		0.6	0.1(0.49)	0.07(0.42)	0.07(0.45)	0.05(0.41)
		0.8	0.12(0.54)	0.07(0.44)	0.08(0.48)	0.05(0.42)
		1	0.09(0.50)	0.03(0.38)	0.05(0.42)	0.02(0.37)
	50	0.2	0.02(0.20)	0.02(0.20)	0.02(0.22)	0.02(0.21)
		0.4	0.02(0.20)	0.02(0.20)	0.02(0.22)	0.01(0.20)
	100	0.2	0.02(0.15)	0.01(0.15)	0.01(0.16)	0.01(0.15)

Negative estimates are corrected. Methods: 1-MOM, 2-corrected MOM, 3-PWM, 4-corrected PWM

For checking the goodness of fit, the following error definitions were used:

$$\mathrm{ASAE} = n^{-1}\sum_{i=1}^{n}\frac{|x_{i:n}-\hat{x}_{i:n}|}{x_{n:n}-x_{1:n}}, \quad \hat{x}_{i:n} = \hat{\sigma}(1-(1-p_{i:n})^{\hat{\gamma}})/\hat{\gamma}, \quad p_{i:n} = \frac{i}{n+1}, \tag{25}$$

and

$$\mathrm{SSQ} = \sum_{i=1}^{n}(F_{\hat{\gamma},\hat{\sigma}}(x_{i:n}) - \hat{F}(x_{i:n}))^2, \tag{26}$$

with $\hat{F}$ being the empirical distribution function. Results are summarized in Table 6 and indicate very good agreement between both corrected methods and ML method, which is considered best for this dataset (according to [1]).

In this example, RMSE (scaled square root of the values given in the last row of the Table 6) of the newly obtained fit is a bit greater than the RMSE obtained when using original MOM and PWM. This is probably due to the fact that there is some contamination in the data that is not incorporated into GPD model. However, corrected MOM and PWM are more adequate in this case, since they are feasible with the sample data.

Table 5 Ratio of the performance of corrected estimates (corr.) to hybrid estimates proposed by Dupuis and Tsao (hyb.): absolute bias corr./absolute bias hib. (RMSE corr./RMSE hyb.)

n	γ	MOM	PWM	MOM	PWM
		Shape par.		**Scale par.**	
15	0.2	0.96 (0.97)	0.92 (0.97)	0.94 (0.97)	0.88 (0.97)
	0.4	0.86 (0.94)	0.71 (0.95)	0.83 (0.94)	0.75 (0.94)
	0.6	0.57 (0.90)	0.11 (0.93)	0.66 (0.89)	0.58 (0.92)
	0.8	0.09 (0.83)	6.61 (0.91)	0.46 (0.82)	0.38 (0.89)
	1	0.41 (0.75)	13.22 (0.88)	0.21 (0.72)	0.16 (0.87)
	2	1.03 (0.42)	29.67 (0.75)	0.63 (0.39)	1.62 (0.73)
50	0.2	0.96 (0.99)	0.77 (0.98)	0.93 (0.99)	0.72 (0.97)
	0.4	0.26 (0.97)	1.88 (0.96)	0.60 (0.96)	0.16 (0.95)
	0.6	2.51 (0.93)	1.64 (0.94)	0.09 (0.92)	0.61 (0.92)
	0.8	2.47 (0.90)	1.74 (0.92)	0.31 (0.89)	1.42 (0.90)
	1	3.21 (0.87)	1.95 (0.90)	0.57 (0.85)	2.16 (0.89)
	2	8.41 (0.66)	7.12 (0.75)	2.69 (0.61)	7.70 (0.72)
100	0.2	0.97 (0.99)	0.83 (0.98)	0.95 (0.99)	0.82 (0.98)
	0.4	0.25 (0.97)	2.76 (0.96)	0.68 (0.96)	0.47 (0.95)
	0.6	3.14 (0.93)	1.95 (0.94)	0.29 (0.93)	0.04 (0.92)
	0.8	2.96 (0.91)	2.14 (0.92)	0.05 (0.89)	0.34 (0.90)
	1	4.00 (0.88)	2.55 (0.90)	0.28 (0.86)	0.65 (0.88)
	2	6.80 (0.73)	50.71 (0.78)	2.18 (0.69)	2.93 (0.76)
		99% q.		**99.9% q.**	
15	0.2	0.99 (1.00)	0.10 (0.58)	0.11 (1.33)	1.00 (1.00)
	0.4	1.11 (1.00)	0.28 (0.66)	1.32 (1.40)	1.01 (1.00)
	0.6	1.03 (1.00)	0.48 (0.71)	4.83 (1.37)	1.01 (1.00)
	0.8	1.03 (1.00)	0.65 (0.76)	1.80 (1.31)	1.01 (1.00)
	1	1.04 (1.00)	0.78 (0.79)	1.39 (1.26)	1.02 (1.00)
	2	1.05 (1.00)	0.96 (0.84)	1.14 (1.19)	1.04 (1.00)
50	0.2	0.99 (1.00)	0.06 (0.47)	0.06 (0.99)	1.01 (1.00)
	0.4	1.05 (1.00)	0.20 (0.58)	0.44 (1.21)	1.02 (1.00)
	0.6	1.06 (0.99)	0.41 (0.68)	4.12 (1.33)	1.03 (1.00)
	0.8	1.07 (0.99)	0.63 (0.76)	2.56 (1.28)	1.05 (1.00)
	1	1.09 (0.99)	0.80 (0.82)	1.56 (1.21)	1.07 (1.00)
	2	1.15 (1.01)	1.00 (0.83)	1.32 (1.22)	1.14 (1.01)
100	0.2	0.97 (1.00)	0.04 (0.40)	0.03 (0.78)	1.01 (1.00)
	0.4	1.06 (0.99)	0.15 (0.52)	0.24 (1.01)	1.03 (1.00)
	0.6	1.09 (0.99)	0.36 (0.64)	1.37 (1.25)	1.07 (0.99)
	0.8	1.12 (0.99)	0.59 (0.74)	4.87 (1.29)	1.10 (0.99)
	1	1.14 (1.00)	0.79 (0.81)	1.84 (1.22)	1.12 (1.00)
	2	1.25 (1.02)	1.06 (0.82)	1.49 (1.27)	1.26 (1.02)

Table 6 The Fish river data: shape ($\hat{\gamma}$), scale ($\hat{\sigma}$) and quantile estimates, estimated upper bound ($\hat{\sigma}/\hat{\gamma}$), number of observations that exceeded the estimated upper bound (d), errors (ASAE, SSQ)

	MOM (Orig.)	MOM (Corr.)	PWM (Orig.)	PWM (Corr.)	EPM	ML
$\hat{\gamma}$	0.705	0.643	0.754	0.661	0.486	0.560
$\hat{\sigma}$	72.092	72.063	74.170	74.125	64.309	65.451
99% q.	98.294	106.301	95.313	106.773	118.146	108.041
99,9% q.	101.490	110.788	97.828	110.946	127.631	114.472
$\hat{\sigma}/\hat{\gamma}$	102.275	112.110	98.366	112.110	132.226	116.918
d	1	0	3	0	0	0
ASAE	0.030	0.034	0.031	0.038	0.030	0.027
SSQ	0.275	0.327	0.287	0.367	0.284	0.246

6 Conclusions and Future Work

The most traditional estimation methods for the generalized Pareto distributions are MOM, PWM and maximum likelihood method (ML). However, all of these methods have problems: MOM and PWM may be inconsistent with the sample data, and ML method often suffers from convergence problems and inefficiency when applied to small samples [5, 11, 13]. For these reasons, several new estimation techniques were proposed, including Bayesian methods (a recent review is given in [5]), robust methods [7, 12, 14], EPM method [3], GPWM [16] and many others. Some recently proposed estimation methods combine several different approaches, for example, ML and Bayesian approach [19], or ML and moment technique [18].

All methods have advantages and disadvantages. For example, robust procedures show good performance and outperform the ML method in cases of contaminated samples, but fail to provide good results in cases with no contamination [12, 13]. Robust and/or likelihood-based techniques usually do not suffer from inconsistency problems, but are much more computationally intensive and sometimes inefficient in small samples [13].

The present paper proposed a way to overcome the infeasibility of GPD estimation methods by introducing a whole new class of estimates, which are all feasible, obtained by correcting the original ones. In case of MOM and PWM estimation methods, simulation experiments demonstrated that corrected shape and scale parameter estimates have smaller bias (in most cases) and RMSE (in all cases) than the estimates obtained with the original methods. The new technique is equally acceptable for estimating high quantiles as the original MOM and PWM.

There are several possibilities for continuing this work.

The technique considered here may be applied to other estimation methods suffering from inconsistency of any type (for example, GPWM [16], according to the study presented in [1]). It can be done by obtaining plausible values for α and β through a simulation study, and then applying this completely determined procedure to real data.

It is also possible to obtain estimates with desired special properties, by combining this procedure with other estimation techniques (such as ML), or using optimization methods. On the other hand, this technique can be also applied to other distributions with finite bound(s) that depend on parameters, for example generalized extreme value distribution.

Acknowledgements This work is supported by the Ministry of Education and Science of the Republic of Serbia, Grant nos. 174012 and TR34007.

The author would like to thank the editors and the referees for their useful comments and suggestions.

References

1. Ashkar, F., Tatsambon, C.N.: Revisiting some estimation methods for the generalized Pareto distribution. J. Hydrol. **346**, 136–143 (2007)
2. Balkema, A., de Haan, L.: Residual life time at great age. Ann. Prob. **2**, 792–804 (1974)
3. Castillo, E., Hadi, A.S.: Fitting the generalized Pareto distribution to data. J. Am. Stat. Assoc. **92**(440), 1609–1620 (1997)
4. Davison, A.C., Smith, R.L.: Models for exceedances over high thresholds. J. Roy. Stat. Soc. Ser. B (Methodological), **52**(3), 393–442 (1990)
5. de Zea Bermudez, P., Kotz, S.: Parameter estimation of the generalized Pareto distribution – Parts 1 and 2. J. Stat. Plann. Infer. **140**, 1353–1373 (2010)
6. Dupuis, D.J.: Estimating the probability of obtaining nonfeasible parameter estimates of the generalized Pareto distribution. J. Stati. Comput. Simulat. **54**, 197–209 (1996)
7. Dupuis, D.J.: Exceedances over high thresholds: a guide to threshold selection. Extremes **1**(3), 251–261 (1998)
8. Dupuis, D.J., Tsao, M.: A hybrid estimator for generalized Pareto and extreme-value distributions. Comm. Stat. Theor. Meth. **27**(4), 925–941 (1998)
9. Embrechts, P., Kluppelberg, C., Mikosch, T.: Modelling Extremal Events. Springer, Berlin (2003)
10. Greenwood, J.A., Landwehr, J.M., Matalas, N.C., Wallis, R.: Probability weighted moments: definition and relation to parameters of several distributions expressible in inverse form. Water Res. Res. **15**(5), 1049–1054 (1979)
11. Hosking, J.R.M., Wallis, J.R. Parameter and quantile estimation for the generalized Pareto distribution. Technometrics **29**(3), 339–349 (1987)
12. Juarez, S., Schucany, W.: Robust and efficient estimation for the generalized Pareto distribution. Extremes **7**, 237–251 (2004)
13. Mackay, E., Challenor, P., Bahaj, A.: A comparison of estimators for the generalised Pareto distribution. Ocean Eng. **38**, 1338–1346 (2011)
14. Peng, L., Welsh, A.H.: Robust estimation of the generalized Pareto distribution. Extremes **4**(1), 53–65 (2001)
15. Pickands, J.: Statistical inference using extreme order statistics. Ann. Stat. **3**, 119–131 (1975)
16. Rasmussen, P.F., Generalized probability weighted moments: application to the generalized Pareto distribution. Water Res. Res. **37**(6), 1745–1751 (2001)
17. Reiss, R.D., Thomas, M.: Statistical analysis of extreme values. Birkhauser, Basel (1997)
18. Zhang, J.: Likelihood moment estimation for the generalised Pareto distribution. Aust. N. Z. J. Stat. **49**(1), 69–77 (2007)
19. Zhang, J., Stephens, M.A.: A new and efficient estimation method for the generalized Pareto distribution. Technometrics **51**(3), 316–325 (2009)

Consistent Sequences of Tests Defined by Bans

Alexander Grusho and Elena Timonina

Abstract Finite probability spaces are important in such problems of operation research as data mining, computer simulation, network and computer security, cryptography and many others. We consider complexity of testing a simple hypothesis $H_{0,n}$ against complex alternative $H_{1,n}$ in finite models. The way to make calculation of tests simpler is to build critical sets dependent on smallest bans (the shortest vectors, which have probability zero). We prove necessary and sufficient conditions when consistent sequence of statistical tests exists and all critical sets of the tests are defined by smallest bans. Existence of such sequences of tests is equivalent to existence of strictly consistent sequence of tests.

1 Introduction

The concept of consistent sequences of statistical tests (CST) has been defined for the first time in the work of Wald and Wolfowitz [1] in 1940. Despite long time since that work not many works have been devoted to a problem of existence of CST though nonexistence of CST means impossibility of reliable detection of necessary property by statistical techniques. For the first time this important property of nonexistence of CST has been pointed by Hoeffding [2].

Let's review some results devoted to CST existence. In the work of Schmetterer [3] dependency between CST and consistent estimations was found. Using these dependencies Schmerkotte [4] proved necessary and sufficient conditions of CST existence for a wide class of probability spaces. To prove these results Schmerkotte

A. Grusho (✉)
Institute of Informatics Problems, RAS, Vaviliva St., 44, build 2, 119333 Moscow, Russia
e-mail: grusho@yandex.ru

E. Timonina
Moscow State University, Leninskie Gory, GSP-2, 119992 Moscow, Russia
e-mail: eltimon@yandex.ru

A. Migdalas et al. (eds.), *Optimization Theory, Decision Making, and Operations Research Applications*, Springer Proceedings in Mathematics & Statistics 31,
DOI 10.1007/978-1-4614-5134-1_20,

[4] used a topology generated by metrics on the space of parameters. Earlier this method was also used by Pfanzagl [5]. In the works [6, 7] we considered sequences of finite probability spaces. It helped to prove similar necessary and sufficient conditions of CST existence without topological structures on space of parameters.

Finite probability spaces are often used as models for description of data mining and computer simulation problems, computer security problems, cryptography and many others.

Stochastic models for finite spaces should be considered from two points of view. From the first point of view it is necessary to consider the productivity for application. From the second point of view it is necessary to consider complexity of calculation. We consider a sequence of finite probability spaces, indexed by natural numbers. For each positive integer n we consider the statistical problem of testing a simple hypothesis $H_{0,n}$ against a complex alternative $H_{1,n}$. Each criterion is defined by a critical set S_n. S_n consists of all elementary events that lead to the acceptance of $H_{1,n}$.

Besides CST existence we consider complexity of calculation of statistical criterions in discrete probability spaces. Any sequence of criteria is defined by the sequence of critical sets. In finite spaces critical sets are finite. In the discrete mathematics for any subset of finite set it is defined a complexity of calculation of membership function for the critical set. Thus complexity of a statistical criterion is defined. Considering sequences of criteria it is possible to speak about asymptotic complexity. In our previous work [8], it has been shown that by changing the time scale it is possible to construct for every CST another CST with asymptotically better computational complexity. Such simplification has been named fictitious and later it has been proved that in a class of monotone sequences of critical sets (in some sense) property of fictitious simplification is impossible.

In finite spaces we searched for tests with minimal complexity of an algorithm for calculation that data belongs to S_n. In previous studies [9, 10] we introduced a definition of a ban for a probability measure on a finite space. A ban means a sequence which has probability zero in a finite space. We have shown that the notion of bans is convenient because it allows to determine the critical set in the simplest way for calculation [9]. Then we have proved necessary and sufficient condition for the existence of a consistent sequence of tests, where all critical sets are defined by bans [10].

Here we generalize the main results of [10] for a sequence of different finite spaces and prove some properties of bans. An interesting result is that the existence of strictly consistent sequence of tests [11] means that we can choose the critical sets defined by bans.

The article is structured as follows. Section 2 introduces definitions and examples. Section 3 defines the necessary and sufficient conditions for the case when the critical sets may be defined in terms of bans. In Sect. 4 we analyze the specific properties of bans and prove that the existence of a strictly consistent sequence of tests means that the critical sets can be chosen depending on the bans. In Conclusion we discuss unsolved problems.

2 Mathematical Models and Examples

Let X_i, $i = 1,2,\dots,n,\dots$, be a sequence of finite sets, $\prod_{i=1}^{n} X_i$ be a Cartesian product of X_i, $i = 1,2,\dots,n$, X^∞ be a set of all sequences where ith element belongs to X_i. Define $\mathscr{A}$ be a σ-algebra on X^∞, generated by cylindrical sets. $\mathscr{A}$ is also Borel σ-algebra in Tichonof product X^∞, where X_i, $i = 1,2,\dots,n,\dots$, has a discrete topology [12].

On $(X^\infty, \mathscr{A})$ a probability measure P_0 is defined. Assume $P_{0,n}$ is a project of P_0 on the first n coordinates of sequences from X^∞. It is clear that for every $B_n \subseteq \prod_{i=1}^{n} X_i$

$$P_{0,n}(B_n) = P_0(B_n \times X_n^\infty),$$

where $X_n^\infty = \prod_{i=n+1}^{\infty} X_i$. Let $D_{0,n}$ be a support of measure $P_{0,n}$:

$$D_{0,n} = \left\{ \underline{x}_n \in \prod_{i=1}^{n} X_i,\, P_{0,n}(\underline{x}_n) > 0 \right\}.$$

Denote $\Delta_{0,n} = D_{0,n} \times X_n^\infty$. The sequence $\Delta_{0,n}$, $n = 1,2,\dots$, is nonincreasing and

$$\Delta_0 = \lim_{n\to\infty} \Delta_{0,n} = \bigcap_{n=1}^{\infty} \Delta_{0,n}.$$

The set Δ_0 is closed and it is a support of P_0.

We also have a set of probability measures $\{P_\theta,\, \theta \in \Theta\}$ on $(X^\infty, \mathscr{A})$. Then as before we define $P_{\theta,n}$, $D_{\theta,n}$, $\Delta_{\theta,n}$, Δ_θ.

If $\overline{\omega}^{(k)} \in \prod_{i=1}^{k} X_i$, then $\widetilde{\omega}^{(k-1)}$ is obtained from $\overline{\omega}^{(k)}$ by dropping the last coordinate.

Definition 1. Ban in measure $P_{0,n}$ is a vector $\overline{\omega}^{(k)} \in \prod_{i=1}^{k} X_i$, $k \le n$, such that

$$P_{0,n}\left(\overline{\omega}^{(k)} \times \prod_{i=k+1}^{n} X_i \right) = 0.$$

If $P_{0,k-1}(\widetilde{\omega}^{(k-1)}) > 0$, then $\overline{\omega}^{(k)}$ is the smallest ban.

If $\overline{\omega}^{(k)}$ is a ban in $P_{0,n}$, then for every $k \le s \le n$ and for every sequence $\overline{\omega}^{(s)}$ starting with $\overline{\omega}^{(k)}$ we have

$$P_{0,s}(\overline{\omega}^{(s)}) = 0. \tag{1}$$

In fact , if $P_{0,k}(\overline{\omega}^{(k)}) = 0$, then $P_0(\overline{\omega}^{(k)} \times X_k^\infty) = 0$ and

$$P_0\left(\overline{\omega}^{(k)} \times \prod_{i=k+1}^{s} X_i \times X_s^\infty \right) = 0.$$

It follows that

$$P_{0,s}\left(\overline{\omega}^{(s)}\right) = P_0\left(\overline{\omega}^{(s)} \times X_s^{\infty}\right) \le P_0\left(\overline{\omega}^{(k)} \times \prod_{i=k+1}^{s} X_i \times X_s^{\infty}\right) = 0.$$

If there exists $\overline{\omega}^{(n)} \in \prod_{i=1}^{n} X_i$ such that $P_{0,n}(\overline{\omega}^{(n)}) = 0$, then there exists the smallest ban. If $P_{0,n-1}(\widetilde{\omega}^{(n-1)}) > 0$, then the assertion is proved.

Otherwise $P_{0,n-1}(\widetilde{\omega}^{(n-1)}) = 0$ and we can repeat the previous arguments. It follows that for every n the set $\overline{D}_{0,n}, \overline{D}_{0,n} \neq \emptyset$, is uniquely determined by least bans in such sense that all elements of $\overline{D}_{0,n}$ are obtained by all possible extensions of smallest bans to the length n. If S_n is a critical set for testing $H_{0,n}$ against $H_{1,n}$ and all vectors in S_n have probability zero in $P_{0,n}$ then S_n is defined by some smallest bans.

We give examples of bans for certain probability distributions.

Example 1. Let $X_i = \{0, 1\}, i = 1, 2, \ldots, \omega_0 = (1, 1, \ldots) \in X^{\infty}, P_0(\omega_0) = 1$. Then, for each n the set of the smallest bans consists of the vectors

$$A_n = \left\{ \overline{\omega}_1^{(1)} = (0), \overline{\omega}_1^{(2)} = (10), \ldots, \overline{\omega}_1^{(n)} = (\underbrace{1\ldots1}_{n-1}0) \right\}.$$

These smallest bans define the set of $2^n - 1$ vectors $\overline{\omega}^{(n)}$ of length n, for which $P_{0,n}(\overline{\omega}^{(n)}) = 0$.

As alternatives let's take a family of Bernoulli schemes of length n with a probability of unity $p = \theta, 0 < \theta < 1$.

By definition a critical set S_n of criterion is defined by bans if it includes all the extensions of the length n of some set of smallest bans. For the set S_n there exists a simple algorithm for computing the membership function for S_n. This algorithm calculates for each smallest ban its presence in the initial section of the vector, resulting in a statistical experiment.

Let us take a specific value $\theta = \frac{1}{2}$ and construct likelihood ratio test. Likelihood ratio function is

$$L(\overline{\omega}^{(n)}) = \frac{1}{2^n I(\overline{\omega}^{(n)} = \overline{\omega}_0^{(n)})}.$$

It is clear that for every $c > \frac{1}{2^n}$

$$P_{0,n}(L(\overline{\omega}^{(n)}) \ge c) = 0.$$

The power of criterion is equal

$$P_{\frac{1}{2},n}(L(\overline{\omega}^{(n)}) \ge c) = \frac{2^n - 1}{2^n} = 1 - \frac{1}{2^n}.$$

Let's compare the likelihood ratio test with the criterion, the critical set S_n of which is determined by the bans

$$P_{0,n}(S_n) = 0, P_{\frac{1}{2},n}(S_n) = 1 - \frac{1}{2^n}.$$

We compare the average number of steps for both criteria. When $H_{1,n}$ is true then the average number of steps to the first deviation from $\overline{\omega}_0^{(n)}$ equals to $2\left[1 - n(\frac{1}{2})^{n-1}\right]$.

However, the average number of operations to calculate the likelihood ratio is fixed and equals to n. In the case of optimization for function calculation $f(\overline{\omega}^{(n)}) = \overline{\omega}_0^{(n)})$ we obtain the previous estimation. Thus we have shown that the criterion defined by bans may possess the same best characteristics as the likelihood ratio test, but can be calculated with the same complexity or easier.

Example 2. Again, let $X = \{0,1\}, i = 1,2,\ldots$. Consider a simple homogeneous Markov chain with transition matrix

$$\begin{pmatrix} p & 1-p \\ q & 1-q \end{pmatrix}$$

and nondegenerate initial distribution. When $q = 1$ any smallest ban is defined by the first appearance of combination $(1,1)$ in a vector.

Note that in the presence of this vector in any fragment of $\overline{\omega}^{(n)}$ imply that $P_{0,n}(\overline{\omega}^{(n)}) = 0$.

In this case a criterion for testing the hypothesis

$$H_0 : 0 < p < 1, q = 1,$$

against the alternative

$$H_1 : 0 < p < 1, 0 < q < 1,$$

can be calculated only on the basis of the values of the ban. Algorithm for computing the data membership in the critical set is defined by an algorithm of a search for the ban in the observed sequence.

3 Conditions for the Existence of Consistent Sequences of Tests Depending on the Ban

Definition 2. Sequence of tests with critical sets S_n is called consistent (CST) [13] if

$$P_{0,n}(S_n) \longrightarrow 0, n \to \infty,$$

and for every $\theta \in \Theta$

$$P_{\theta,n}(S_n) \longrightarrow 1, n \to \infty.$$

Remind that the supports of P_0 and P_θ can be defined by the equalities

$$\Delta_0 = \lim_{n\to\infty} \Delta_{0,n},\ \Delta_{0,n} = D_{0,n} \times X_n^\infty,$$

$$\Delta_\theta = \lim_{n\to\infty} \Delta_{\theta,n},\ \Delta_{\theta,n} = D_{\theta,n} \times X_n^\infty,$$

and for every set $B_n \subseteq \prod_{i=1}^n X_i$, $\theta \in \Theta$, we have

$$P_{\theta,n}(B_n) = P_\theta(B_n \times X_n^\infty).$$

Theorem 1. *There exists CST for which all critical sets are defined by bans iff for every $\theta \in \Theta$*

$$P_\theta(\Delta_0) = 0.$$

Proof. Let in CST for each n the critical set S_n is defined by bans. Then $S_n \subseteq \overline{D}_{0,n}$. Denote $\Sigma_n = S_n \times X_n^\infty$. Then for each n

$$(\overline{D}_{0,n} \times X_n^\infty) \bigcap \Delta_{0,n} = \emptyset.$$

The sequence of sets $\Delta_{0,n}$ does not increase and, consequently, for each n

$$(\overline{D}_{0,n} \times X_n^\infty) \bigcap \Delta_0 = \emptyset.$$

The sequence $\overline{\Delta}_{0,n} = \overline{D}_{0,n} \times X_n^\infty$, so

$$\Lambda = \lim_{n\to\infty} \overline{\Delta}_{0,n} = \bigcup_{n=1}^\infty \overline{\Delta}_{0,n}.$$

Clearly, $\Lambda \bigcap \Delta_0 = \emptyset$. From the condition of consistency it follows that for every $\theta \in \Theta$

$$\lim_{n\to\infty} P_\theta(\Sigma_n) = 1.$$

Then

$$\lim_{n\to\infty} P_\theta(\overline{\Delta}_{0,n}) = P_\theta(\Lambda) = 1.$$

If there is a measurable set $A \subseteq \Delta_0$ such that $P_\theta(A) > 0$, then it follows from $\Lambda \bigcap A = \emptyset$ that

$$P_\theta(\Lambda \bigcup A) = P_\theta(\Lambda) + P_\theta(A) > 1.$$

Then for every $\theta \in \Theta$ we see that the probability $P_\theta(\Delta_0) = 0$.

Consider the $\overline{D}_{0,n}$ and $D_{\theta,n}$. By definition $P_{\theta,n}(D_{\theta,n}) = 1$ and assume that a set of bans for the sequence $P_{0,n}$ is not empty. We put

$$S_n = \overline{D}_{0,n},\ \Sigma_n = S_n \times X_n^\infty.$$

For all n the probability $P_{0,n}(S_n) = 0$. Sequence Σ_n does not decrease, and denote its limit by Σ. By the total probability formula we have

$$P_\theta(\Delta_\theta) = P_\theta(\Delta_\theta \bigcap \Sigma) + P_\theta(\Delta_\theta \bigcap \overline{\Sigma}).$$

From the construction of critical sets

$$\overline{\Sigma} = \overline{\bigcup_{n=1}^{\infty} \overline{\Delta}_{0,n}} = \Delta_0.$$

In addition, we have

$$P_\theta(\Delta_\theta \bigcap \overline{\Sigma}) = P_\theta(\Delta_\theta \bigcap \Delta_0) = P_\theta(\Delta_0) - P_\theta(\overline{\Delta}_\theta \bigcap \Delta_0).$$

On the right side of this equality the first probability is equal to 0 by the condition of the theorem, and the second is equal to 0, by definition Δ_θ. It follows that $P_\theta(\Delta_\theta \bigcap \Sigma) = 1$. Therefore, $P_\theta(\Sigma) = 1$. Hence, we find that

$$\lim_{n\to\infty} P_\theta(\Sigma_n) = 1.$$

This proves that the constructed sequence of tests is consistent.

It remains to show that in the conditions of the theorem the set of bans cannot be empty. Assume the contrary. Then for each n the set $\overline{D}_{0,n} = \emptyset$. This implies that for every n the set $D_{0,n} = \prod_{i=1}^{n} X_i$. Passing to the limit we get that $\Delta_0 = X^\infty$. This contradicts the condition that for every $\theta \in \Theta$ the probability $P_\theta(\Delta_0) = 0$. □

4 Properties of the Bans

It is easy to see that the smallest bans in $\prod_{i=1}^{n} X_i$ form a prefix code. In fact, any smallest ban cannot be part of another smallest ban by (1). This means that if $|X_i| = m$ then the lengths of smallest bans must satisfy an analogue of Kraft inequality [14].

We use the same idea to prove the following equations.

Theorem 2. *Let ν_h, $h = \overline{1;n}$, be the number of the smallest bans of length h for $P_{0,n}$. Then, for all h, $h = \overline{1;n}$, $|X_i| = m_i$, $i = \overline{1;n}$, we have*

$$\nu_1 \prod_{i=2}^{h} m_i + \nu_2 \prod_{i=3}^{h} m_i + \cdots + \nu_{h-1} m_h + \nu_h + |D_{0,h}| = \prod_{i=1}^{h} m_i. \tag{2}$$

Proof. It is clear that for $h = 1$ and for every point $\overline{\omega}^{(1)}$ of X_1 we have $P_{0,1}(\overline{\omega}^{(1)}) = 0$ or $P_{0,1}(\overline{\omega}^{(1)}) > 0$. In the first case the point belongs to the smallest bans and in the second case it belongs to $D_{0,1}$. Then $\nu_1 + |D_{0,1}| = m_1$. When $h \le n$ the number of vectors of the length h is equal to $\prod_{i=1}^{h} m_i$. If there is the smallest ban $\overline{\omega}_n^{(k)}$, $k < h$, of the length k, then the number of vectors which have probability zero and begin with $\overline{\omega}^{(k)}$ equals to $\prod_{i=k+1}^{h} m_i$. If the total number of the smallest bans of the length k equals to ν_k, then at the length h we have $\nu_k \prod_{i=k+1}^{h} m_i$ vectors of probability zero and which have in the beginning some smallest ban of the length k. Then summing all together we get

$$\nu_1 \prod_{i=2}^{h} m_i + \nu_2 \prod_{i=3}^{h} m_i + \cdots + \nu_{h-1} m_h + \nu_h + |D_{0,h}| = \prod_{i=1}^{h} m_i.$$

□

In the case $m_i = m$ for all i we have

$$\nu_1 m^{n-1} + \cdots + \nu_{n-1} m + \nu_n + |D_{0,n}| = m^n$$

for all $n = 1, 2, \ldots$.

In [11, 15] we introduced and studied the notion of strictly consistent sequence of tests for $X_i = X$, $i = 1, 2, \ldots$.

Definition 3. The sequence of tests with critical sets $S_n, n = 1, 2, \ldots$, is called strictly consistent (SCST) [11] iff

$$P_{0,n}(S_n) \longrightarrow 0, \, n \to \infty,$$

and for every $\theta \in \Theta$

$$P_{\theta,n}(S_n) \longrightarrow 1, \, n \to \infty.$$

and

$$\bar{S}_n \times X^{\infty} \supseteq \bar{S}_{n+1} \times X^{\infty},$$

where

$$\bar{S}_n = X^n \setminus S_n, \, n = 1, 2 \ldots$$

In our model the sequence of tests $S_n, n = 1, 2, \ldots$, is strictly consistent if it is consistent and the sequence $S_n \times X_n^{\infty}$ is not decreasing.

Theorem 3. *Strictly consistent sequence of tests exists iff there is a CST for which all critical sets are defined by smallest bans.*

Proof. Assume that there exists CST determined by smallest bans. If the set of all smallest bans is finite, then according to (2), starting with some h all ν_h equal to 0. Let it be for all $h \geq h_0$. We have

$$\nu_1 \prod_{i=2}^{h} m_i + \cdots + \nu_{h_0} \prod_{i=h_0+1}^{h} m_i + |D_{0,h}| = \prod_{i=1}^{h} m_i,$$

so

$$\prod_{i=h_0+1}^{h} m_i \left(v_1 \prod_{i=2}^{h_0} m_i + \cdots + v_{h_0} + |D_{0,h_0}| \right) - |D_{0,h_0}| \prod_{i=h_0+1}^{h} m_i + |D_{0,h}| = \prod_{i=1}^{h} m_i.$$

Then

$$\prod_{i=1}^{h} m_i - |D_{0,h_0}| \prod_{i=1}^{h_0+1} m_i + |D_{0,h}| = \prod_{i=1}^{h} m_i.$$

It means that

$$D_{0,h} = D_{0,h_0} \times \prod_{i=1}^{h_0+1} X_i,$$

and $\Delta_0 = \Delta_{0,h_0} \times X_{h_0}^{\infty}$.

By the theorem 1 $P_\theta(\Delta_0) = 0$ for all θ. It follows that $P_\theta(\overline{D}_{0,h_0} \times X_{h_0}^{\infty}) = 1$ for all θ. Then

$$S_n = \overline{D}_{0,h_0} \times \prod_{i=h_0+1}^{n} X_i, \, n \geq h_0,$$

satisfy equations $P_{0,n}(S_n) = 0$ and $P_{\theta,n}(S_n) = 1$. This sequence of tests is CST. It is clear that $S_n \times X_n^{\infty}$ is not decreasing sequence of sets. We can also define S_i, $i < h_0$, as projections of S_{h_0} to the first coordinates in $\prod_{k=1}^{i} X_k$. Then this sequence satisfies definition of strictly CST.

Let the set of smallest bans is countable. We have $P_{0,n}(\overline{D}_{0,n}) = 0$ for all $n = 1, 2, \ldots$. According to the theorem 1

$$\lim(\overline{D}_{0,n} \times X_n^{\infty}) = \overline{\Delta}_0,$$

and for every $\theta \in \Theta$

$$\lim_{n\to\infty} P_\theta(\overline{D}_{0,n} \times X_n^{\infty}) = 1.$$

That means that the sequence of tests with critical sets $S_n = \overline{D}_{0,n}$ is CST.

All smallest bans in $\overline{D}_{0,h}$ are also in $\overline{D}_{0,h+1}$, $h = 1, 2 \ldots$. Then all conditions of SCST are fulfilled for tests with critical sets S_n, $n = 1, 2, \ldots$.

Let SCST exists with critical sets S_n and $S_n \times X_n^{\infty}$ is an open set. Then

$$\lim_{n\to\infty} (S_n \times X_n^{\infty}) = S$$

is an open set and $P_0(S) = 0$. That means that there exists a closed set $\overline{S}$ and $P_0(\overline{S}) = 1$.

It follows from CST that $\forall \theta \in \Theta, P_\theta(\overline{S}) = 0$. The set Δ_0 is the support of P_0. By the definition it is intersection of all closed sets $\mathscr{D}$ with $P_0(\mathscr{D}) = 1$. It follows that $\Delta_0 \subseteq \overline{S}$. Then $\forall \theta \in \Theta, P_\theta(\Delta_0) \leq P_\theta(\overline{S}) = 0$. Now we use the theorem 1 to conclude that there exists CST defined by the smallest bans. □

5 Consistent Sequences of Tests for Finite Markov Chains

Let all measures be finite homogeneous Markov chains which are defined by initial positive distributions $\overline{P}_0$, $\overline{P}_\theta$ and matrixes $\mathbf{P}_0 = \left\|P_{ij}^0\right\|$, $\mathbf{P}_\theta = \left\|P_{ij}^\theta\right\|$, We say that P_{ij}^θ contradicts P_{ij}^0 if $P_{ij}^0 = 0$ but $P_{ij}^\theta > 0$.

Theorem 4. *There exists CST for which all critical sets are defined by bans iff for every $\theta \in \Theta$ and every ergodic class $\mathbf{P}_0$ there exists (ij) in it for whichP_{ij}^θ contradicts P_{ij}^0.*

6 Conclusion

An open question in ban theory is detection of bans in P_0. It is a simple problem in homogeneous Markov chains but sometimes the problem may be hard.

Here is an example. Let X be an alphabet, A^* be a set of all finite words in alphabet X. $\widetilde{A}$ is a rich language $\widetilde{A} \subseteq A^*$, A^∞ is a set of the infinite texts in $\widetilde{A}$ and A_n are projections of A^∞ to the first n coordinates. Let $P_{0,n}(\overline{\omega}^{(n)}) > 0$ if $\overline{\omega}_0^{(n)} \in A_n$ and $P_{0,n}(\overline{\omega}^{(n)}) = 0$ if $\overline{\omega}_0^{(n)} \notin A_n$. To determine all smallest bans here we should calculate that the word belongs to the language. It is not difficult to build a language in which such a problem is hard.

The idea to search bans by statistical methods seems to be perspective. Here the problem consists of appearing of a false ban, and the possibility of acceptance of the false decision connected with it. It is obvious that theory transferring onto this case will need some additional restrictions on probability measures.

Other direction of researches is generalization of a considered approach onto a case of a complex zero hypothesis.

In practical sense the offered approach is convenient for applying in systems of consecutive control. Really, in the considered case the probability of acceptance of an alternative hypothesis when the zero hypothesis is true, is equal to zero. It means an absence of false alarms in the monitoring system.

There are some other open problems. Nevertheless when we find some bans it is much easier to solve statistical problems.

Acknowledgements This work was supported by Russian Foundation for Basic Research, project 10-01-00480.

References

1. Wald, A., Wolfowitz, J.: On a test whether two samples are from the same population. Ann. Math Stat. **11**, 147–162 (1940)
2. Hoeffding, W.: The role of assumptions in statistical decission. In: Proceedings of the 3rd Berkeley Symposium on Mathematical Statistics and Probability, vol. 1, pp. 105–114. University of California Press, CA (1956)
3. Schmetterer, L.: Einfiihrung in die mathematische Statistik. Springer, Wien-New York (1966)
4. Schmerkotte, H.: Interdependence between consistent tests and estimates. Z. Wahrscheinlichkeitstheorie verw. Geb. **16**, 293–306 (1970)
5. Pfanzagl, J.: On the existence of consistent estimates and tests. Z. Wahrscheinlichkeitstheorie verw. Geb. **10**, 43–62 (1968)
6. Grusho, A., Grusho, N., Timonina, E.: Theorems of non-existence of the consistent sequences of tests in some discrete problems. Discrete Math. **20**(2), 25–31 (2008)
7. Grusho, A., Chentsov, V., Timonina, E.: Existence of consistent sequences of statistical tests in the discrete statistical problems at complex null hypothesis. Informatics Appl. 2(2), 64–66 (2008)
8. Grusho, A., Grusho, N., Timonina, E.: Complexity and consistency of statistical criteria. In: Systems and Means of Informatics. Special Issue. Mathematical and Computer Modeling in Applied Problems, pp. 32–39. IPI RAS, Moscow (2008)
9. Grusho, A., Grusho, N., Timonina, E.: Problems of modeling in the analysis of covert channels. In: Proceedings of the 5th International Conference on Mathematical Methods, Models, and Architectures for Computer Network Security (MMM-ACNS 2010), pp. 118–124. Springer, Berlin (2010)
10. Grusho, A., Timonina, E.: Prohibitions in discrete probabilistic statistical problems. Discrete Math. Appl. **21**(3), 275–281 (2011)
11. Grusho, A., Timonina, E.: Some relations between discrete statistical problems and properties of probability measures on topological spaces. Discrete Math. Appl. **16**(6), 547–554 (2006)
12. Bourbaki, N.: Topologie Générale. Russian translation. Science, Moscow (1968)
13. Lehmann, E.L.: Testing Statistical Hypotheses (Springer Texts in Statistics), 2nd edn. Springer, Berlin (1997)
14. Fano, R.: Transmission of Information: A Statistical Theory of Communications. Russian translation. Mir, Moscow (1965)
15. Grusho, A., Kniazev, A., Timonina, E.: Detection of illegal information flow. In: Proceedings of the 3rd International Workshop on Mathematical Methods, Models, and Architectures for Computer Network Security (MMM-ACNS 2005), pp. 235–244. Springer, Berlin (2005)

Impact Assessment Through Collaborative Asset Modeling: The STORM-RM Approach

Theodoros Ntouskas, Panayiotis Kotzanikolaou, and Nineta Polemi

Abstract Existing Risk Management (RM) methodologies are mainly expert driven and require a large number of interviews with the security experts, which makes rather inefficient to take into account the knowledge from all the organization's participants. In this paper we extend the STORM-RM multi-criteria group decision-making methodology. More specifically, we propose specific asset and user models, which make use of the AHP multi-criteria decision-making methodology in order to identify the organization's assets and calculate their potential security impacts.

Key words Impact assessment • Asset modeling • AHP • Multi-criteria decision making

1 Introduction

Existing information risk management (RM) methodologies (e.g. OCTAVE [12], CRAMM [4], EBIOS [6], ISO-15408-1 [7], Mehari [10], MAGERIT [9], Austrian IT Security Handbook [2], BSI-Standard 100-3 [3], Dutch A & K Analysis [5]) are mainly expert driven. Although the input required for the risk assessment is based on information provided by selected participants from the organization hosting the Information System in question (such as users, technical personnel and managers), usually information gathering is a time- and resource-consuming process, which requires the active involvement of security experts in various interviews.

T. Ntouskas (✉) • P. Kotzanikolaou • N. Polemi
Department of Informatics, University of Piraeus, Karaoli and Dimitriou 80, 185 34 Piraeus, Greece
e-mail: tdouskas@unipi.gr; pkotzani@unipi.gr; dpolemi@unipi.gr

A. Migdalas et al. (eds.), *Optimization Theory, Decision Making, and Operations Research Applications*, Springer Proceedings in Mathematics & Statistics 31,
DOI 10.1007/978-1-4614-5134-1_21,

In a previous work [11] a new risk management methodology, the STORM-RM methodology was proposed. The STORM-RM methodology treats information risk management as collaborative decision-making problem, by combining the Analytic Hierarchy Process (AHP) [14] in various phases of the risk assessment, with well-known security management standards (ISO27001 [8] and AS/NZS 4360 [1]). The main goal of STORM-RM is to provide a user-driven methodology and to reduce the active involvement of security experts in the various phases of an assessment. Since risk assessment is a complicated process depending on multiple criteria, designing a user-driven methodology requires the design of well-defined procedures that will enable the collaboration between the various participants. In our previous work [11] the main phases and steps of the STORM-RM methodology were presented. Here we specifically define the first phase of methodology, the Cartography phase, by defining models for the representation of the dependencies between participating users (*User Group* model) and assets (*Asset Group* model). We also define in detail the calculation of the impact values for each asset required in the Impact Assessment phase, based on the User and Asset Group models.

The rest of the paper is organized as follows: Section 2 defines in detail the STORM-RM Organization Cartography Phase and its basic steps. Section 3 presents the STORM-RM Impact Assessment Phase and the functions that handle the opinions of various organization's participants and finally, and Sect. 4 draws conclusions and directions for further research.

2 The STORM-RM Organization Cartography Phase

In contrast to decision-maker-driven methodologies, the STORM-RM methodology is user driven. The various participants (users) of the organization will collaborate, in order to evaluate the information security risks of the organization. However, the collaboration of multiple users with different organization views and different knowledge on ICT security issues introduces increased complexity and high uncertainty on the accuracy of the results. It would be a rather strong and unrealistic assumption to expect that the users will understand all the steps of a risk assessment methodology, the required level of precision and abstraction in their expected input and finally they will successfully coordinate their participation in the risk assessment.

The goal of this phase is to *reduce the complexity* of the risk assessment and *minimize the involvement* of each participant, only to those phases of the methodology and those systems of the organization that they are expected to understand. Since not all the opinions of all the participants in the risk assessment are equally important, this step of the methodology also aims to define the proper weights of the various actors for each phase of the risk assessment.

In order to minimize the involvement of the participants to the most appropriate phases of the risk assessment, the STORM-RM Cartography phase defines models that capture the roles between the various participants, the relation between the various assets and more importantly, the relations between the participants and the assets of the system. Once the relations between users and assets are defined, then the participants will only be involved into those steps of each phase of the methodology (impact, threat and vulnerability analysis), where their opinions will be the most beneficial. Bellow, we will first describe the basic structures used to describe these relations, mainly the User Group Model and the Asset Group Model. These models are pre-defined by the STORM-RM methodology and are utilized during the Cartography phase. Then, we will describe the basic steps that are executed during the Cartography phase, which include the assignment of users and assets to inter-related groups.

We note that the initial steps of the Cartography phase are executed under the co-ordination of the Security Officer of the organization. Then, the organization will be able to run the rest of the phases in a user self-driven approach, according to the decisions taken during this phase.

2.1 Modeling Users and Assets in STORM-RM

In order to model the roles of the participants as well as their relation with particular assets, the STORM-RM cartography phase defines the *User Group Model* and the *Asset Group Model*, as described below.

2.1.1 User Group Model

The User Group model is used in order to categorize the users of the organization (at least the ones participating in the risk assessment) into a group which will determine the role of the user to all the phases of the analysis. The STORM-RM User Group Model defines two levels of user groups, as shown in Fig. 1.

Note that although the STORM-RM pre-defines the above model, it is possible to modify this according to the specific details of the organization running the assessment. For example, the Application administrators sub-group can be further divided into different sub-groups for different end-user applications. In the same way, the Organization Users can be further divided into multiple groups, according to the different end-user services provided by the organization; or the External Users can be omitted, if this sub-group is not applicable for a particular organization. The proposed model, however, covers a wide range of organizations.

2.1.2 Asset Group Model

The Asset Group Model is an abstract model used in order to categorize all the ICT assets of the organization and capture their dependencies, as shown in Fig. 2. These dependencies will be used in the following phases of the methodology.

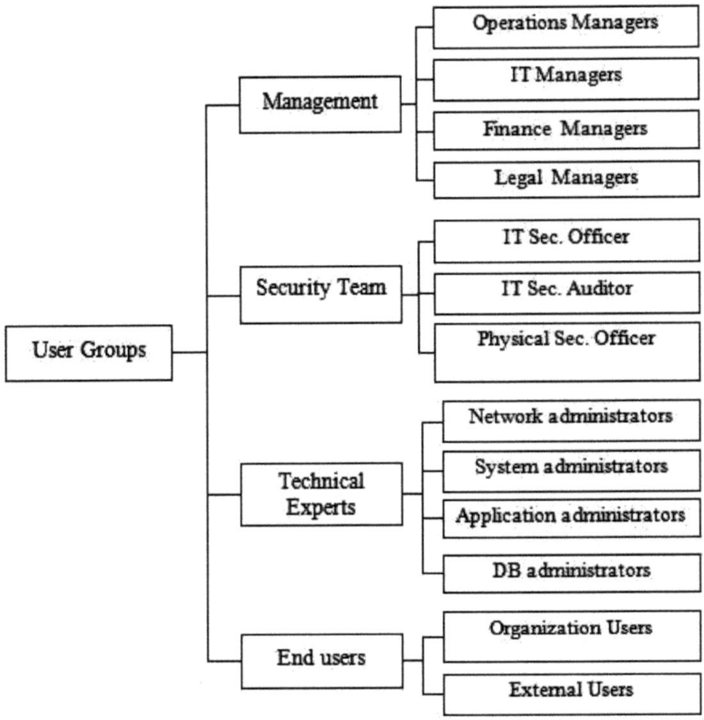

Fig. 1 The user group model of the STORM-RM methodology

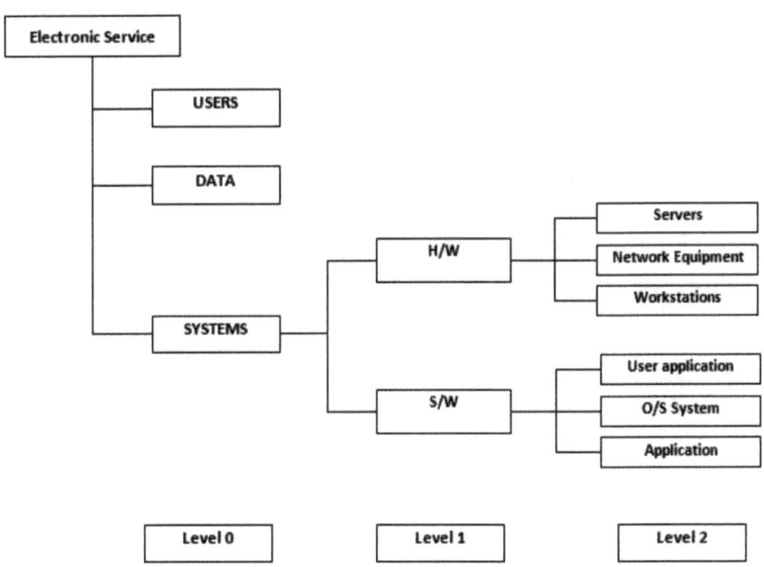

Fig. 2 The asset group model of the STORM-RM methodology

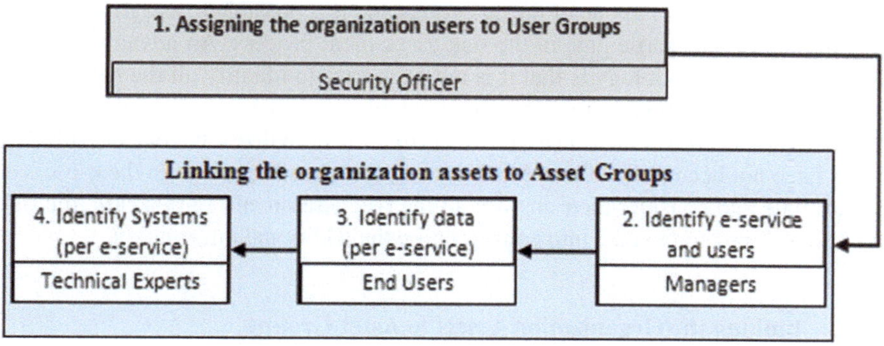

Fig. 3 The basic steps of the Cartography phase and their participants

The STORM-RM methodology defines as a key element of the asset model, the Electronic Services of the organization. Since all the ICT assets are used as supporting tools for the *Electronic Services*, it is natural to assume that all the ICT assets will support one or more Electronic Services. The STORM-RM methodology defines the Asset Group model in three levels. Each Electronic Service is provided by a combination of users, data and IT Systems (Level 0). We use the abstraction of a *System* in our model, in order to represent the combination of one or more hardware and software ICT assets (Level 1). These in turn can be further analyzed in various sub-types of ICT assets (Level 2). For example, the hardware assets can be divided into servers, workstations, network equipment, etc. The output data of this Asset Group Model can be later used in order to capture the dependencies between business-wise impacts and IT systems. Finally, it also includes the users that are linked with a particular electronic service. These users can be members of various groups in the User Group model. In this way it will be possible to find the appropriate actors to participate in the following phases of the risk assessment, for each particular asset.

2.2 *Identifying the Users and Assets of the Organization*

Based on the User Group and the Asset Group models defined in the STORM-RM method, in the Cartography phase the key users of the organization that will participate in the risk assessment, as well as the assets of the organizations will be identified. Figure 3 describes the basic steps of the Organization Cartography phase, along with the participants involved in each step.

2.2.1 Assigning the Organization Users to User Groups

During this step, the Security Officer will assign the users of the organization to one of the sub-groups defined in the User Group model. It is not necessary to assign all the organization users to one of the categories and it is possible to assign only those key users that will participate in the risk assessment process. An advantage of the STORM-RM methodology is that it is not necessary to identify all the participants at the beginning (although a proper selection will reduce the time needed to perform the assessment). If in a later phase the Security Officer realizes that one or more key users have not been utilized in the assessment, it is possible to assign these users at a later time and consider their answers in the risk assessment. In this case, the risk will be re-evaluated to take into consideration the additional information.

2.2.2 Linking the Organization Assets to Asset Groups

During this step, the various participants (already assigned to the User Groups) will collaborate in order to link the information assets to Asset Groups. This is executed in three sub-steps:

1. In the first sub-step, the participants belonging to the Management group are responsible to identify the Electronic Services belonging to the areas of their responsibility, as well as the key users which are involved in each Electronic Service. Although managers are not aware of the specific implementation details, they have an overall view of their electronic services and key personnel related with these e-services.
2. After the Electronic Services and the key users of each service have been identified, the end users who have been assigned to each electronic service will identify the required Data that are related to each electronic service. Again, the end users belonging to each business unit will be the most appropriate to identify the above assets, since they consist part of their everyday work.
3. Finally, after the electronic services have been identified, the Technical Experts (administrators) will identify the Systems (with details for their Hardware/Software assets) which are related to the provision of each electronic service. Although the administrators may not be aware of the business view of each electronic service, they will be aware of the technical details related to the provision of each electronic service. Note that if the hardware assets of the organization are hosted in more than one location, information related to the physical location of each hardware asset will also be provided in this step.

At the end of this phase, the assets that will be assessed and the users of the organization will be mapped to Asset Groups, based on the Asset Group model described in Fig. 2.

3 The STORM-RM Impact Assessment Phase

As in the Cartography phase, in order to minimize the involvement of the participants, each participant will only assess the impact of those assets of the organization that he is familiar with. In addition, in order to increase the accuracy of the impact

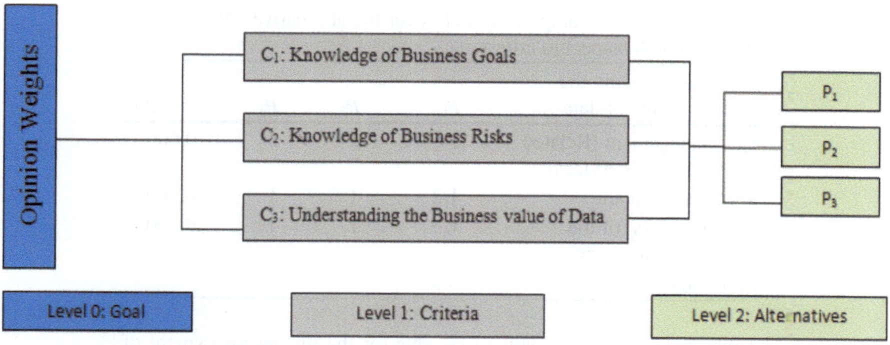

Fig. 4 AHP model for the selection of opinion weights for the Impact assessment of data assets

assessment, the STORM-RM methodology enables the participation of various users in the assessment of each asset. According to the User Group of each participant, a different view of each asset will be considered.

3.1 *Assigning and Weighting User Groups to Asset Categories*

Each asset will be assessed for the impact caused due to the violation of a security property. The Data assets are assessed for their possible unavailability, disclosure and modification, while the Systems are evaluated for their significance for the provision of electronic services to which they are connected. It is important to mention that the H/W and S/W assets will not be assessed for their security importance and their Impact value will derived from the Impact value of the System that they are connected to according to the Asset Model (Fig. 2).

So in order to reduce the complexity derived from the involvement of different users with different views, the Impact Assessment phase utilizes the AHP [14] methodology. For each asset category the STORM-RM [11] defines which User Groups will participate in the impact assessment and what is the weight of each User Group involved.

3.1.1 Data Assets

The opinions of the related Managers (P_1), Security Team (P_2) and End Users (P_3) are considered. Their weights are based on the AHP methodology (Fig. 4) taking into account different criteria such as Knowledge of Business Goals, Knowledge of Business Risks, Understanding the Business values of Data.

Considering that the opinions of the Managers are three times more important than the opinions of the other participants, since the manager will be aware of the business impacts of the loss of security of a Data asset, and consequently of the related electronic service, Table 1 shows the results of the pairwise comparison with respect to the criterion C_3*: Understanding the Business value of Data*. However, the

Table 1 Pairwise comparison matrix for the alternatives (Participants P_1,P_2,P_3) with respect to criterion C_3

C_3: Understanding the business value of data	P_1	P_2	P_3	Weights
P_1 : Management (Related with the e-services)	1	3	3	0.600
P_2 : Security team	1/3	1	1	0.200
P_3 : End users (of the particular e-services.	1/3	1	1	0.200
Consistency ratio:	0.00			

Table 2 Final results (opinion weights for the impact assessment of data assets) shown as normalised priorities

Criteria	C_1	C_2	C_3	
Criteria priorities	0.333	0.333	0.333	Weights
P_1 : Management (Related with the e-services)	0.574	0.685	0.600	0.620
P_2 : Security Team	0.286	0.179	0.200	0.222
P_3 : End users (of the particular e-services.	0.140	0.136	0.200	0.159

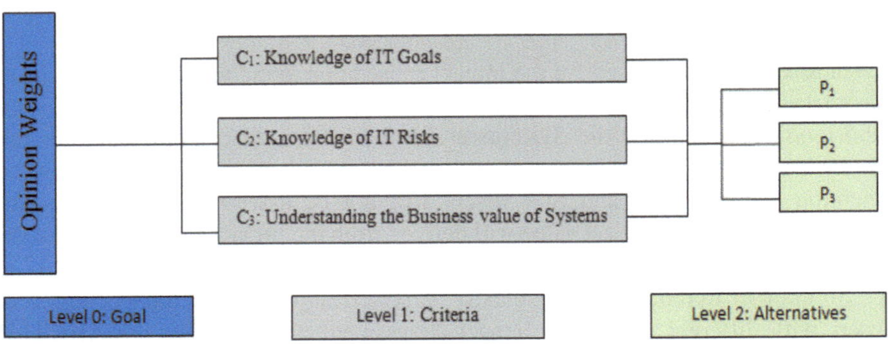

Fig. 5 AHP model for the selection of opinion weights for the impact assessment of systems

opinions of the End-Users related to the particular electronic service should also be considered, as well as the opinion of the security team who has an overall view of the organization. The final opinion weights are shown in Table 2.

3.1.2 Systems

In a similar way, the opinions of the Technical Experts, IT Manager and Security Team are considered for the Systems taking into account different criteria (e.g. Knowledge of IT Goals, Knowledge of IT Risks, Understanding the Business value of Systems) as shown in Fig. 5 and their weights of participation are shown in Table 3.

Table 3 Final results (opinion weights for the impact assessment of Systems) shown as normalised priorities

Criteria	C_1	C_2	C_3	
Criteria priorities	0.333	0.333	0.333	Weights
P_1 : Technical experts (Related with the H/W - S/W)	0.480	0.501	0.405	0.462
P_2 : IT manager	0.405	0.310	0.480	0.398
P_3 : Security team	0.115	0.189	0.115	0.140

3.2 *Calculating the Impact of Each Asset*

After the participation and the weights of each User Group have been assigned for each asset category of Level 0 (i.e. Data Assets or Systems) and the involved participants have provided their opinions, the impact of each asset should be evaluated.

Each participant is asked through dynamically generated questionnaires, to provide his opinion for the possible impact(s) of each asset he has been assigned to in the previous phase. The data assets are assessed for their impact due to possible unavailability ($I_{\text{un}}(A)$), disclosure ($I_{\text{dis}}(A)$) and modification ($I_{\text{mod}}(A)$).

The use of the System abstraction in the asset group model will be exploited in order to minimize the burden of the impact assessment phase for the H/W and S/W assets. Instead of performing a time consuming impact assessment for each H/W and S/W asset, we will only define a *Correlation Factor* CF for each System defined within an Electronic Service.

These different ways of the calculation of the Impact values of Data Assets and Systems (and as a result of H/W and S/W) are described in detail below.

3.2.1 Calculating the Impact of Each Data Asset

The impact scale and the possible consequences asked in this phase are based on STORM-RM scale [11] (1 = Very Low, 2 = Low, 3 = Medium, 4 = High, 5 = Very High consequences). Since multiple users from different User Groups will provide their opinion, the final Impact value for each Data asset, is evaluated as follows.

Let G_j, $j \in [1,m]$ denote the jth User Group and let $U_{j,i}$ denote the ith user of the User Group j. We assume that each Data asset A may have one or more Impact values, according to its category. Without loss of generality, we examine the unavailability impact of an asset A.

Since in the STORM-RM, the impact of an asset is based on the opinions of various users belonging to different groups, the impact value of Data asset A shall consider the opinion of each related User Group. We denote as $op_{G_j}(I_{\text{un}}(A))$ the opinion of the User Group G_j for the impact value regarding the unavailability of the Data asset A.

In turn, the value $op_{G_j}(I_{\mathrm{un}}(A))$ is computed, based on the opinions of each participating user belonging to the User Group G_j. We denote the view of a user $U_{j,i}$ for the Impact value of the Data asset A as $op_{U_{j,i}}(I_{\mathrm{un}}(A))$. Since a user may have a different opinion for the unavailability impact of a Data asset, we consider that $op_{U_{j,i}}(I_{\mathrm{un}}(A))$ is the maximum reply of the user $U_{j,i}$, for all the possible examined consequences. For example, if $U_{j,i}$ assesses that the unavailability of the asset A will have a value equal to 4 for the consequence type 'loss of reputation' and a value equal to 5 for the consequence 'economic loss', then $op_{U_{j,i}}(I_{\mathrm{un}}(A)) = \max(4,5) = 5$. Based on the user opinions, the group opinion of the User Group G_j for the unavailability impact of the Data asset A is computed as:

$$op_{G_j}(I_{\mathrm{un}}(A)) = \sum_{i=1}^{n} \frac{op_{U_{j,i}}(I_{\mathrm{un}}(A))}{n}. \tag{1}$$

Let w_j denote the weight of the User Group G_j (as computed in Sect. 3.1). Then the estimated unavailability impact of the Data asset A is computed as:

$$I(A) = \sum_{j=1}^{m} w_j \cdot op_{G_j}(I(A)). \tag{2}$$

3.2.2 Calculating the Impact of Each System

In order to calculate the Impact value of each System and as a result of all the connected assets (i.e. H/W and S/W assets based on the Asset Model), the STORM-RM methodology defines the Correlation Factor (CF) which expresses the correlation between the System and the Electronic service(s) that it supports. More specifically we define three different CF values between a System (S) and an Electronic Service (E):

- The unavailability correlation factor ($CF_{\mathrm{un}}(S,E)$) : This implies the correlation between the unavailability impact caused to the electronic service E, due to the unavailability caused to one or more H/W or S/W assets of the system S.
- The disclosure correlation factor ($CF_{\mathrm{dis}}(S,E)$) : This implies the correlation between the disclosure impact caused to the electronic service E, due to the disclosure caused to one or more H/W or S/W assets of the system S
- The modification correlation factor ($CF_{\mathrm{mod}}(S,E)$) : This implies the correlation between the modification impact caused to the electronic service E, due to the modification caused to one or more H/W or S/W assets of the system S.

The appropriate users define values for these CF of each System that he has been assigned to in the previous phase (Cartography Phase). In particular, the users define the percentage of significance (1 = Very Low (or correlation nearly 0%), 2 = Low (or correlation nearly 25%), 3 = Medium (or correlation nearly 50%), 4 = High (or correlation nearly 75%), 5 = Very High (or correlation nearly 100%)) of each

System for the availability ($CF_{\text{un}}(S,E)$), confidentiality ($CF_{\text{dis}}(S,E)$) and integrity ($CF_{\text{mod}}(S,E)$) of the correlated Electronic Service (s). The calculation of the group and final CF values of each System are performed with the same way with the calculation of Impact values of Data assets (as it described in Sect. 3.2.1). The assets of Level 3 (i.e. H/W and S/W assets) that are connected with each System (S) inherit these CF values as their Impact values for possible unavailability, disclosure and modification.

3.3 *Implied Values and Finalization*

Based on the previous step, the impact values of all the assets have been evaluated. However, these values are isolated and do not consider the interdependencies between the different assets. In this step the impact evaluation of each asset is finalized based on the interdependencies between the assets. In this way the indirect implied impacts of each asset are calculated. The implied asset values are computed based on the following rules (applied in the particular order):

Implied values. If the unavailability of an asset implies the unavailability on another asset, then the worst case will be considered for the unavailability impact of the asset in question. For example, if the unavailability impact for a data asset has been evaluated as 4.8 and the unavailability of an electronic service that depends on this data asset has been evaluated as 5, then the final unavailability impact of the data asset will be 5.

Assets belonging to multiple asset groups. If an asset belongs to more than one Asset Groups, then the procedure described in Sect. 3.2 will have been performed more than one time for the particular asset. In this case, the worst impact value(s) of the particular asset will be considered as the final impact value(s).

4 Conclusions

Risk management methodologies should take into consideration the views of many users with different roles, in order to increase the accuracy of the results. This, however, increases the complexity of the risk assessment process, turning it into an expert-driven process. In this paper we have presented the STORM-RM methodology which attempts to reduce this complexity, by properly applying the AHP methodology in the organization cartography and impact assessment phases. In this way, the roles and the weights of the participants are fully defined and in this way the risk assessment is shifted to a more user-driven process. In order to fully exploit the advantages of multi-criteria approach such as the AHP, we plan to consider its applicability in the rest of the phases of the STORM-RM, such as the threat/vulnerability assessment and countermeasures selection. The STORM-RM methodology will be implemented as service at the S-PORT system [13] and will be tested by three Greek commercial Ports (Piraeus Port Authority S.A., Thessaloniki Port Authority S.A, Municipal Port Fund Mykonos).

Acknowledgements This work has been performed in the framework of the GSRT/SYNERGASIA/ S-Port project (09SYN-72-650) (http://s-port.unipi.gr).

References

1. AS/NZS 4360. Risk management standards australia, Strathfield (1999)
2. Austrian IT Security Handbook, Austrian federal chancellery (2004)
3. BSI-Standard 100-3. Risk analysis based on it-grundschutz (2005)
4. CRAMM. Ccta risk analysis and management method, cramm version 5.2 information security toolkit (2003)
5. Dutch A&K Analysis (1996)
6. Ebios. Expression des besoins et identification des objectifs de securite (2004)
7. ISO/IEC:15408-1. Information technology - security techniques - evaluation criteria for it security – part 1: Introduction and general model (2005)
8. ISO/IEC:27001. Information technology - security techniques - information security management systems - requirements (2005)
9. MAGERIT. Methodology for information systems risk analysis and management. Public Administration Ministry (2005)
10. Mehari. Méthode harmonisée d' analyse de risque (2007)
11. Theodoros Ntouskas and Nineta Polemi. STORM-RM: A collaborative and multicriteria risk management methodology. To appear in Int. J. Multicriteria Decision Making.
12. OCTAVE. Octave method implementation guide version 2.0. Carnegie Mellon University, June (2001)
13. S-PORT. S-port project.
14. Saaty, T.L.: Decision making with the analytic hierarchy process. Int. J. Service Sci. **1**, 83–98 (2008)

Testing the Homoskedasticity/Heteroskedasticity of the Errors Using the White Test: Pattern Classification by k-Variances and Informational Criteria

Daniel Ciuiu

Abstract In this paper we will test the homoskedasticity/heteroskedasticity of the errors for a linear regression model using the White homoskedasticity test. In the case of heteroskedasticity we use the k-variance algorithm to classify the data such that all the classes are homoskedastic. The informational criteria analogues to the Akaike and Schwartz criteria are used to choose the best classification.

Key words Homoskedasticity • k-variance • Informational criteria

1 Introduction

Consider n points in $\mathbb{R}^{k+1}$, $X^{(1)},\ldots,X^{(n)}$, where $X^{(i)}=\left(X_1^{(i)},X_2^{(i)},\ldots,X_k^{(i)},Y_i\right)$. The equation of the regression hyper-plane, used in [1] to classify patterns, is (see [7])

$$H: Y = A_0+\sum_{i=1}^{k} A_i X_i \text{ such that} \tag{1}$$

$$\sum_{i=1}^{n} u_i^2 \text{ is minimum,} \tag{1'}$$

D. Ciuiu (✉)
Technical University of Civil Engineering, Bucharest, Bd. Lacul Tei No. 122-124, Sector 2, Bucharest, Romania

Romanian Institute for Economic Forecasting, Calea 13 Septembrie No. 13, Sector 5, Bucharest, Romania
e-mail: dciuiu@yahoo.com

A. Migdalas et al. (eds.), *Optimization Theory, Decision Making, and Operations Research Applications*, Springer Proceedings in Mathematics & Statistics 31,
DOI 10.1007/978-1-4614-5134-1_22,

where the residues u_i are given by the formula

$$u_i = Y_i - A_0 - \sum_{j=1}^{k} A_j X_j^{(i)}. \tag{1''}$$

For the estimation of A_i from (1) we have to solve the following linear system (see [7]):

$$\sum_{j=0}^{k} \overline{X_i \cdot X_j} \cdot A_j = \overline{X_i \cdot Y},\ i = \overline{0,k}, \tag{2}$$

where $\overline{X_0 \cdot X_i} = \overline{X_i}$ and $\overline{X_0^2} = 1$.

The polynomial model is in fact the multilinear model (1) with the explanatory variables $X_1 = X, X_2 = X^2$ and so on (see [2]).

For the obtained estimators of A_i using (2) and for the residues u_i we have the following assumptions (see [4, 10]):

1. The estimators of A_i are linear.
2. The estimators of u_i have the expectation 0 and the same variance (homoskedasticity).
3. The estimators of u_i are normal.
4. The random variables u_i are independent.

From the above assumptions and from Gauss—Markov theorem, we obtain the following properties (see [4, 10]):

1. The estimators of A_i are consistent.
2. The estimators of A_i are unbiased.
3. The estimators of A_i have the minimum variance.
4. The estimators of A_i have the maximum likelihood.

Denoting by VT the total variance of the resulting variable Y, by VTM the total variance explained by the model, and by VTR the total variance of the residues we obtain

$$\begin{cases} VT = \sum\limits_{t=1}^{n} \left(Y_t - \overline{Y}\right)^2 \\ VTM = \sum\limits_{t=1}^{n} \left(\widehat{Y}_t - \overline{Y}\right)^2, \\ VTR = \sum\limits_{t=1}^{n} u_t^2 \end{cases} \tag{3}$$

where $\widehat{Y}_t = \widehat{A}_0 + \sum_{i=1}^{k} \widehat{A}_i X_{i;t}$, and $\widehat{A}_i$ is the estimator of A_i by the less squares method.

We know (see [4, 10]) that if the regression contains the term A_0 and the parameters A_i are estimated by the less squares method then

$$VT = VTM + VTR. \tag{3'}$$

The coefficient of determination is

$$R^2 = \frac{VTM}{VT} = 1 - \frac{VTR}{VT}. \tag{4}$$

Remark 1. If we divide VT and VTR from 3 by n we obtain the variance of the resulting variable Y, respectively, the variance of the residues. If we use these substitutions we obtain $R^2 = 1 - \frac{\mathrm{Var}(u)}{\mathrm{Var}(Y)}$.

This is used in our $C++$ program, where we also use $\frac{1}{n} \cdot X'X$ instead of $X'X$ and $\frac{1}{n} \cdot X'Y$ instead of $X'Y$ (see (2) and [4]). If we denote by $\widehat{A}$ the estimator of the linear regression coefficients, the variance of the residues is computed using the formula (see [4, 10])

$$\mathrm{Var}(u) = \mathrm{Var}(Y) - \frac{2}{n} \cdot \widehat{A}'X'Y + \frac{1}{n} \cdot \widehat{A}'X'X\widehat{A}. \tag{5}$$

To compute $\mathrm{Var}(Y)$ we compute, in our $C++$ program, $\frac{1}{n} \cdot X'X$ by adding a new component (having the index $k+1$) to X: the values of Y. The second moment of Y is the component $k+1, k+1$ of the matrix, and $X'Y$ from [4] and [10] divided by n is the last column of the obtained matrix with the first k elements.

A test for the homoscedasticity of the errors is the White test (see [4, 10]). For this test we consider the following linear regression:

$$u^2 = \alpha_0 + \alpha_1 Z_1 + \cdots + \alpha_p Z_p, \tag{6}$$

where the explanatory variables $Z_1, Z_2, \ldots, Z_p$ are as follows.

If the number of explanatory variables in (1) is 1 and we denote by $X = X_1$, the above explanatory variables are $Z_1 = X$ and $Z_2 = X^2$, and $p = 2$.

If $k = 2$, the above explanatory variables are $Z_1 = X_1$, $Z_2 = X_2$, $Z_3 = X_1^2$, $Z_4 = X_2^2$ and $Z_5 = X_1 \cdot X_2$, and $p = 5$.

If $k > 3$, but small enough we have (see [4, 10]) the above explanatory variables $Z_1 = X_1$, $Z_2 = X_2, \ldots, Z_k = X_k$, $Z_{k+1} = X_1^2$, $Z_{k+2} = X_2^2, \ldots, Z_{2k} = X_k^2$, $Z_{2k+1} = X_1 \cdot X_2, \ldots, Z_{2k+C_k^2} = X_{k-1} \cdot X_k$, and $p = 2k + C_k^2$ (as for $k = 3$).

If $k > 3$, but large enough we have the above explanatory variables $Z_1 = X_1$, $Z_2 = X_2, \ldots, Z_k = X_k$, $Z_{k+1} = X_1^2$, $Z_{k+2} = X_2^2, \ldots, Z_{2k} = X_k^2$, and $p = 2\ k$ (as for $k = 2$).

Next we take into account that $n \cdot R^2$ has the distribution χ_p^2, where R^2 is the coefficient of determination for the regression (6) (see [4, 10]). Therefore we accept the homoskedasticity if $n \cdot R^2 < \chi_{p;\varepsilon}^2$ (ε is the first degree error of the test).

An informational criterion for (1) is the Akaike criterion, AIC (see [4]):

$$AIC = \frac{VTR}{n} \cdot e^{\frac{2(k+1)}{n}} = \frac{\sum_{t=1}^{n} u_t^2}{n} \cdot e^{\frac{2(k+1)}{n}}, \tag{7}$$

or, in logarithmic expression,

$$\ln AIC = \ln\left(\frac{\sum_{t=1}^{n} u_t^2}{n}\right) + \frac{2(k+1)}{n}. \tag{7'}$$

Another informational criterion for (1) is the Schwartz criterion, *BIC*:

$$BIC = \frac{VTR}{n} \cdot n^{\frac{k+1}{n}} = \frac{\sum_{t=1}^{n} u_t^2}{n} \cdot n^{\frac{k+1}{n}}, \tag{8}$$

or, in logarithmic expression,

$$\ln BIC = \ln\left(\frac{\sum_{t=1}^{n} u_t^2}{n}\right) + \frac{k+1}{n} \cdot \ln n. \tag{8'}$$

For the $ARMA_{p,q}$ models these criteria are as follows (see [6]). The Akaike criterion is

$$AIC = -2\ln L_z\left(\widehat{\beta}, \widehat{\sigma_a^2}\right) + 2(p+q), \tag{9}$$

where $\widehat{\beta}$ is the vector of estimated parameters, $\widehat{\sigma_a^2}$ is the estimation of the variance of the corresponding white noice a_t, and $L_z\left(\widehat{\beta}, \widehat{\sigma_a^2}\right)$ is the maximum likelihood.

The Schwartz criterion is

$$\begin{aligned} BIC = & (n-p-q)\ln\frac{n\widehat{\sigma_a^2}}{n-p-q} + n\ln\left(1+\ln\sqrt{2\pi}\right) \\ & + (p+q)\ln\frac{n\left(\widehat{\sigma_z^2}-\widehat{\sigma_a^2}\right)}{p+q}, \end{aligned} \tag{10}$$

where $\widehat{\sigma_z^2}$ is the sample variance of the initial time series z_t.

Apparently these criteria are different in the time series case, but we can approximate them by

$$\begin{cases} AIC \approx n\ln\widehat{\sigma_a^2} + 2(p+q) \\ BIC \approx MDL = n\ln\widehat{\sigma_a^2} + (p+q)\ln n \end{cases}, \tag{11}$$

where *MDL* is the Minimum Description Length of Rissanen (see [6]).

Comparing the formulae of informational criteria in the case of linear regression and of the time series, it is pointed out in [2] that

$$\begin{cases} AIC = \alpha + 2 \cdot n_{\text{par}} \\ BIC = \alpha + n_{\text{par}} \cdot \ln m \end{cases}, \tag{12}$$

where n_{par} is the number of the estimated parameters and m is the number of degrees of freedom if we do not take into account the constraints (in the case of regression the number of degrees of freedom is $n-k-1$, and in the case of time series this number is $n-p-q$).

For the Goldfeld—Quandt homoskedasticity test we obtain the Akaike criterion (see [2])

$$AIC = -2\ln(1-\varepsilon - frstatfmax) + 2 \cdot nrcls \cdot (k+1), \tag{13}$$

and the Schwartz criterion *BIC*

$$BIC = -2\ln(1-\varepsilon - frstatfmax) + nrcls \cdot (k+1) \cdot \ln(n_1+n_2), \tag{14}$$

where $frstatfmax$ is the maximum cumulative distribution function applied to the involved Snedecor—Fisher statistics for the Goldfeld—Quandt test with the error ε, $nrcls$ is the number of classes, k is the number of explanatory variables, and n_i are from the Goldfeld—Quandt test on the class for which $frstatfmax$ is obtained.

In the next section we will present a similar approach where we use the White homoskedasticity test instead of Goldfeld—Quandt homoskedasticity test.

2 Informational Criteria and Classification

Suppose now we have n points in $\mathbb{R}^{k+1}$: k explanatory variables and one resulting variable. Consider also $nrcls$ classes in which we classify the n points (for only one class we have $nrcls = 1$). Denoting by p the number of explanatory variables in the regression of u^2 we take into account that m in the case of linear regression is n, i.e. the size of available data. Therefore we define the informational criteria in the case of White test when all the $nrcls$ classes are homoskedastic as

$$AIC = -2\ln\left(\chi^2_{p;\varepsilon} - \max\ nR^2\right) + 2 \cdot nrcls \cdot p, \textit{ and} \tag{15}$$

$$BIC = -2\ln\left(\chi^2_{p;\varepsilon} - \max\ nR^2\right) + nrcls \cdot p \cdot \ln\ n, \tag{16}$$

where $\max\ nR^2$ is the maximum of the values $n_i \cdot R_i^2$, obtained for the $nrcls$ classes (i is the index of the class).

As in [2], after we have computed the value of the informational criterion, we compute the maximum number of classes. In the case of Akaike informational criterion, the maximum number of classes is

$$\max nrcls = \frac{AIC + 2\ln\left(\chi^2_{p;\varepsilon}\right)}{2 \cdot p}. \tag{17}$$

In the case of Schwartz informational criterion, the maximum number of classes is

$$\max nrcls = \frac{BIC + 2\ln\left(\chi^2_{p;\varepsilon}\right)}{p \cdot \ln n}. \tag{17'}$$

Even when the only one class containing all the n points is homoskedastic we compute the informational criteria and we check to see if for another classification (in more classes) we obtain a smaller value of the information criterion. In both cases (homoskedasticity, respectively, heteroskedasticity) we choose the classification with all the classes homoskedastic, and with the minimum value of the informational criterion.

To compute $\chi^2_{p;\varepsilon}$ by a numerical method we have to solve the Cauchy problem

$$\begin{cases} y'(x) = \frac{1}{h_p(y)} \\ y(x_0) = y_0 \end{cases}, \tag{18}$$

where h_p is the pdf of the χ^2_p distribution. The values of x_0 and y_0 are chosen such that $H_p(y_0) = x_0$ (H_p is the cdf of the χ^2_p distribution) as follows. First we have to avoid for $p \geq 3$ zero at denominator. For this we find first the linear regression of the centil $\chi^2_{p;0.1}$ in terms of p:

$$\chi^2_{p;0.1} = 1.6 \cdot p + 4.5. \tag{19}$$

After this we take $y_0 = 1.6 \cdot p + 4.5$ and we compute

$$x_0 = \int_0^{y_0} h_p(t)\mathrm{d}t \tag{19'}$$

by a numeric method. The method used in our $C++$ program may be one of the rectangle method, the trapezoid method or the Simpson method (see [5]). The method to solve the Cauchy problem can be the Euler method, the modified Euler method or the Runge—Kutta method. In our examples we will use the most precise methods, i.e. the Simpson method to compute the integral, respectively, the Runge—Kutta method to solve the Cauchy problem (see [5]). Of course, we have to compute $\chi^2_{p;\varepsilon} = y(1-\varepsilon)$.

To compute the value of $\chi^2_{p;\varepsilon}$ for $p \geq 3$ by the Monte Carlo method we simulate 10000 random variables having the distribution χ^2_p. Once we order these values ascending we take $\chi^2_{p;\varepsilon}$ to be the value from the position equal to the integer part of $10000 \cdot (1-\varepsilon)$. In our $C++$ program we can generate the involved normal variables using the central limit method, the Box—Muler method, or the Butcher 1 method (see [9]). We choose in examples the Box—Muler method, because it is the most rapid.

If the number of explanatory variables is $k = 1$ we have $p = 2$. In this case we take into account that the distribution χ_2^2 coincide with the exponential distribution with $\lambda = 0.5$. Therefore $\chi_{2;\varepsilon}^2 = -2\ln\varepsilon$.

The k-means algorithm (see [3]) is an algorithm to classify n points in k classes as follows.

1. First k points are allocated each to different classes.
2. Next $n-k$ points are allocated to the class with the nearest gravity center. The gravity center of this class is recomputed.
3. All the points are checked in order to see whether another class has the gravity center closer than that of the current class. In this case the point is moved to the closest class, and we recompute the gravity centers of the involved classes. The algorithm ends when no point is moved.

In the above algorithm the distance between the point and the class is considered the distance between the point and the gravity center of the class. In [1] another distance is considered: the residue of the linear regression corresponding to the class.

But heteroskedasticity is not produced by expectation, or by the residues. It is produced by the variance of the residues. Therefore we consider (in our algorithm, and in our $C++$ program) as distance the absolute value of the difference between the square of the residue and the variance of the residues of the class.

Therefore the k-variances algorithm is as follows:

1. First $nrcls \cdot (p+2)$ points are allocated to different classes as follows. First $p+2$ to class 1, next $p+2$ to class 2 and so on.
2. Next $n - nrcls \cdot (p+2)$ points are allocated to the class with the minimum distance between the square of the residue (corresponding to the class) and the variance of residues for the class. We recompute the variance of the residues for this class.
3. All the points are checked to see whether for some other class the mentioned distance is less. Once we find the best class, we move the point and we recompute the variances of the residues for the involved classes. The algorithm ends when no point is moved.

After we have found the best classification of the n points into $nrcls$ classes, we check if all these classes are homoskedastic. In the affirmative case we compute the value of the involved informational criterion, and the maximum number of classes using (17) or (17').

3 Applications

Example 1. Consider the resulting variable Y representing the interest rates applied by banks on loans, and the explanatory variable X representing the reference interest. The monthly data are from January 2006 to December 2010, according to The Statistical Section of National Bank of Romania, June and December 2007 and 2008, and December 2006, 2009 and 2010. They are introduced in Table 3, Appendix.

The results of our $C++$ program for the linear regression are presented in Table 1. The first column contains the number of classes (starting with the case of the lack of the classification, where this number is one), the second column contains the linear regressions for each class and the third one contains the variance of the residues for these regressions. The fourths column contains the linear regressions of u^2 from (6), the fifths column contains the corresponding values of $n \cdot R^2$, and the sixths column contains "Yes" if the class is homoskedastic, and "No" in the contrary case.

The values $n \cdot R^2$ from the fifths column in the above table must be compared with $\chi^2_{p;\varepsilon}$, with $p = 2$, because we are in the case of linear regression. When we compute the centil $\chi^2_{2;0.05}$ we take into account that the χ^2_2 distribution is identical to exponential distribution with $\lambda = 0.5$, as we have mentioned in section 2. Denoting by α the above value $\chi^2_{2;0.05}$, we obtain from $F(\alpha) = 1 - e^{-0.5\alpha} = 0.95$ the value $\alpha = -2\ln(0.05) = 5.99146$.

For a classification having at least one heteroskedastic class (including the case of only one heteroskedastic class, as in Table 1) we can not apply formulae (17) and (17') to compute the maximum number of classes. In this case, regardless the number of considered classes, the maximum number of classes is such that each class has at least four points, in order to have not degenerative regression in (6). Therefore this number is the upper integer part of $\frac{n}{4}$, i.e. $15 = \frac{60}{4}$, and it remains the same until we find a classification with all *nrcls* homoskedastic (in the case of Table 1 we have $nrcls = 2$).

As we can see in the above table, comparing the values of $n \cdot R^2$ with the above centil, we have both classes with homoskedastic errors in the case of two classes. The first class (23 months) is from October 2008 to August 2010, i.e. during the world economic crisis. The other class (the second one: 37 months) is from January 2006 to September 2008, and from September 2010 to December 2010.

The values of the informational criteria are $AIC = 4.7607$, respectively, $BIC = 13.13808$, and the recomputed maximum number of classes using (17) and (17') is two. Therefore we have not to check the classification in three or more classes.

Because in this example we have considered only one explanatory variable (the reference interest) we can also consider the parabolic model $Y = a_0 + a_1X + a_2X^2$. We obtain the regression $Y = 22.21213 - 2.78566X + 0.22209X^2$, and $n \cdot R^2 = 7.93794$. The centil $\chi^2_{4;0.05}$ is 9.48773 estimated by the Runge—Kutta method, 9.60415 estimated by the Monte Carlo method, considering AIC as informational criterion, respectively, 9.49132 estimated by the Monte Carlo method, considering BIC as informational criterion. Therefore the only one class is homoskedastic, and we proceed to compute the AIC and BIC informational criteria.

The value of AIC informational criterion is 7.12377 in the case of Runge—Kutta method, respectively, 6.9789 in the case of Monte Carlo method. The value of BIC informational criterion is 15.50114 in the case of Runge—Kutta

Table 1 The linear model $Y = a_0 + a_1 \cdot X$

No. of classes	Linear regressions	Variance of residues	The regression of u^2	$n \cdot R^2$	Homoskedastic? Yes/no
1	$Y = 7.53166 + 0.87154X$	1.32345	$u^2 = -10.31827 + 2.45342X - 0.12341X^2$	18.17567	No
2	$\begin{cases} Y = 8.25489 + 0.9398X \\ \text{(23 points);} \\ Y = 10.01352 + 0.463X \\ \text{(37 points).} \end{cases}$	$\begin{pmatrix} 0.2095 \\ 0.06613 \end{pmatrix}$	$\begin{cases} u^2 = -0.38925 + 0.12552X - \\ 0.00631X^2; \\ u^2 = 0.68681 - 0.15319X + \\ 0.00927X^2. \end{cases}$	$\begin{pmatrix} 0.2046 \\ 0.94015 \end{pmatrix}$	$\begin{pmatrix} \text{Yes} \\ \text{Yes} \end{pmatrix}$

method, respectively, 15.49651 in the case of Monte Carlo method. Applying the formulae (17) and (17'), we obtain the maximum number of classes one in all the four above cases. Therefore we need no classification in more classes for the parabolic model.

Example 2. Consider the resulting variable Y representing the non-government loans, and the explanatory variables X_1 representing the interest rates applied by banks on loans, and X_2 representing the CPI (Consumer Price Index). The monthly data are from January 2006 to December 2010, according to The Statistical Section of National Bank of Romania, June and December 2007 and 2008, and December 2006, 2009 and 2010. They are introduced in Table 3, Appendix.

The results of our $C++$ program for the linear regression are presented in Table 2. The columns in this table are the same as in Table 1, except for the last column which was omitted due to lack of space.

The centil $\chi^2_{5;0.05}$ that is compared to the values of $n_i \cdot R_i^2$ from the above table is 11.0704 if we use the Runge—Kutta method, 11.1691 if we use the Monte Carlo method, we generate the normal variables by the Box—Muler method and we choose AIC as informational criterion, and 11.13233 if we use the Monte Carlo method, we generate the normal variables by the Box—Muler method and we choose BIC as informational criterion.

The first classification with all classes homoskedastic is that with four classes, as above. This classification is as follows. First class contains 8 months from January to May 2006 and from April to June 2008, the second class contains 11 months from August 2006 to April 2007, December 2008 and June 2009, and the third class contains 19 months from June to July 2006, from May 2007 to March 2008, from July to August 2008, October 2008 and from January to March 2009. The fourths class contains the main period of the economic crisis in Romania: September and November 2008, **from April to May 2009 and from July 2009 to December 2010**.

The values of AIC criterion are 43.87938 in the case of Runge—Kutta method and 42.83391 in the case of Monte Carlo method. The values of BIC criterion are 85.76628 in the case of Runge—Kutta method and 85.0498 in the case of Monte Carlo method. The obtained maximum number of classes using (17) and (17') is four in all the above cases. Therefore we need no classification in more than four classes for the above multilinear model.

4 Conclusions

The main idea for conception of informational criteria like Akaike and Schwartz is to build criteria increasing on the errors of the model, but, in the same time, increasing on the number of parameters. It is obvious that we obtain lower errors of the model by increasing the number of parameters, but, sometimes this number can make the time to estimate the parameters become prohibitively.

Table 2 The linear model $Y = a_0 + a_1 \cdot X_1 + a_2 \cdot X_2$

No. of classes	Linear regressions	Variance of residues	The regression of u^2	$n \cdot R^2$
1	$Y = -98.48476 + 12.74889X_1 + 13.33178X_2$	707.86376	$u^2 = -31017.96198 + 4175.14838X_1 + 673.35478X_2 - 139.83287X_1^2 - 54.97945X_2^2 - 8.85949X_1 \cdot X_2$	14.1507
2	$Y = 2967.15794 - 206.33315X_1 - 12.27067X_2$ (16 points); $Y = -36.53382 + 11.03718X_1 + 8.00996X_2$ (44 points).	$\begin{pmatrix} 47.98414 \\ 670.34682 \end{pmatrix}$	$u^2 = -350206.25780 + 50029.50496X_1 + 5655.91819X_2 - 1786.45360X_1^2 - 21.85218X_2^2 - 404.32048X_1 \cdot X_2$; $u^2 = 464.18035 + 469.18593X_1 - 341.16314X_2 - 30.20067X_1^2 - 27.53264X_2^2 + 35.72408X_1 * X_2$.	$\begin{pmatrix} 1.70278 \\ 17.78457 \end{pmatrix}$
3	$Y = 1135.09285 - 76.40167X_1 + 1.61391X_2$ (9 points) $Y = -109.2573 + 13.57222X_1 + 10.35688X_2$ (28points) $Y = 232.80765 - 2.08736X_1 + 0.21743X_2$ (23 points)	$\begin{pmatrix} 12.44351 \\ 30.88103 \\ 7.76398 \end{pmatrix}$	$u^2 = -674939.88707 + 95157.23710X_1 + 12885.36367X_2 - 3353.45363X_1^2 - 48.51287X_2^2 - 913.51134X_1 \cdot X_2$ $u^2 = -7533.88938 + 839.84606X_1 + 439.3864X_2 - 21.53135X_1^2 - 7.55135X_2^2 - 26.62723X_1 \cdot X_2$ $u^2 = -120.77408 + 79.4336X_1 - 183.26748X_2 - 3.6862X_1^2 + 8.02518X_2^2 + 6.42587X_1 \cdot X_2$	$\begin{pmatrix} 6.73814 \\ 12.8306 \\ 11.59561 \end{pmatrix}$
4	$Y = -3067.64589 + 223.74645X_1 + 2.51537X_2$ (8 points) $Y = -256.80688 + 24.51834X_1 + 4.38354X_2$ (11 points) $Y = -102.37709 + 12.80934X_1 + 11.17288X_2$ (19 points) $Y = 233.13557 - 2.10605X_1 + 0.20708X_2$ (22 points)	$\begin{pmatrix} 13.17047 \\ 6.28951 \\ 24.49809 \\ 8.10776 \end{pmatrix}$	$u^2 = -4986520.22732 + 712268.89332X_1 - 31187.48175X_2 - 25435.02921X_1^2 - 62.93884X_2^2 + 2236.18172X_1 \cdot X_2$ $u^2 = 2846.02162 - 366.66489X_1 - 2.68627X_2 + 11.71176X_1^2 + 0.27026X_2^2 - 0.06685X_1 \cdot X_2$ $u^2 = -3222.25862 + 342.59082X_1 + 233.91188X_2 - 8.02190X_1^2 - 2.85015X_2^2 - 14.68337X_1 \cdot X_2$ $u^2 = -102.46536 + 76.20586X_1 - 180.82867X_2 - 3.57488X_1^2 + 7.85773X_2^2 + 6.38697X_1 \cdot X_2$	$\begin{pmatrix} 6.4664 \\ 4.30588 \\ 5.31184 \\ 10.92665 \end{pmatrix}$

In the case of the Goldfeld—Quandt homoskedasticity test a natural order is obtained for the n points, and the classification was reduced to the separation of the classes (see [2]). But, in the present paper case (the White homoskedasticity test), no natural order is obtained. Therefore we have to use algorithms like those used in artificial inteligence. For instance we use the proposed k-variances algorithm from section 2.

In the case of Example 1 we notice that in the period of the world economic crisis the coefficient of X and the variance of the residues are greater than in the other period: 0.9398 and 0.2095, respectively, 0.463 and 0.06613. In the case of the coefficient of X (the reference interest) the information about the evolution of the USA economic crisis was incompleted. Therefore the banks had to take into account more the signals from the central bank in their credit policy. We notice also that the coefficient of X during the economic crisis is closed to those from the case of only one heteroskedastic class (0.87154).

For this example we can use two methods to avoid the heteroskedasticity: to increase the degree of the polynomial (the parabolic model is homoskedastic) and to increase the number of classes (k-variance classification algorithm).

In Example 2 we notice that the coefficient of X_1 (the interest rates applied by banks on loans) is negative only for the fourths class. This can be explained by the fact that the credits were given very easy (as those only with the ID card). The first two classes have big coefficients for X_1: 223.74643 and 24.51834. These classes contain the majority of the year 2006, before Romania joined EU. In this period many Romanians made household credits even they did not need really a home: they have bought flats hopping they could sell them to EU citisens in 2007.

In both examples from previous section we notice that the first classification with all the classes homoskedastic is also optimal from the point of view of an informational criterion as AIC or BIC. Theoreticaly, it is possible (because $\lim_{x\to 0} \ln x = -\infty$) to obtain a classification with all classes homoskedastic that would require increasing the number of classes. We need only a value of $n_i \cdot R_i^2$ less than the involved χ^2 centil, but closed to it. An open problem is to find economic data having this property.

Appendix

Acknowledgements This paper is supported by the Sectorial Operational Programme Human Resources Development (SOP HRD) financed from the European Social Fund and by the Romanian Government under the contract number SOP HRD/89/1.5/S/62988.

Table 3 The reference interest, the non-government credit (billions lei), the interest rates applied by banks on loans and the CPI from January 2007 to December 2010

Month/year	Reference interest	Non-government credit	Interest rates applied by banks on loans	CPI
January/2006	7.5	61.627	14	1.03
February/2006	7.5	62.4039	14	0.24
March/2006	8.47	65.6752	14	0.21
April/2006	8.5	68.1239	14	0.42
May/2006	8.5	72.3104	14	0.6
June/2006	8.5	76.4558	14	0.15
July/2006	8.50	79.4007	14	0.11
August/2006	8.75	82.1613	14	−0.07
September/2006	8.75	85.2889	14	0.05
October/2006	8.75	89.0168	14	0.21
November/2006	8.75	91.9023	14	1.09
December/2006	8.75	93.2834	14	0.74
January/2007	8.75	92.4949	13.68	4.01
February/2007	8.75	95.4817	13.72	3.81
March/2007	8.08	98.9642	13.68	3.66
April/2007	8	102.6061	14	3.77
May/2007	7.5	106.4999	13.68	3.81
June/2007	7.25	109.0313	13.28	3.8
July/2007	7.25	114.6615	13.14	3.99
August/2007	6.1	122.0958	12.97	4.96
September/2007	6.48	129.0622	12.92	6.03
October/2007	6.87	133.3196	13.02	6.84
November/2007	7	141.1176	13.04	6.67
December/2007	7.5	148.1807	13.05	6.57
January/2008	7.5	154.2675	13.16	7.26
February/2008	8	158.3409	13.49	7.97
March/2008	9	164.6068	13.75	8.63
April/2008	9.03	168.7341	14.36	8.62
May/2008	9.5	171.8343	14.4	8.46
June/2008	9.75	178.1803	14.4	8.61
July/2008	9.75	178.6922	14.6	9.04
August/2008	10	183.6299	14.9	8.02
September/2008	10.25	194.1741	15.29	7.3
October/2008	10.25	193.0636	16.67	7.39
November/2008	10.25	195.131	17.45	6.74
December/2008	10.25	198.0557	17.47	6.3
January/2009	10.25	206.4357	17.87	6.71
February/2009	10.25	206.8901	18.11	6.89
March/2009	10.14	202.617	18.15	6.71
April/2009	10.07	200.5538	18.08	6.45
May/2009	10.02	199.0795	17.73	5.95
June/2009	9.71	198.0563	17.46	5.86
July/2009	9.5	197.9049	17	5.06

(continued)

Table 3 (continued)

Month/year	Reference interest	Non-government credit	Interest rates applied by banks on loans	CPI
August/2009	9	198.6828	16.68	4.96
September/2009	8.53	198.9147	16.5	4.94
October/2009	8.5	201.2144	16.6	4.3
November/2009	8	200.8716	16.57	4.65
December/2009	8	199.8871	16.58	4.74
January/2010	8	199.285	16.3	5.2
February/2010	7.5	199.1671	15.6	4.49
March/2010	7.25	199.4041	14.99	4.2
April/2010	7	200.3224	14.23	4.28
May/2010	6.5	203.1121	14.26	4.42
June/2010	6.25	210.8089	13.9	4.38
July/2010	6.25	206.6989	13.89	7.14
August/2010	6.25	207.6677	13.59	7.58
September/2010	6.25	207.9305	13.42	7.77
October/2010	6.25	206.8363	13.18	7.88
November/2010	6.25	207.9248	12.93	7.73
December/2010	6.25	209.298	12.66	7.96

References

1. Ciuiu, D.: Pattern classification using polynomial and linear regression. In: Proceedings of the International Conference Trends and Challenges in Applied Mathematics, 20–23 June 2007. Technical University of Civil Engineering, Bucharest, Romania, pp. 153–156 (2007)
2. Ciuiu, D.: Informational criteria for the homoskedasticity of the errors. Rom. J. Econ. Forecast. **XIII**(2), 231–244 (2010)
3. Dumitrache, I., Constantin, N., Drăgoicea, M.: Reţele Neurale. Identificarea şi Managementul proceselor (English: Neural Networks. Identification and Management of the Processes). Matrix Rom, Bucureşti (1999)
4. Jula, D.: Introducere în Econometrie (English: Introduction to Econometrics). Professional Consulting, Bucureşti (2003)
5. Păltineanu, G., Matei, P., Mateescu, G.D.: Analiză Numerică (English: Numerical Analysys). Conspress, Bucureşti (2010)
6. Popescu, Th.: Serii de Timp. Aplicaţii în Analiza Sistemelor (English: Time Series. Application to Analysis of the Systems). Efitura Tehnică, Bucureşti (2000)
7. Saporta, G.: Probabilités, Analyse des Donées et Statistique. Editions Technip, Paris (1990)
8. The Statistical Section of National Bank of Romania. Monthly Bulletin from June 2007, December 2007, June 2008, December 2008, December 2009 and December 2010. www.bnr.ro (Romanian). Accessed on December 2010
9. Văduva, I.: Modele de Simulare (English: Simulation Models). Editura Universităţii Bucureşti (2004)
10. Voineagu V. et al.: Teorie şi Practică Econometrică (English: Econometric Theory and Practice). Meteor Press, Bucureşti (2007)

An Innovative Decision Making e-key Application For the Identification of Fish Species

George Minos, Vassilis Kostoglou, and Emmanouil Tolis

Abstract The most important tool for ichthyologists, as well as biologists, fishery biologists and other relevant scientists is an identification key, that is an information system providing them the capability to identify specimens accurately or to find information on correct names, biology and distribution of species. Dichotomous identification keys organize fishes based on their similarities and differences. This research work focuses on the development and implementation of a new innovative information system which is able to identify correctly fish species. The developed system is a fully interactive fish identification e-key which can be used in both forms; locally and remotely via Internet, and more specifically the Telnet service. This new dichotomous classification e-key provides the capability to identify any species in a compact and easy-to-use environment which gives the user excellent operation capabilities and complete information about all included fish species. Moreover, the application provides the capability to search for a sporadic fish species and to show a list which includes all the fish species that exist to the application's database until that time.

Key words Information system • Decision-making application • Identification key • E-key • Fish • Species

G. Minos (✉)
Department of Aquaculture and Fisheries Technology, Alexander Technological Educational Institute of Thessaloniki, P.O. Box 157, 63200, N. Moudania, Greece
e-mail: gminos@aqua.teithe.gr

V. Kostoglou • E. Tolis
Department of Informatics, Alexander Technological Educational Institute of Thessaloniki, P.O. Box 141, 57400, Thessaloniki, Greece
e-mail: vkostogl@it.teithe.gr; emtolis@yahoo.gr

A. Migdalas et al. (eds.), *Optimization Theory, Decision Making, and Operations Research Applications*, Springer Proceedings in Mathematics & Statistics 31,
DOI 10.1007/978-1-4614-5134-1_23,

1 Introduction

There are about 28,000 living species of fishes which makes very difficult their correct classification [9]. The identification of the various fish species is based on morphometric characters (measurable structures such as fin length, head length, etc.), meristic characters (countable structures such as number of scales in the lateral line, number of vertebrae, etc.), anatomical characters of the skeleton and the soft anatomy or characters that include any fixed, describable differences among taxa such as color (presence of stripes, spots) photophores (number and position) and sexually dimorphic structures [1, 9].

To classify different species a dichotomous identification key is used, which is an extremely important tool in science. The primary aims of an identification key are to enable species to be identified correctly and to summarize what is known on their biology and geographical distribution. In order to identify a fish with the use of a dichotomous key, the user works through a series of questions and illustrations which eventually lead him to the species matching best the characteristics he has set.

Systematics (or taxonomy) is the biological science responsible for the classification of living organisms in a hierarchically organized system representing the evolutionary kinship of the various systematic groups. In classification, the use of morphological, anatomical, physiological and other characteristics is made to decide the existing relationships [1]. The basic systematic unit (taxon) is the species followed (in ascending order) by the genus, family, order, class, superclass, subphylum and the phylum.

Like other animals and plants, fishes are known by common names and scientific names. While common names differ from country to country, scientific names are universal. Aristotle was the first to classify the animals known in his days, but the first generally acknowledged scientific classification of animals and plants was by Carl Linnaeus who introduced the binomial system, in which every species was given two Latin or Greek names. Since the scientific name consists of two parts, the first italicized word, with the initial letter capitalized, is the genus while the second italicized word is the specific (species) name.

The existing identification keys are divided into two categories: (1) printed keys and (2) electronic keys (e-keys). The former are printed in the form of a book and they mostly still keep this form so far. Nevertheless, e-keys have been developed in recent years. Examples of printed identification keys are books dealing with fishes from the Mediterranean Sea [7, 13], Atlantic Ocean [2, 6, 12], Indian and Pacific Ocean [3–5] and Greek seas [10]. The main disadvantage of the printed identification keys is that it is easy to make a few wrong decisions when navigating through the test. So, when someone is deadlocked or makes a wrong selection, it is not easy to go to a previous selection (family, suborder, etc.) because there is not an area that shows the history of the selections. In a case like this, the reader has to find the previous selections that he made and the page in which they were. Furthermore, when the selections are too many, it leads to confusion for the reader. Also, it must be mentioned the case where new dichotomous keys must be created and printed which will be used to identify new organisms (fish species) that will appear in the area in the future.

For this reason, fish identification keys evolved as e-keys. There are many examples of fish identification e-keys because, as technology boomed several keys of this type were developed to facilitate ichthyologists, students or persons who needed such tools. All fish identification e-keys are based on printed keys (books). No new keys have been created, but the book contents have been digitized. Some examples of identification keys in World Wide Web are in *FishBase* [8] per Food and Agriculture Organization (FAO) area, per order or per family or quick identification by image and also identification by morphometrics. An important tool is the *Marine Species Identification Portal* (http://species-identification.org) while the *Fish Identification Site* (http://svrsh2.kahaku.go.jp/fishis) helps to identify fishes utilizing countable characters such as numbers of fin rays, scales, pores, gill rakers, body rings and vertebrae. Other identification e-keys are for specific state in the USA (www.theanglingchannel.com/fish-identification-encyclopedia-resources.html) like the *Identification Key to Native Freshwater Fishes of Peninsular Florida* (www.flmnh.ufl.edu/fish/southflorida/everglades/marshes/fishkeyedu.html). The common feature of these keys is that they are web applications. This means that they are uploaded as webpages into a web site and when someone wants to use them, he just has to visit the specific website.

There are few mobile applications for iPad, iPod and iPhone (http://itunes.apple.com/us/genre/ios/id36?mt=8) to identify fish species. The *Marine Fishes - Identification Guide* is based on the book entitled *Marine Fishes of Brazil - A Practical Identification Guide* and is limited on a specific number (200) of marine fish species from Brazil. In the application *Fish*, the number of fishes is very limited, since it is a Fish Guide reporting only a part of the common fish swimming in streams, lakes and rivers across the North Woods, US. In the application *FishID - Know every fish, fish every spot and spot the best catch*, appear only eight saltwater fishes. There are also other mobile fish-related applications for iPhone and iPad that are not identification keys. The *Sharks Magazine* contains information on sharks, *MarineLife - Genus trait Handbook* on marine life species, *Fish Alkhaleej* on common fish in Arabian countries, *Fish Complete Reference* and *Fishes of the World - eFishesW* information on fish species and *Marine Fish Encyclopedia* on common marine aquarium fish.

The available Android applications (https://market.android.com/) are fewer in number than the iTunes ones. None of them gives the ability for the species identification. Some give the description and illustrations for the most common fishes (e.g. *North American Fish Guide, FL SW Fishing Regulations, Saltwater Pocket Fisherman, The Pocket Fisherman-Freshwater edition*). Other Android applications are useful only for fishing on where and how to catch fish, such as *My Fishing Advisor, Fishing Status, Tide Prediction, Fishing Calendar* and *Fish Cast 2012*.

A dichotomous key is a tool that allows the user to determine the identity (specific name) of a fish. These keys consist of a series of "either or" choices that lead the user to the correct name. "Dichotomous" means "divided into two parts". Therefore, dichotomous keys always give two choices in each step.

Technically, there are two types of e-keys. The more simple e-keys (with static content) are developed with HyperText Markup Language (HTML) and contain a set of information which is stored and divided into a number of pages. These pages compose a webpage. The above set of information is about key's selections, data on fish species, fish images and all the necessary information which compose a fish identification key. A simple e-key is not flexible because it does not provide updating capabilities (with which the user can add new fish species). Furthermore, a simple key lacks proper organization because it does not contain any database which can provide organized information storage. The more complex e-keys are developed under both HTML and a scripting language. The scripting language is usually either PHP (Hypertext Pre Processor) or ASP (Active Server Pages). By using a scripting language the developer is able to create a webpage with dynamic content, ensuring also that all the necessary for the operation of the e-key information is being stored into a database. This e-key does not simply show to the user a set of information which is divided into a number of webpages, but every time the user makes a selection, a set of information is being recovered from the database to be shown to the browser. As e-keys of this type are using databases, they could provide updating capabilities. An extensive literature review of the existing fish identification e-keys revealed that there are no e-keys providing complete and correct update capability.

The information system that has been designed and developed in the present work constitutes the first fish identification e-key for all the Mediterranean fish species of the Greek seas. The designed system is fully interactive with the user and can be used in both forms; locally and remotely via the World Wide Web. As the information system constitutes a desktop application, it provides an easy and user-friendly environment which gives the user multifaceted fish identification capabilities and an effective search function for all included fish species. Furthermore, its navigation function is a strong and useful advantage. Finally, its additional function which shows information about the fish systematic taxonomy is innovative.

2 Methodology

The information system has been developed with the Java object-oriented programming language. The system was necessary to have a database to connect to, so that to recover from this all necessary data and also to present this data as information to the end user. The database includes full information about every level of the fish systematic taxonomy, as well as images of every one of the 511 existing fish species. The *Relational Database Management System* (RDBMS) that was used to manage the database of the information system is MySQL. It was selected in order to provide to the information system the capability to connect to a single database, common to all users as it is uploaded to a Web Server. The above feature is usual for MySQL and it does not exist in all other RDBMSs, either they are free or not. MySQL can also be used locally at the personal computer of a user. In this case, each user's personal computer takes Web Server's role. So, each personal computer keeps the database

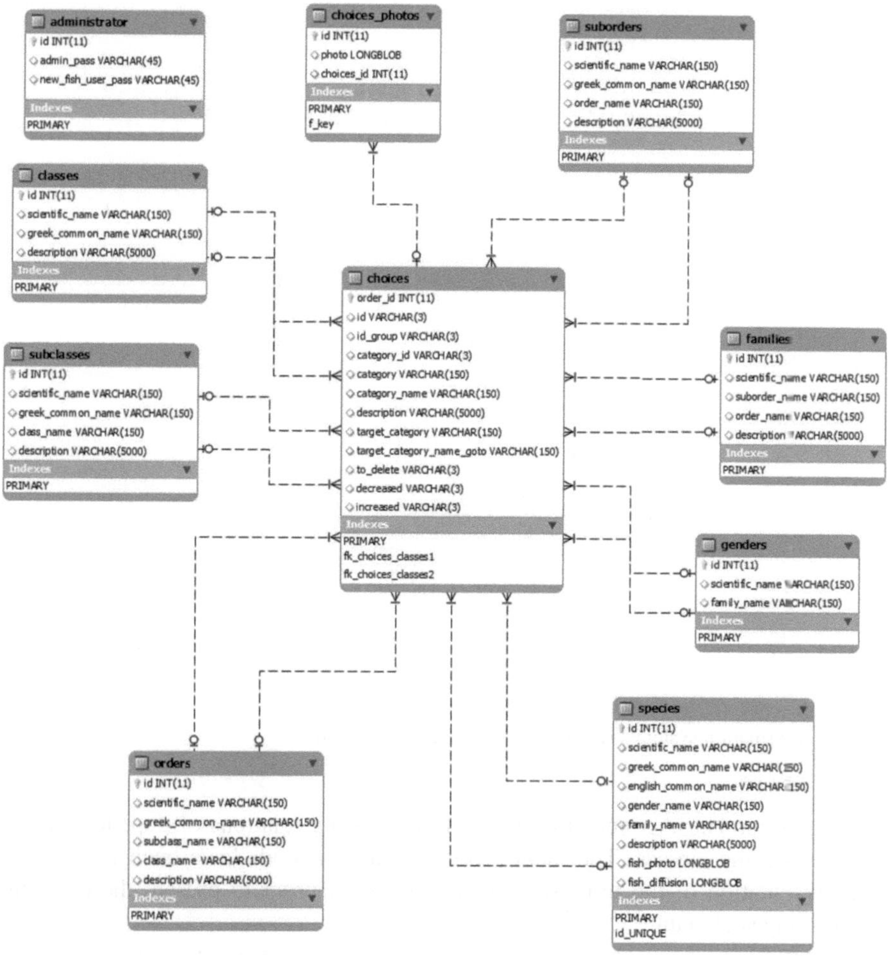

Fig. 1 Information system's database structure through an ER diagram

stored in it. Regarding the fish information included in the new information system, all data and images of the main relevant published identification keys [10, 11] were digitized and transferred.

Figure 1 presents the structure of the database through an Entity-Relationship (ER) diagram. The diagram includes the main table containing information about all possible selections of the user, as well as all the other tables with the corresponding information. These tables have direct link with the main table, as every level of the fish systematic taxonomy hierarchy participates in the fish identification key selections.

A relevant class diagram is illustrated in Fig. 2. This diagram belongs to the object-oriented programming diagrams and generally reflects the structure of a system. It contains all the existing system units, called classes, and the connections

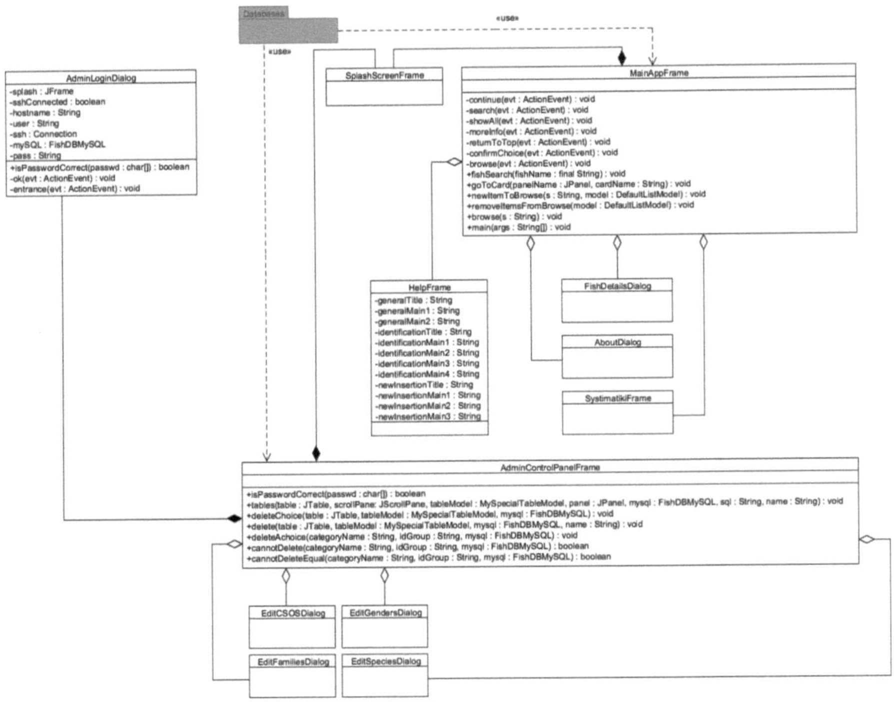

Fig. 2 Information system's structure through a class diagram

among them. These connections present the existing relations of dependence and use among the classes. The main properties and methods have been included in the presentation of the most significant classes, in order to present the elements composing such a unit.

Figure 3 illustrates the overall flow control of the information system through an *activity diagram*, presenting the software process as a flow of work through a series of actions. The diagram represents graphically the workflows of stepwise activities and actions with support for choice, iteration and concurrency. The activity diagram depicts all the main application's processes and the choices the user can make while using the information system. More specifically, it depicts all the activities (rounded rectangles), the control flows between them (arrows), the decision and merge nodes (diamonds), the object node (rectangle), the initial node (filled circle) and the activity final node (filled circle with border).

Two programming tools have been used for the development of the information system: Netbeans IDE 6.9.1 for the programming part, and MySQL Workbench 5.2.31 CE for database design and management. Apart from the above basic tools, a Java library was also used. This library is the Ganymed SSH-2 for Java library and was used in order to support the development of the information system. This library implements the SSH-2 network protocol and gives to a Java program the capability to connect with an SSH Server.

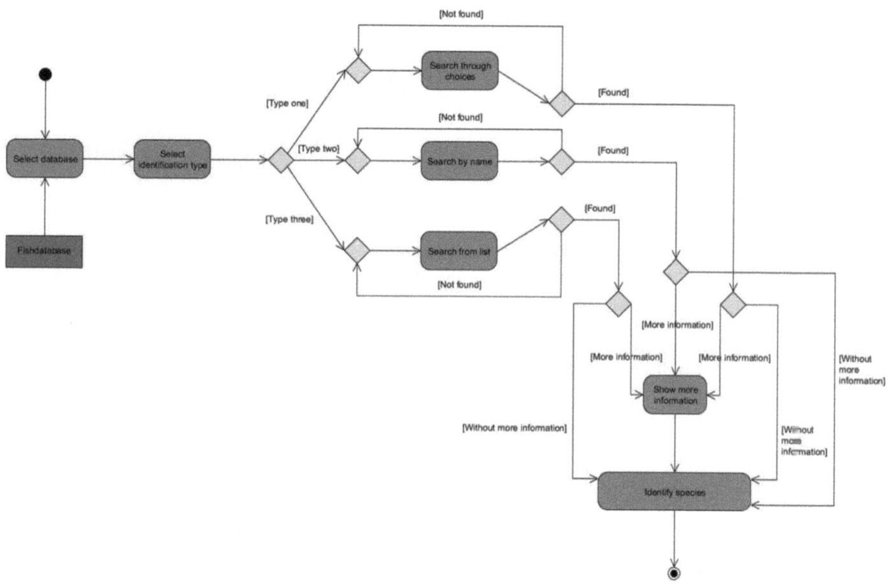

Fig. 3 Information system's overall flow of control through an activity diagram

3 The Developed Information System

The information system provides the user the capability to identify a fish species by making some selections relevant to species' external morphology. When starting the application the user has to choose one of the two databases (local or online) and type a password. After that, the main menu of the e-key appears at the upper part of the screen (Fig. 4).

The screen is divided into three horizontal parts. The upper part and the bottom part are static. Conversely, the central part is dynamic. In the upper part of the screen there are application's two main functions: (1) the *Show all species* function and (2) the *Search species* function. The bottom part constitutes a status bar which contains information about the connected database and a bar which informs the user about the progress of a specific search procedure.

The central part of the e-key (Fig. 4) consists of the following sections: (1) the table at the center which contains the texts of the current dichotomous selection and two relevant actions, to confirm the selection or to return to the beginning, (2) the area below the table which shows more information to help user's selection, (3) the navigation at the left side and (4) the area at the right side which shows more information about the current level in which user is based on his selections. The upper section of the center part (above the table) contains a title which informs the user about the application's function that is being executed at that time.

If the user intends to use the fish identification function, he has to select one of the two selections in the central part (area 1) and press the button *Confirm your table*

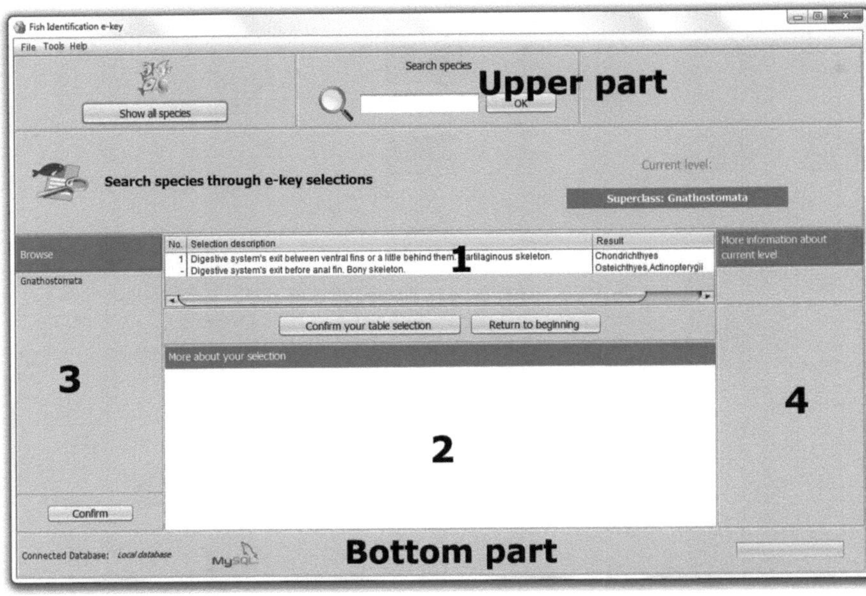

Fig. 4 Main screen of the fish identification e-key

selection. Anytime he wants to return to the selections in the start of the key, he can just press the button *Return to beginning*. Every time the user makes a selection, he can read its full description (area 2). Also, when it is necessary, this area shows an image which describes optically user's selection. Furthermore, every user's selection is being recorded in the navigation list (area 3), so that he can anytime go back to a previous step. This list is very useful, as it presents the whole route till the final successful fish identification.

The left selection of the upper part (area 1—*Show all species*) leads to a table including all the fish species that are stored at the selected database until that time. This area also contains two selections (*More information about the species* and *Back to e-key*). The former requires the selection of a row from the fishes' table and presents all the stored information about the selected species including names, picture, description and geographical distribution (Fig. 5). The latter restores the fish identification screen at the central part of the screen.

4 The User Interface

The information system's user interface has been designed with the aim to be simple and friendly in its use. It has been designed to correspond fully to the needs of an ichthyologist without significant experience in the use of computers.

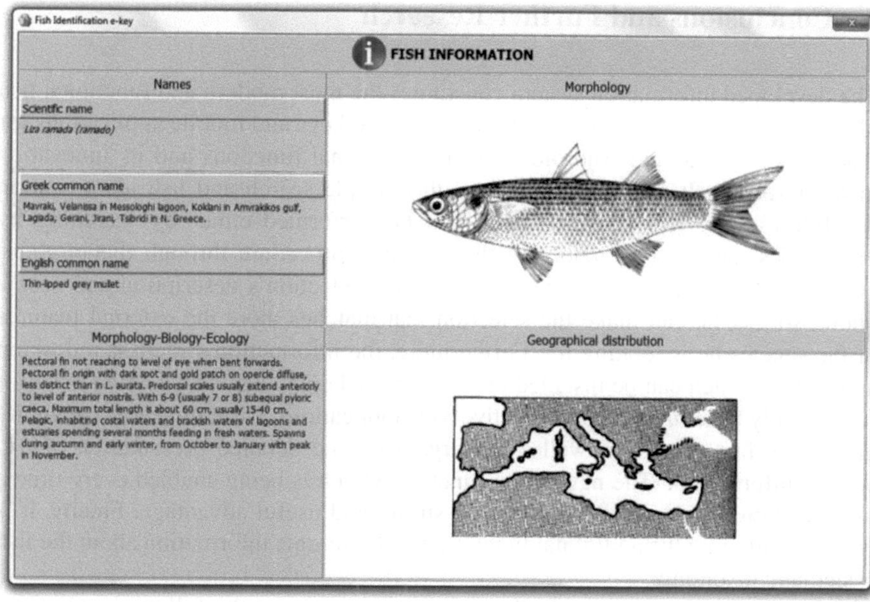

Fig. 5 Fish identification - Window with species information

In the design phase of the system, emphasis has been given to the segregation of the main screen into distinguished parts enabling the user to see all the information he needs in a well legible way. Additionally, the use of titled buttons corresponding to the various operations enables the user to perform them immediately and very easily.

In the user's interface section, concerning the main operation of the identification of a fish species, the main window has been separated in four horizontal parts (Fig. 2). The first part contains the three main operations (show, search and insertion) in order to be continuously available to the user. The second part informs the user about the current operation, as well as for relevant useful information. The third horizontal part changes according to the selected operation, presenting to the user either his search results or his selections for the specification of a certain species. The last part is the system's status bar informing the user about technical issues of the application. All the operations are supported by corresponding buttons and full navigation capabilities.

The developed information system has been tested extensively through successive pilot trials with 10 users. Users' feedback was satisfactory as the information system covered their needs. It is worth noting that most users appreciated the navigation operation and noted that it has been proved very useful and easy to use.

5 Conclusions and Further Research

The developed information system constitutes the most modern and functional fish identification e-key, compared with the existing e-keys and mobile applications for iPad, iPod, iPhone and Android. Both its additional functions and its innovation make it special. It provides the user with multiple search and fish identification capabilities. The fish species search can be performed either by searching by the species name, or by applying the selection procedure through dichotomous questions. In this procedure the user reads the selection's description and sees a photo, so that he can make the selection that matches more the external features of the species he is looking for. Furthermore, the information system is a desktop application which can be installed on any personal computer. This feature makes it functionally faster than other respective web applications. Also, its search capability gives very fast results, as well as an organized and detailed presentation of fish species information. The navigation function, which is being enabled every time a user is trying to identify a species, is a strong and useful advantage. Finally, it is worth mentioning the additional function which presents information about the fish systematic taxonomy.

Some proposed issues for further relevant research are the following: (1) Conversion of the information system to a mobile application for use by mobile phones, smart phones or tablets with touchscreen utilizing popular operating systems, such as iOS (iPhone, iPad), Android or Windows Phone, (2) Application's extension to a wider geographical area (Mediterranean Sea, Atlantic Ocean, etc.) and/or to a specific fish fauna (e.g. fishes in fresh waters), and (3) the most innovative new feature would be the addition of updating capabilities which will allow users to add to the existing database new fish species. Current research of the authors of the present article focuses on the requirement analysis and implementation of this specific issue.

References

1. Bristow, P.: The Illustrated Encyclopedia of Fishes. Chancellor Press, London (1992)
2. Carpenter, K.: The living marine resources of the Western Central Atlantic, 1–3. FAO Species Identification Guide for Fishery Purposes and American Society of Ichthyologists and Herpetologists, Special Publication No. 5. FAO, Rome (2002)
3. Carpenter, K, Niem, V.: FAO species identification guide for fishery purposes. The living marine resources of the Western Central Pacific, 2–6. FAO, Rome (1998–2001)
4. Fischer, W., Bianchi, G.: FAO species identification guide for fishery purposes. Western Indian Ocean (Fishing Area 51), 1–6. Food and Agricultural Organization of the United Nations, Rome (1984)
5. Fischer, W., Whitehead, P.: FAO species identification sheets for fishery purposes. Eastern Indian Ocean (fishing area 57) and Western Central Pacific (fishing area 71), 1–4. FAO, Rome (1974)

6. Fischer, W., Bianchi, G., Scott, W.: FAO species identification sheets for fishery purposes. Eastern Central Atlantic. Fishing Area 34 and part of 47, 1–6. Department of Fisheries and Oceans Canada, by arrangement with the Food and Agriculture Organisation of the United Nation, Ottawa (1981)
7. Fischer, W., Bauchot, M., Schneider, M.: Fiches FAO d'identification des especes pour les besoins de la peche. (Revision 1). Mediterranee et mer Noire. Zone de peche 37, II. Vertebres. FAO, Rome (1987)
8. Froese, R., Pauly, D.: Fishbase. World Wide Web electronic publication. www.fishbase.org, version 12/2011 (2011)
9. Helfman, G., Collette, B., Facey, D., Bowen, B.: The diversity of fishes. Biology, evolution, and ecology. 2nd edition. Wiley-Blackwell, Chichester (2009)
10. Kaspiris, P.: The Fishes of Greece (Identification Keys). TYPOffset Manoudi-Stanidi, Patras, Greece (in Greek) (2000)
11. Minos, G.: Fish Biology and Systematics. 2nd Part. ATEI of Thessaloniki, Department of Aquaculture & Fisheries Technology, Greece (in Greek) (2011)
12. Perlmutter, A.: Guide to Marine Fishes. Bramhall House, New York (1961)
13. Whitehead, P., Bauchot, M., Hureau, J., Neilsen, J., Tortonese, E.: Fishes of the North-eastern Atlantic and the Mediterranean, vol. I-III. Unesco, Paris (1984)

Primal-Dual Algorithms for $P_*(\kappa)$ Linear Complementarity Problems Based on Kernel-Function with Trigonometric Barrier Term

Mohamed El Ghami

Abstract Recently, El Ghami et al. [Journal of Computational and Applied Mathematics, May, 2011, doi:10.1016/j.cam.2011.05.036.] investigated a new kernel function which differs from the self-regular kernel functions. The kernel function has a trigonometric Barrier Term. In this paper we generalize the analysis presented in the above paper for $P_*(\kappa)$ Linear Complementarity Problems (LCPs). It is shown that the interior-point methods based on this function for large-update methods, the iteration bound is improved significantly. For small-update interior point methods the iteration bound is the best currently known bound for primal-dual interior point methods. The analysis for LCPs deviates significantly from the analysis for linear optimization. Several new tools and techniques are derived in this paper.

Key words Interior-point • Kernel function • Primal-dual method • Large update, Small update • Linear complementarity problem

1 Introduction

In this paper we consider the following linear complementarity problem:

$$\begin{aligned} s &= Mx+q, \\ xs &= 0, \\ x,s &\geq 0, \end{aligned} \tag{1}$$

M.E. Ghami (✉)
Høgskolen i Nesna, 8700 Nesna, Norway
e-mail: mohamede@hinesna.no

A. Migdalas et al. (eds.), *Optimization Theory, Decision Making, and Operations Research Applications*, Springer Proceedings in Mathematics & Statistics 31,
DOI 10.1007/978-1-4614-5134-1_24,

where $M \in \mathbf{R}^{n\times n}$ is a $P_*(\kappa)$ matrix and q,x,s are vectors of $\mathbf{R}^n$, and xs denotes the componentwise product (Hadamard product) of vectors x and s. Linear complementarity problems have many applications in mathematical programming and equilibrium problems. Indeed, it is known that by exploiting the first-order optimality conditions of the optimization problem, any differentiable convex quadratic program can be formulated into a monotone linear complementarity problem, i.e. $P_*(0)$ *LCP*, and vice versa [16]. Variational inequality problems are widely used in the study of equilibrium in economics, transportation planning, and game theory, and have a close connection to the *LCPs*. The reader can refer to Sect. 5.9 in [5] for the basic theory, algorithms, and applications.

The primal-dual *IPM* for linear optimization (*LO*) problems was first introduced in [9, 12] and extended to various class of problems, e.g., [3, 14]. Kojima et al. [9] and Monteiro et al. [12] first proved the polynomial computational complexity of the algorithm for *LO* problem independently, and since then many other algorithms have been developed based on the primal-dual strategy. Kojima et al. [10] proved the existence of the central path for any $P_*(\kappa)$ *LCP*, generalized the primal-dual interior-point algorithm in [9] to $P_*(\kappa)$ *LCP* and proved the same complexity results. Miao [11] extended the Mizuno-Todd-Ye predictor-corrector method to $P_*(\kappa)$ *LCPs*. His algorithm uses the l_2-neighborhood of the central path and has $O((1+\kappa)\sqrt{n}L)$ iteration complexity. Illés and Nagy [8] give a version of the Mizuno-Todd-Ye predictor-corrector interior point algorithm for the $P_*(\kappa)$ *LCP* and show that the complexity of the algorithm is $O\left((1+\kappa)^{\frac{3}{2}}\sqrt{n}L\right)$. They choose τ and τ' neighborhood parameters in such a way that at each iteration a predictor step is followed by one corrector step. For larger value of κ the values of τ and τ' decrease fast, therefore the constant in the complexity results is increasing.

Most of the polynomial-time interior point algorithms for *LO* are based on the use of the logarithmic barrier function [9,15]. Peng et al. [14] introduced self-regular barrier functions for primal-dual interior-point methods (*IPMs*) for *LO*, semidefinite optimization (SDO), second order cone optimization (SOCO) and also extended to $P_*(\kappa)$ *LCPs*. Recently in [2, 7] the authors proposed a new primal-dual *IPM* for *LO* based on a new class of kernel functions which are not logarithmic and not necessarily self-regular barrier functions.

In this paper we propose a new large-update primal-dual *IPM* which generalizes the results obtained in [7] to $P_*(\kappa)$ *LCPs*. We use a new search direction based on kernel functions which are neither logarithmic nor self-regular barrier. The new analysis which is derived in this paper is different from the one used in early papers [8, 10, 11, 14]. Furthermore, our analysis provides a simpler way to analyze primal dual *IPMs*.

We use the following notational conventions. Throughout the paper, $\|\cdot\|$ denotes the 2-norm of a vector. The nonnegative orthant and positive orthant are denoted as $\mathbf{R}^n_+$ and $\mathbf{R}^n_{++}$, respectively. If $z \in \mathbf{R}^n_+$ and $f : \mathbf{R}_+ \to \mathbf{R}_+$, then $f(z)$ denotes the vector in $\mathbf{R}^n_+$ whose ith component is $f(z_i)$, with $1 \le i \le n$. Finally, for $x \in \mathbf{R}^n$, $X = \mathrm{diag}(x)$ is the diagonal matrix from vector x, and $J = \{1,2,\ldots,n\}$ is the index set.

This paper is organized as follows. In Sect. 2 we recall basic concepts and the notion of the central path. In Sect. 3 we review known results relevant for the development of the analysis. Section 4 contains new results to compute the feasible step size and the study of the amount of decrease of the proximity function during an inner iteration. Section 5 combines the results from Sect. 3 and the derived results in Sect. 4 to show the bound for the total number of iterations of the algorithm. Finally, concluding remarks are given in Sect. 6.

2 Preliminaries

In this section we introduce the definition of $P_*(\kappa)$ matrix and its proprieties [10].

Definition 1. Let Y be an open convex subset of $\mathbf{R}^n$ and $\kappa \geq 0$. A matrix $M \in \mathbf{R}^{n\times n}$ is called a $P_*(\kappa)$-matrix on Y if and only if

$$(1+4\kappa)\sum_{i\in J_+(x)} x_i\,(Mx)_i + \sum_{i\in J_-(x)} x_i\,(Mx)_i \geq 0,$$

for all $x \in Y$, where

$$J_+(x) = \{i \in J : x_i\,(Mx)_i \geq 0\} \quad \text{and} \quad J_-(x) = \{i \in J : x_i\,(Mx)_i < 0\}.$$

Further, M is called a P_*-matrix if it is a $P_*(\kappa)$-matrix for some $\kappa \geq 0$.

Note that the class of P_*-matrices is the union of all $P_*(\kappa)$-matrices for $\kappa \geq 0$, and contains the class of positive semi-definite matrices, i.e. symmetric matrices M satisfying $\sum_{i\in J} x_i(Mx)_i \geq 0$ for all $x \in \mathbf{R}^n$, by choosing $\kappa = 0$. The class of P_* matrices also contains matrices with all positive principal minors. In the following we recall some results which are essential in our analysis.

Proposition 1 (Lemma 4.1 in [10]). *If $M \in \mathbf{R}^{n\times n}$ is a P_* matrix, then*

$$M' = \begin{pmatrix} -M & I \\ S & X \end{pmatrix}$$

is a nonsingular matrix for any positive diagonal matrices X, $S \in \mathbf{R}^{n\times n}$.

We use the following corollary of Proposition 1 to prove that the modified Newton system has a unique solution.

Corollary 1. *Let $M \in \mathbf{R}^{n\times n}$ be a P_* matrix and x, $s \in \mathbf{R}^n_{++}$. Then for all $a \in \mathbf{R}^n$ the system*

$$-M\triangle x + \triangle s = 0,$$

$$S\triangle x + X\triangle s = a,$$

has a unique solution $(\triangle x, \triangle s)$.

The basic idea of primal-dual interior-point methods is to replace the second equation in (1) by the nonlinear equation $xs = \mu e$, where e is the all-one vector, and $\mu > 0$. Thus we have the following parameterized system:

$$\begin{aligned} s &= Mx + q, \\ xs &= \mu e, \\ x &\geq 0, \;\; s \geq 0, \end{aligned} \tag{2}$$

where $\mu > 0$. We assume that there exists strictly positive x and s that satisfy (1).

Since M is a $P_*(\kappa)$ matrix and (1) is strictly feasible, then the parameterized system (2) has a unique solution $(x(\mu), s(\mu))$ for each $\mu > 0$. $(x(\mu), s(\mu))$ is called μ-center of (2), the set of μ-centers ($\mu > 0$) defines a homotopy path, which is called the *central path* of (2). If $\mu \to 0$ the limit of the central path exists. This limit satisfies the complementarity condition and belongs to the solution set of (1) [10].

Let (x,s) be a strictly feasible point and $\mu > 0$. We define the vector

$$v := \sqrt{\frac{xs}{\mu}}. \tag{3}$$

Note that the pair (x,s) coincides with the μ-center $(x(\mu), s(\mu))$ if and only if $v = e$. Let $\Psi(v)$ be a smooth, strictly convex function defined for all $v > 0$, which is minimal at $v = e$, with $\Psi(e) = 0$. Following [1,2,4,6,14] we define search directions Δx, Δs by

$$\begin{aligned} -M\Delta x + \Delta s &= 0, \\ s\Delta x + x\Delta s &= -\mu v \nabla \Psi(v). \end{aligned} \tag{4}$$

Since M is a P_* matrix, the system (4) uniquely defines $(\Delta x, \Delta s)$ for any $x > 0$ and $s > 0$. Note that $\Delta x = 0$, $\Delta s = 0$, if and only if $v = e$, because the right-hand sides in (4) vanish if and only if $\nabla \Psi(v) = 0$, and this occurs if and only if $v = e$.

Let (x,s) be a strictly feasible point. We define the vector p by

$$p := \sqrt{\frac{x}{s}}. \tag{5}$$

Introducing the following notations

$$\bar{M} := PMP \text{ and } P := \operatorname{diag}(p),\, V := \operatorname{diag}(v) \text{ where } v = \sqrt{\frac{xs}{\mu}}, \tag{6}$$

and

$$d_x := \frac{v\Delta x}{x}, \qquad d_s := \frac{v\Delta s}{s}, \tag{7}$$

Algorithm 24.1: Generic Primal-Dual Algorithm for LCP

Input:
a proximity function $\Psi(v)$;
a threshold parameter $\tau \geq 1$;
an accuracy parameter $\varepsilon > 0$;
a barrier update parameter $\theta, 0 < \theta < 1$;
begin
$x := x^0$; $s := s^0$; $\mu := \mu^0$;
while $n\mu \geq \varepsilon$**do**
begin
$\mu := (1-\theta)\mu$;
while $\Psi(v) > \tau$**do**
begin
Solve $(\Delta x, \Delta s)$ from (4)
$x := x + \alpha\Delta x$;
$s := s + \alpha\Delta s$;
$v := \sqrt{\frac{xs}{\mu}}$;
end
end
end

system (4) can be reformulated as

$$\begin{aligned} -\bar{M}d_x + d_s &= 0, \\ d_x + d_s &= -\nabla\Psi(v). \end{aligned} \tag{8}$$

From the solution d_x and d_s, the vectors Δx and Δs can be computed from (7).

Note that the vectors d_x and d_s are not orthogonal. So our analysis in this paper will deviate significantly from the analysis used for *LO* in [7].

The algorithm considered in this paper is described in Fig. 24.1.

The inner while loop in the algorithm is called *inner iteration* and the outer while loop *outer iteration*. So each outer iteration consists of an update of the barrier parameter and a sequence of one or more inner iterations. We assume that (1) is strictly feasible, and the starting point (x^0, s^0) is strictly feasible for (1). Choose τ and $v^0 = \sqrt{\frac{x^0 s^0}{\mu^0}}$ initial strictly feasible point such that $\Psi(v^0) \leq \tau$ where τ is threshold value in Fig. 24.1. We then decrease μ to $\mu := (1-\theta)\mu$, for some $\theta \in (0,1)$. In general this will increase the value of $\Psi(v)$ above τ. To get this value smaller again, and coming closer to the current μ-center, we solve the scaled search directions from (8), and unscaled these directions by using (4). By choosing an appropriate step size α, we move along the search direction, and construct a new pair (x_+, s_+) with

$$x_+ = x + \alpha\triangle x \quad s_+ = s + \alpha\triangle s. \tag{9}$$

Table 1 Examples of kernel functions studied in early paper [6] with complexity results.

i	Kernel functions ψ_i	Large-update	Small-update
1	$\frac{t^2-1}{2} - \log t$	$O\left((1+2\kappa)n\log\frac{n}{\varepsilon}\right)$	$O\left((1+2\kappa)\sqrt{n}\log\frac{n}{\varepsilon}\right)$
2	$\frac{t^2-1}{2} + \frac{t^{1-q}-1}{q(q-1)} - \frac{q-1}{q}(t-1)$	$O\left((1+2\kappa)qn^{\frac{q+1}{2q}}\log\frac{n}{\varepsilon}\right)$	$O\left((1+2\kappa)q\sqrt{n}\log\frac{n}{\varepsilon}\right)$
3	$\frac{1}{2}\left(t-\frac{1}{t}\right)^2$	$O\left((1+2\kappa)n^{\frac{2}{3}}\log\frac{n}{\varepsilon}\right)$	$O\left((1+2\kappa)\sqrt{n}\log\frac{n}{\varepsilon}\right)$
4	$\frac{t^2-1}{2} + e^{\frac{1}{t}-1} - 1$	$O\left((1+2\kappa)\sqrt{n}\log^2 n\log\frac{n}{\varepsilon}\right)$	$O\left((1+2\kappa)\sqrt{n}\log\frac{n}{\varepsilon}\right)$
5	$\frac{t^2-1}{2} - \int_1^t e^{\frac{1}{\xi}-1}d\xi$	$O\left((1+2\kappa)\sqrt{n}\log^2 n\log\frac{n}{\varepsilon}\right)$	$O\left((1+2\kappa)\sqrt{n}\log\frac{n}{\varepsilon}\right)$
6	$\frac{t^2-1}{2} + \frac{t^{1-q}-1}{q-1},\ q>1$	$O\left((1+2\kappa)qn^{\frac{q+1}{2q}}\log\frac{n}{\varepsilon}\right)$	$O\left((1+2\kappa)q^2\sqrt{n}\log\frac{n}{\varepsilon}\right)$
7	$t-1+\frac{t^{1-q}-1}{q-1},\quad q>1$	$O\left(qn\log\frac{n}{\varepsilon}\right)$	$O\left((1+2\kappa)q^2\sqrt{n}\log\frac{n}{\varepsilon}\right)$

If necessary, we repeat the procedure until we find iterates such that $\Psi(v)$ no longer exceed the threshold value τ, which means that the iterates are in a small enough neighborhood of $(x(\mu), s(\mu))$. Then μ is again reduced by the factor $1-\theta$ and we apply the same procedure targeting at the new μ-centers. This process is repeated until μ is small enough, i.e. until $n\mu \leq \varepsilon$. At this stage we have found an ε-solution of (1). Just as in the *LO* case, the parameters τ, θ, and the step size α should be chosen in such a way that the algorithm is 'optimized' in the sense that the number of iterations required by the algorithm is as small as possible. Obviously, the resulting iteration bound will depend on the kernel function underlying the algorithm, and our main task becomes to find a kernel function that minimizes the iteration bound.

Table 1 gives some examples of kernel functions that have been analyzed in [6] with the complexity results for the corresponding algorithms.

The aim of this paper is to investigate a new kernel function studied first in linear optimization case in [7], namely

$$\psi(t) = \frac{t^2-1}{2} + \frac{6}{\pi}\tan(h(t)), \quad \text{with} \quad h(t) = \frac{\pi(1-t)}{4t+2}, \tag{10}$$

and to show that the interior-point methods for linear complementarity based on these function have favorable complexity results.

Note that the growth term of our kernel function is quadratic as all kernel functions in Table 1. However, this function (10) deviates from all other kernel functions [6] since its barrier term is trigonometric as $\frac{6}{\pi}\tan\frac{\pi(1-t)}{4t+2}$. In order to study the new kernel function, several new arguments had to be developed for the analysis.

3 Properties of the New Proximity Function

This section is started by technical lemma, and then some properties of the new kernel function introduced in this paper are derived.

3.1 Some Technical Results

The first three derivatives of ψ are given by

$$\psi'(t) = t + \frac{6h'(t)}{\pi}\left(1+\tan^2(h(t))\right), \tag{11}$$

$$\psi''(t) = 1 + \frac{6}{\pi}\left(1+\tan^2(h(t))\right)\left(h''(t)+2h'(t)^2\tan(h(t))\right). \tag{12}$$

$$\psi'''(t) = \frac{6}{\pi}\left(1+\tan^2(h(t))\right)k(t), \tag{13}$$

with

$$k(t) := 6h''(t)h'(t)\tan(h(t)) + h'''(t) + 2h'(t)^3\left(3\tan^2(h(t))+1\right). \tag{14}$$

The next lemma serves to prove that the new kernel function (10) is eligible.

Lemma 1 (Lemma 2 in [7]). *Let ψ be as defined in (10) and $t>0$. Then,*

$$\psi''(t) > 1, \tag{15a}$$

$$t\psi''(t)+\psi'(t) > 0, \tag{15b}$$

$$t\psi''(t)-\psi'(t) > 0, \tag{15c}$$

$$\text{and} \quad \psi'''(t) < 0. \tag{15d}$$

It follows that $\psi(1) = \psi'(1) = 0$ and $\psi''(t) \geq 0$, proving that ψ is defined by $\psi''(t)$.

$$\psi(t) = \int_1^t \int_1^\xi \psi''(\zeta)\,\mathrm{d}\zeta\,\mathrm{d}\xi. \tag{16}$$

The second property (15b) in Lemma 1 is related to Definition 2.1.1 and Lemma 2.1.2 in [14]. This property is equivalent to convexity of the composed function $z \mapsto \psi(e^z)$ and this holds if and only if $\psi(\sqrt{t_1t_2}) \leq \frac{1}{2}(\psi(t_1)+\psi(t_2))$ for any $t_1, t_2 \geq 0$. Following [1], we therefore say that ψ is exponentially convex, or shortly, e-convex, whenever $t>0$.

Lemma 2. *Let ψ be as defined in (10), one has*

$$\psi(t) < \frac{1}{2}\psi''(1)(t-1)^2, \quad \text{if} \quad t>1.$$

Proof. By Taylor's theorem and $\psi(1) = \psi'(1) = 0$, we obtain

$$\psi(t) = \frac{1}{2}\psi''(1)(t-1)^2 + \frac{1}{6}\psi'''(\xi)(\xi-1)^3,$$

where $1<\xi<t$ if $t>1$. Since $\psi'''(\xi)<0$, the lemma follows. □

Lemma 3. *Let ψ be as defined in (10), one has*

$$t\psi'(t) \geq \psi(t), \quad \text{if } t \geq 1.$$

Proof. Defining $g(t) := t\psi'(t) - \psi(t)$ one has $g(1) = 0$ and $g'(t) = t\psi''(t) \geq 0$. Hence $g(t) \geq 0$ and the lemma follows. □

Following [2], we now introduce a norm-based proximity measure $\delta(v)$, according to

$$\delta(v) := \frac{1}{2}\|\psi'(v)\| = \frac{1}{2}\sqrt{\sum_{i=1}^{n} \psi'(v_i)^2} = \frac{1}{2}\|d_x + d_s\|, \tag{17}$$

in terms of $\Psi(v)$. Since $\Psi(v)$ is strictly convex and attains its minimal value zero at $v = e$, we have

$$\Psi(v) = 0 \quad \Leftrightarrow \quad \delta(v) = 0 \quad \Leftrightarrow \quad v = e.$$

3.2 Relations Between Proximity Measure and Norm-Based Proximity Measure

For the analysis of the algorithm in Sect. 4 we need to establish relations between $\Psi(v)$ and $\delta(v)$. A crucial observation is that the inverse function of $\psi(t)$, for $t \geq 1$, plays an important role in this relation.

The next theorem, which is one of main results in [2], gives a lower bound on $\delta(v)$ in term of $\Psi(v)$. This is due to the fact that $\psi(t)$ satisfies (15d). The theorem is a special case of Theorem 4.9 in [2], and is therefore stated without proof.

We denote by $\rho : [0,\infty) \to [1,\infty)$ and $\rho : [0,\infty) \to (0,1]$ the inverse functions of $\psi(t)$ for $t \geq 1$, and $-\frac{1}{2}\psi'(t)$ for $t \leq 1$, respectively. In other words

$$s = \psi(t) \quad \Leftrightarrow \quad t = \rho(s), \quad t \geq 1, \tag{18}$$

$$s = -\frac{1}{2}\psi'(t) \quad \Leftrightarrow \quad t = \rho(s), \quad t \leq 1. \tag{19}$$

Theorem 1 (Theorem 4.9 in [2]). *Let ρ be as defined in (18). One has*

$$\delta(v) \geq \frac{1}{2}\psi'\left(\rho\left(\Psi(v)\right)\right).$$

Corollary 2. *Let ρ be as defined in (18). Thus we have*

$$\delta(v) \geq \frac{\Psi(v)}{2\rho\left(\Psi(v)\right)}.$$

Proof. Using Theorem 1, i.e., $\delta(v) \geq \frac{1}{2}\psi'(\rho(\Psi(v)))$, we obtain from Lemma 3

$$\delta(v) \geq \frac{\psi(\rho(\Psi(v)))}{2\rho(\Psi(v))} = \frac{\Psi(v)}{2\rho(\Psi(v))}.$$

This proves the corollary. □

Theorem 2. *If* $\Psi(v) \geq 1$, *then*

$$\delta(v) \geq \frac{1}{6}\Psi^{\frac{1}{2}}. \tag{20}$$

Proof. The inverse function of $\psi(t)$ for $t \in [1,\infty)$ is obtained by solving t from

$$\psi(t) = \frac{t^2-1}{2} + \frac{6}{\pi}\tan\frac{\pi(1-t)}{4t+2} = s, \qquad t \geq 1.$$

We derive an upper bound for t, as this suffices for our goal. One has from (16) and $\psi''(t) \geq 1$,

$$s = \psi(t) = \int_1^t \int_1^\xi \psi''(\zeta)\,\mathrm{d}\zeta\,\mathrm{d}\xi \geq \int_1^t \int_1^\xi \mathrm{d}\zeta\,\mathrm{d}\xi = \frac{1}{2}(t-1)^2,$$

which implies

$$t = \rho(s) \leq 1 + \sqrt{2s}. \tag{21}$$

Assuming $s \geq 1$, we get $t = \rho(s) \leq \sqrt{s} + \sqrt{2s} \leq 3s^{\frac{1}{2}}$. Omitting the argument v, and assuming $\Psi(v) \geq 1$, we have $\rho(\Psi(v)) \leq 3\Psi(v)^{\frac{1}{2}}$. Now, using Corollary 2, we have

$$\delta(v) \geq \frac{\Psi(v)}{2\rho(\Psi(v))} \geq \frac{1}{6}\Psi(v)^{\frac{1}{2}}.$$

This proves the lemma. □

Note that if $\Psi(v) \geq 1$, substitution in (20) gives

$$\delta(v) \geq \frac{1}{6}. \tag{22}$$

3.3 Growth Behavior of the Barrier Function

Note that at the start of each outer iteration of the algorithm, just before the update of μ with the factor $1-\theta$, we have $\Psi(v) \leq \tau$. Due to the update of μ the vector v is divided by the factor $\sqrt{1-\theta}$, with $0 < \theta < 1$, which in general leads to an increase in the value of $\Psi(v)$. Then, during the subsequent inner iterations, $\Psi(v)$ decreases until it passes the threshold τ again. Hence, during the course of the algorithm the largest values of $\Psi(v)$ occur just after the updates of μ. In this section we derive an estimate for the effect of a μ-update on the value of $\Psi(v)$. We start with an important theorem which is valid for all kernel functions $\psi(t)$ that are strictly convex (15a), and satisfies (15c).

Theorem 3 (Theorem 3.2 in [2]). *Let $\rho : [0,\infty) \to [1,\infty)$ be the inverse function of ψ on $[0,\infty)$. Then for any positive vector v and any $\beta > 1$ we have:*

$$\Psi(\beta v) \leq n\psi\left(\beta\rho\left(\frac{\Psi(v)}{n}\right)\right). \tag{23}$$

Corollary 3. *Let $0 < \theta < 1$ and $v_+ = \frac{v}{\sqrt{1-\theta}}$. Then*

$$\Psi(v_+) \leq n\psi\left(\frac{\rho\left(\frac{\Psi(v)}{n}\right)}{\sqrt{1-\theta}}\right). \tag{24}$$

Proof. Substitution of $\beta = \frac{1}{\sqrt{1-\theta}}$ into (23), the corollary is proved. □

Suppose that the barrier update parameter θ and threshold value τ are given. According to the algorithm, at the start of each outer iteration we have $\Psi(v) \leq \tau$. By Theorem 3, after each μ-update the growth of $\Psi(v)$ is limited by (24). Therefore we define

$$L = L(n,\theta,\tau) := n\psi\left(\frac{\rho\left(\frac{\tau}{n}\right)}{\sqrt{1-\theta}}\right). \tag{25}$$

Obviously, L is an upper bound of $\Psi(v_+)$, the value of $\Psi(v)$ after the μ-update.

4 Analysis of the Algorithm

In this section, we show how to compute a feasible step size α of a Newton step with the decrease of the barrier function. Since d_x and d_s are not orthogonal the analysis in this paper is different from that of LO case. After a damped step, with step size α, using (3) and (7) we have

$$x_+ = x + \alpha\Delta x = \frac{x}{v}\left(v + \alpha d_x\right), \quad s_+ = s + \alpha\Delta s = \frac{s}{v}\left(v + \alpha d_s\right).$$

Thus we obtain

$$v_+^2 = \frac{x_+ s_+}{\mu} = \left(v + \alpha d_x\right)\left(v + \alpha d_s\right). \tag{26}$$

In the sequel we use the following notation:

$$v := \min_{i\in J} v_i, \quad \delta := \delta(v), \quad \sigma_+ := \sum_{i\in J_+} d_{x_i}d_{s_i}, \quad \sigma_- := -\sum_{i\in J_-} d_{x_i}d_{s_i}. \tag{27}$$

Since M is a $P_*(\kappa)$ matrix, we have

$$(1+4\kappa)\sum_{i\in J_+}\Delta x_i(M\Delta x)_i + \sum_{i\in J_-}\Delta x_i(M\Delta s)_i \geq 0,$$

where $J_+ = \{i\in J : \Delta x_i(M\Delta x)_i \geq 0\}$, $J_- = J - J_+$. Using the first equation in (4) we have for $\Delta x \in \mathbf{R}^n$, $M\Delta x = \Delta s$, and

$$(1+4\kappa)\sum_{i\in J_+}\Delta x_i\Delta s_i + \sum_{i\in J_-}\Delta x_i\Delta s_i \geq 0.$$

From (7) it follows that $d_xd_s = \frac{v^2\Delta x\Delta s}{xs} = \frac{\Delta x\Delta s}{\mu}$ with $\mu > 0$, and

$$(1+4\kappa)\sum_{i\in J_+} d_{x_i}d_{s_i} + \sum_{i\in J_-} d_{x_i}d_{s_i} = (1+4\kappa)\sigma_+ - \sigma_- \geq 0. \tag{28}$$

The next lemma gives an upper bound of σ_+ and σ_-

Lemma 4. *One has*

$$\sigma_+ \leq \delta^2, \quad \textit{and} \quad \sigma_- \leq (1+4\kappa)\delta^2.$$

Proof. By definition of σ_+, σ_- and δ, we have

$$\sigma_+ = \sum_{i\in J_+} d_{x_i}d_{s_i} \leq \frac{1}{4}\sum_{i\in J_+}(d_{x_i}+d_{s_i})^2 \leq \frac{1}{4}\sum_{i\in J}(d_{x_i}+d_{s_i})^2 = \frac{1}{4}\|d_{x_i}+d_{s_i}\|^2 = \delta^2.$$

Since M is a $P_*(\kappa)$ matrix, using (28), we get

$$(1+4\kappa)\sigma_+ - \sigma_- \geq 0.$$

Thus

$$\sigma_- \leq (1+4\kappa)\sigma_+ \leq (1+4\kappa)\delta^2.$$

This proves the lemma. □

The following lemma gives an upper bound for $\|d_x\|$ and $\|d_s\|$.

Lemma 5. *One has*

$$\sum_{i=1}^{n}(d_{x_i}^2 + d_{s_i}^2) \leq 4(1+2\kappa)\delta^2, \quad \|d_x\| \leq 2\sqrt{1+2\kappa}\,\delta, \quad \textit{and} \quad \|d_s\| \leq 2\sqrt{1+2\kappa}\,\delta.$$

Proof. From the definitions (17) and (27), we have

$$\delta = \frac{1}{2}\left\|d_x + d_s\right\|, \quad \text{and} \quad \sum_{j\in J} d_{x_i} d_{s_i} = \sigma_+ - \sigma_-,$$

then

$$2\delta = \|d_x + d_s\| = \sqrt{\sum_{i=1}^{n}\left(d_{x_i} + d_{s_i}\right)^2} = \sqrt{\sum_{i=1}^{n}\left(d_{x_i}^2 + d_{s_i}^2\right) + 2\left(\sigma_+ - \sigma_-\right)}.$$

Using (28), and Lemma 4, we get

$$2\delta \geq \sqrt{\sum_{i=1}^{n}\left(d_{x_i}^2 + d_{s_i}^2\right) + 2\left(\frac{1}{1+4\kappa}\sigma_- - \sigma_-\right)} = \sqrt{\sum_{i=1}^{n}\left(d_{x_i}^2 + d_{s_i}^2\right) - \frac{8\kappa}{1+4\kappa}\sigma_-}.$$

Then, we get

$$4\delta^2 + \frac{8\kappa}{1+4\kappa}\sigma_- \geq \sum_{i=1}^{n}\left(d_{x_i}^2 + d_{s_i}^2\right).$$

Using again Lemma 4, we have

$$4\left(1+2\kappa\right)\delta^2 \geq 4\delta^2 + \frac{8\kappa}{1+4\kappa}\sigma_- \geq \sum_{i=1}^{n}\left(d_{x_i}^2 + d_{s_i}^2\right).$$

Thus

$$\|d_x\| \leq \sqrt{\sum_{i=1}^{n}\left(d_{x_i}^2 + d_{s_i}^2\right)} \leq 2\sqrt{1+2\kappa}\,\delta.$$

Using the same argument, we can prove that

$$\|d_s\| \leq 2\sqrt{1+2\kappa}\,\delta.$$

Thus the lemma follows. □

Our aim is to find an upper bound for

$$f(\alpha) := \Psi\left(v_+\right) - \Psi\left(v\right) := \Psi\left(\sqrt{\left(v+\alpha d_x\right)\left(v+\alpha d_s\right)}\right) - \Psi\left(v\right),$$

where $\Psi : \mathbf{R}^n \to \mathbf{R}$ is given by

$$\Psi(v) = \sum_{i=1}^{n} \psi(v_i). \tag{29}$$

To do this, the next four technical lemmas are needed. It is clear that $f(\alpha)$ is not necessarily convex in α. To simplify the analysis we use a convex upper bound for $f(\alpha)$. Such a bound is obtained by using that $\psi(t)$ satisfies the condition (15b). Hence, $\psi(t)$ is e-convex. This implies

$$\Psi(v_+) = \Psi\left(\sqrt{(v+\alpha d_x)(v+\alpha d_s)}\right) \leq \frac{1}{2}\left[\Psi(v+\alpha d_x) + \Psi(v+\alpha d_s)\right].$$

Thus we have $f(\alpha) \leq f_1(\alpha)$, where

$$f_1(\alpha) := \frac{1}{2}\left[\Psi(v+\alpha d_x) + \Psi(v+\alpha d_s)\right] - \Psi(v)$$

is a convex function of α, since $\Psi(v)$ is convex. Obviously, $f(0) = f_1(0) = 0$. Taking the derivative of $f_1(\alpha)$ to α, we get

$$f_1'(\alpha) = \frac{1}{2}\sum_{i=1}^{n}\left(\psi'(v_i+\alpha d_{xi})\, d_{xi} + \psi'(v_i+\alpha d_{si})\, d_{si}\right).$$

This gives, using last equation in (8) and (17),

$$f_1'(0) = \frac{1}{2}\nabla\Psi(v)^T(d_x+d_s) = -\frac{1}{2}\nabla\Psi(v)^T\nabla\Psi(v) = -2\delta(v)^2. \tag{30}$$

Differentiating once more, we obtain

$$f_1''(\alpha) = \frac{1}{2}\sum_{i=1}^{n}\left(\psi''(v_i+\alpha d_{xi})\, d_{xi}^2 + \psi''(v_i+\alpha d_{si})\, d_{si}^2\right). \tag{31}$$

From this stage on we can apply word-by-word the same arguments as in [6] to obtain the following results that are therefore stated without proof.

The following lemma gives an upper bound of $f_1(\alpha)$ in terms of δ and $\psi''(t)$.

Lemma 6 (Lemma 4.3 in [6]). *One has*

$$f_1''(\alpha) \leq 2(1+2\kappa)\delta^2\,\psi''\left(v - 2\alpha\sqrt{1+2\kappa}\,\delta\right).$$

Putting

$$\delta_\kappa := \sqrt{1+2\kappa}\,\delta, \tag{32}$$

we have

$$f_1''(\alpha) \leq 2\delta_\kappa^2\,\psi''(v - 2\alpha\delta_\kappa), \tag{33}$$

Since $f_1(\alpha)$ is convex, we will have $f_1'(\alpha) \leq 0$ for all α less than or equal to the value where $f_1(\alpha)$ is minimal, and vice versa. In this respect the next result is important.

Lemma 7 (Lemma 4.4 in [6]). *One has $f_1'(\alpha) \le 0$ if α satisfies the inequality*

$$-\psi'(v - 2\alpha\delta_\kappa) + \psi'(v) \le \frac{2\delta_\kappa}{(1+2\kappa)}. \tag{34}$$

The next lemma uses the inverse function $\rho : [0,\infty) \to (0,1]$ of $-\frac{1}{2}\psi'(t)$ for $t \in (0,1]$, as defined in (19).

Lemma 8 (Lemma 4.5 in [6]). *The largest value of the step size α satisfying (33) is given by*

$$\bar{\alpha} := \frac{1}{2\delta_\kappa}\left[\rho(\delta) - \rho\left(\frac{1+\sqrt{1+2\kappa}}{1+2\kappa}\delta_\kappa\right)\right]. \tag{35}$$

Moreover

$$\bar{\alpha} \ge \frac{1}{(1+2\kappa)\,\psi''\left(\rho\left(\frac{1+\sqrt{1+2\kappa}}{1+2\kappa}\delta_\kappa\right)\right)}. \tag{36}$$

For future use we define

$$\widetilde{\alpha} := \frac{1}{(1+2\kappa)\,\psi''\left(\rho\left(\frac{1+\sqrt{1+2\kappa}}{1+2\kappa}\delta_\kappa\right)\right)}, \tag{37}$$

as the default step size. By Lemma 8 this step $\widetilde{\alpha}$ satisfies (34). By (36) we have $\bar{\alpha} \ge \tilde{\alpha}$. We recall without proof the following lemma from [13].

Lemma 9 (Lemma 3.12 in [13]). *Let $h(t)$ be a twice differentiable convex function with $h(0) = 0$, $h'(0) < 0$ and let $h(t)$ attain its (global) minimum at $t^* > 0$. If $h''(t)$ is increasing for $t \in [0,t^*]$, then*

$$h(t) \le \frac{th'(0)}{2}, \quad 0 \le t \le t^*.$$

Lemma 10. *If the step size α satisfies (34), then*

$$f(\alpha) \le -\alpha\delta^2. \tag{38}$$

Proof. Let $h(\alpha)$ be defined by

$$h(\alpha) := -2\alpha\delta^2 + \alpha\delta_\kappa\psi'(v) - \frac{1}{2}\psi(v) + \frac{1}{2}\psi(v - 2\alpha\delta_\kappa).$$

Then

$$h(0) = f_1(0) = 0, \quad h'(0) = f_1'(0) = -2\delta^2, \quad h''(\alpha) = 2\delta_\kappa^2\psi''(v - 2\alpha\delta_\kappa).$$

Due to Lemma 6, $f_1''(\alpha) \le h''(\alpha)$. As a consequence, $f_1'(\alpha) \le h'(\alpha)$ and $f_1(\alpha) \le h(\alpha)$. Taking $\alpha \le \bar{\alpha}$, with $\bar{\alpha}$ as defined in Lemma 8, we have

$$\begin{aligned} h'(\alpha) &= -2\delta^2 + 2\delta_\kappa^2 \int_0^\alpha \psi''(v - 2\xi\delta_\kappa)\, d\xi \\ &= -2\delta^2 - \delta_\kappa\left(\psi'(v - 2\alpha\delta_\kappa) - \psi'(v)\right) \le 0. \end{aligned}$$

Since $h''(\alpha)$ is increasing in α, using Lemma 9, we may write

$$f_1(\alpha) \le h(\alpha) \le \frac{1}{2}\alpha h'(0) = -\alpha\delta^2.$$

Since $f(\alpha) \le f_1(\alpha)$, the proof is complete. □

Theorem 4. *Let ρ be defined in (19) and $\widetilde{\alpha}$ in (37). Then*

$$f(\widetilde{\alpha}) \le -\frac{\delta^2}{(1+2\kappa)\,\psi''\left(\rho\left(\frac{1+\sqrt{1+2\kappa}}{\sqrt{1+2\kappa}}\,\delta\right)\right)} \le -\frac{\delta^{\frac{1}{2}}}{2593\,(1+2\kappa)}, \tag{39}$$

Proof. By combining (36) in Lemma 8 and results in Lemma 10, using also (32). Thus the first inequality in (39) follows.

To obtain the inverse function $t = \rho(s)$ of $-\frac{1}{2}\psi'(t)$ for $t \in (0,1]$, we need to solve t from the equation $-\left(t + \frac{6h'(t)}{\pi}\left(1+\tan^2(h(t))\right)\right) = 2s$. This implies

$$1+\tan^2(h(t)) = \frac{-\pi}{6h'(t)}(2s+t) = \frac{2\pi\,(2t+1)^2}{18\pi}(2s+t) \le 2s+1 \quad \text{for} \quad t \le 1.$$

Hence, putting $t = \rho\left(\frac{1+\sqrt{1+2\kappa}}{\sqrt{1+2\kappa}}\,\delta\right)$, which is equivalent to $\frac{2\left(1+\sqrt{1+2\kappa}\right)}{\sqrt{1+2\kappa}}\delta = -\psi'(t)$. Using that $\frac{1+\sqrt{1+2\kappa}}{\sqrt{1+2\kappa}} \le 2$ for all $\kappa \ge 0$, we get

$$\tan(h(t)) \le 2\sqrt{\delta}. \tag{40}$$

By (40), thus we have

$$\begin{aligned} \widetilde{\alpha} &= \frac{1}{(1+2\kappa)\,\psi''\left(\rho\left(\frac{1+\sqrt{1+2\kappa}}{\sqrt{1+2\kappa}}\,\delta\right)\right)} \\ &= \frac{1}{(1+2\kappa)\left(1+\frac{6}{\pi}\left(1+\tan^2(h(t))\right)\left(h''(t)+2h'(t)^2\tan(h(t))\right)\right)} \\ &\ge \frac{1}{(1+2\kappa)\left(1+\frac{6}{\pi}(1+4\delta)\left(h''(t)+4h'(t)^2\sqrt{\delta}\right)\right)}. \end{aligned}$$

Since $h''(t) = \frac{6\pi}{(2t+1)^3} \le 6\pi$, and $h'(t)^2 = \frac{9\pi^2}{4(2t+1)^4} \le \frac{9\pi^2}{4}$ for all $0 \le t \le 1$. Then we have

$$\widetilde{\alpha} \ge \frac{1}{(1+2\kappa)\left(1+\frac{6}{\pi}(1+4\delta)\left(6\pi+9\pi^2\sqrt{\delta}\right)\right)}$$
$$= \frac{1}{(1+2\kappa)\left(1+18(1+4\delta)\left(2+3\pi\sqrt{\delta}\right)\right)}.$$

Also using (22) (i.e., $6\delta \ge 1$) we get,

$$\widetilde{\alpha} \ge \frac{1}{(1+2\kappa)\left((6\delta)^{\frac{3}{2}}+18(6\delta+4\delta)\left(2\sqrt{6\delta}+3\pi\sqrt{\delta}\right)\right)}$$
$$= \frac{1}{(1+2\kappa)\left(\left(6^{\frac{3}{2}}+180\left(2\sqrt{6}+3\pi\right)\right)\delta^{\frac{3}{2}}\right)} \ge \frac{1}{2593(1+2\kappa)\delta^{\frac{3}{2}}}.$$

Hence

$$f(\widetilde{\alpha}) \le -\frac{\delta^2}{2593(1+2\kappa)\delta^{\frac{3}{2}}} = -\frac{\delta^{\frac{1}{2}}}{2593(1+2\kappa)}.$$

Thus, the theorem follows. □

Substitution in (20) gives

$$f(\tilde{\alpha}) \le -\frac{\delta^{\frac{1}{2}}}{2593(1+2\kappa)} \le -\frac{\Psi^{\frac{1}{4}}}{2593\sqrt{6}(1+2\kappa)} \le -\frac{\Psi^{\frac{1}{4}}}{6532(1+2\kappa)}.$$

5 Iteration Complexity

In this section we derive the complexity bounds for large-update methods and small-update methods.

5.1 *Upper Bound for the Total Number of Iterations*

Let K denote the number of inner iterations. An upper bound for the total number of iterations is obtained by multiplying (the upper bound for) the number K by the number of barrier parameter updates, which is bounded above by $\frac{1}{\theta}\log\frac{n}{\varepsilon}$ (cf. [15] Lemma II.17, page 116).

Lemma 11 (Proposition 2.2 in [13]). *Let $t_0, t_1, \ldots, t_K$ be a sequence of positive numbers such that*

$$t_{k+1} \leq t_k - \kappa t_k^{1-\gamma}, \qquad k = 0, 1, \ldots, K-1,$$

where $\kappa > 0$ and $0 < \gamma \leq 1$. Then $K \leq \left\lfloor \frac{t_0^\gamma}{\kappa\gamma} \right\rfloor$.

Lemma 12. *If K denotes the number of inner iterations, we have*

$$K \leq \frac{26128(1+2\kappa)}{3} \Psi_0^{\frac{3}{4}} \leq 8710(1+2\kappa) \Psi_0^{\frac{3}{4}}.$$

Proof. The definition of K implies $\Psi_{K-1} > \tau$ and $\Psi_K \leq \tau$ and

$$\Psi_{k+1} \leq \Psi_k - \kappa (\Psi_k)^{1-\gamma}, \qquad k = 0, 1, \ldots, K-1,$$

with $\kappa = \frac{1}{6532(1+2\kappa)}$ and $\gamma = \frac{3}{4}$. Application of Lemma 11, with $t_k = \Psi_k$ yields the desired inequality. □

Using $\psi_0 \leq L$, where the number L is as given in (25), and Lemma 12 we obtain the following upper bound on the total number of iterations:

$$\frac{8710(1+2\kappa) L^{\frac{3}{4}}}{\theta} \log \frac{n}{\varepsilon}. \tag{41}$$

5.2 Large-Update

We just established that (41) is an upper bound for the total number of iterations, using

$$\psi(t) = \frac{t^2-1}{2} + \frac{6}{\pi} \tan \frac{\pi(1-t)}{4t+2} \leq \frac{t^2-1}{2}, \quad \text{for} \quad t \geq 1,$$

and (21), by substitution in (25) we obtain

$$L \leq n \frac{\left(\frac{\rho\left(\frac{\tau}{n}\right)}{\sqrt{1-\theta}} \right)^2 - 1}{2} \leq \frac{n}{2(1-\theta)} \left(\theta + 2\sqrt{2\frac{\tau}{n}} + \frac{2\tau}{n} \right) = \frac{\left(\theta n + 2\sqrt{2\tau n} + 2\tau\right)}{2(1-\theta)}.$$

Using (41), thus the total number of iterations is bounded above by

$$\frac{K}{\theta} \log \frac{n}{\varepsilon} \leq \frac{8710(1+2\kappa)}{\theta \left(2(1-\theta)^{\frac{3}{4}}\right)} \left(\theta n + 2\sqrt{2\tau n} + 2\tau\right)^{\frac{3}{4}} \log \frac{n}{\varepsilon}.$$

A large-update methods uses $\tau = O(n)$ and $\theta = \Theta(1)$. The right-hand side expression is then $O\left((1+2\kappa)n^{\frac{3}{4}}\log\frac{n}{\varepsilon}\right)$, as easily may be verified.

5.3 *Small-Update Methods*

For small-update methods one has $\tau = O(1)$ and $\theta = \Theta\left(\frac{1}{\sqrt{n}}\right)$. Using Lemma 2, with $\psi''(1) = \frac{2\pi+9}{9}$, we then obtain

$$L \leq \frac{n(2\pi+9)}{18}\left(\frac{\rho\left(\frac{\tau}{n}\right)}{\sqrt{1-\theta}} - 1\right)^2.$$

Using (21), then

$$L \leq \frac{n(2\pi+9)}{18}\left(\frac{1+\sqrt{\frac{2\tau}{n}}}{\sqrt{1-\theta}} - 1\right)^2.$$

Using $1-\sqrt{1-\theta} = \frac{\theta}{1+\sqrt{1-\theta}} \leq \theta$, this leads to $L \leq \frac{(2\pi+9)}{18(1-\theta)}\left(\theta\sqrt{n}+\sqrt{2\tau}\right)^2$. We conclude that the total number of iterations is bounded above by

$$\frac{K}{\theta}\log\frac{n}{\varepsilon} \leq \frac{8710(1+2\kappa)(2\pi+9)^{\frac{3}{4}}}{\theta(18(1-\theta))^{\frac{3}{4}}}\left(\theta\sqrt{n}+\sqrt{2\tau}\right)^{\frac{3}{2}}\log\frac{n}{\varepsilon}.$$

Thus, the right-hand side expression is then $O\left((1+2\kappa)\sqrt{n}\log\frac{n}{\varepsilon}\right)$.

6 Concluding Remarks

In this paper we extended the results obtained for kernel-function-based IPMs in [7] for LO to $P_*(\kappa)$ linear complementarity problems. The observation that the vectors d_x and d_s are not in general orthogonal implies that the analysis in [7] does not hold. The analysis in this paper is new and different from the one using for *LO*. Several new tools and techniques are derived in this paper. The proposed function has a trigonometric barrier term but the function is not logarithmic and not self-regular. We proved that the iteration bound of a large-update interior-point method based on the kernel function considered in this paper is $O\left((1+2\kappa)n^{\frac{3}{4}}\log\frac{n}{\varepsilon}\right)$, which improves the classical iteration complexity with a factor $n^{\frac{1}{4}}$. For small-update methods we obtain the best known iteration bound, namely $O\left((1+2\kappa)\sqrt{n}\log\frac{n}{\varepsilon}\right)$.

The resulting iteration bounds for $P_*(\kappa)$ linear complementarity problems depend on the parameter κ. For $\kappa = 0$, the iteration bounds are the same as the bounds that were obtained in [7] for linear optimization.

References

1. Bai, Y.Q., Ghami, M.El, Roos, C.: A new efficient large-update primal-dual interior-point method based on a finite barrier. SIAM J. Optim. **13**(3), 766–782 (2003)
2. Bai, Y.Q., Ghami, M.El., Roos, C.: A comparative study of kernel functions for primal-dual interior-point algorithms in linear optimization. SIAM J. Optim. **15**(1), 101–128 (2004)
3. Cho, E.M.: Log-barrier method for two-stagequadratic stochastic programming. Appl. Math. Comput. **164**, 45–69 (2005)
4. Cho, G.M., Kim, M.-K.: A new Large-update interior point algorithm for $P_*(\kappa)$ LCPs Based on kernel functions. Appl. Math. Comput. **182**, 1169–1183 (2006)
5. Cottle, R., Pang, J.S., Stone, R.E.: The Linear Complementarity Problem. Academic, Boston (1992)
6. Ghami, M.El., Steihaug, T.: Kernel-function based primal-dual algorithms for $P_*(\kappa)$ linear complementarity problems. RAIRO-Oper. Res. **44**(3), 185–205 (2010)
7. Ghami, M.El., Guennoun, Z.A., Steihaug, S., Bouali T.: Primal-Dual Interior-Point Methods for Linear Optimization Based on a Kernel Function with Trigonometric Barrier Term. J. Comput. Appl. Math. **236**(15), 3613–3623 (2012).
8. Illés, T., Nagy, M.: The Mizuno-Todd-Ye predictor-corrector algorithm for sufficient matrix linear complementarity problem. Alkalmaz. Mat. Lapok **22**(1), 41–61 (2005)
9. Kojima, M., Megiddo, N., Noma, T., Yoshise, A.: A primal-dual interior point algorithm for linear programming. In: Megiddo, N. (ed.) Progress in Mathematical Programming, Interior Point Related Methods, vol. 10, pp. 29–47. Springer, New York (1989)
10. Kojima, M., Megiddo, N., Noma, T., Yoshise, A.: In: A Unified Approach to Interior Point Algorithms for Linear Complementarity Problems. LNCS, vol. 538. Springer, Berlin (1991)
11. Miao, J.: A quadratically convergent $o((1+k)\sqrt{n}l$-iteration algorithm for the $p_*(k)$-matrix linear complementarity problem. Math. Program. **69**, 355–368 (1995)
12. Monteiro, R.D.C., Adler, I.: Interior path following primal-dual algorithms. Part I: Linear programming. Math. Program. **44**, 27–41 (1989)
13. Peng, J., Roos, C., Terlaky, T.: Self-regular functions and new search directions for linear and semidefinite optimization. Math. Program. **93**, 129–171 (2002)
14. Peng, J., Roos, C., Terlaky, T.: Self-Regularity: A New Paradigm for Primal-Dual Interior-Point Algorithms. Princeton University Press, Princeton (2002)
15. Roos, C., Terlaky, T., Vial, J.Ph.: Theory and Algorithms for Linear Optimization. An Interior-Point Approach. Springer, New York (2006)
16. Wright, S.J.: Primal-Dual Interior-Point Methods. SIAM, Philadelphia (1997).

An Approximation Algorithm for the Three Depots Hamiltonian Path Problem

Aristotelis Giannakos, M'hand Hifi, Rezika Kheffache, and Rachid Ouafi

Abstract In this paper, we propose an approximation algorithm for solving the three depots Hamiltonian path problem (3DHPP). The problem studied can be viewed as a variant of the well-known Hamiltonian path problem with multiple depots (cf., Demange [Mathématiques et Informatique, Gazette, 102 (2004)] and Malik et al. [Oper. Res. Lett. **35**, 747–753 (2007)]). For the 3DHPP, we show the existence of a $\frac{3}{2}$-approximation algorithm for a broad family of metric cases which also guarantees a ratio $r < 2$ in the general metric case. The proposed algorithm is mainly based on extending the construction scheme already used by Rathinam et al. [Oper. Res. Lett. **38**, 63–68 (2010)]. The aforementioned result is established for a variant of the three-depot problem, that is, when costs are symmetric and satisfy the triangle inequality.

Key words Approximation algorithms • Hamiltonian path problem • Traveling salesman problem

A. Giannakos (✉) • M. Hifi
Picardie University, 33 Rue Saint Leu 80039 Amiens Cedex 1 France
e-mail: aristotelis.giannakos@u-picardie.fr; hifi@u-picardie.fr

R. Kheffache
Mouloud Mammeri University, Faculty of Science Tizi Ouzon, Algeria
e-mail: kheffache.rezika@yahoo.fr

R. Ouafi
University of Technology, Bab Ezzouar, Algeria
e-mail: rouafi@usthb.dz

A. Migdalas et al. (eds.), *Optimization Theory, Decision Making, and Operations Research Applications*, Springer Proceedings in Mathematics & Statistics 31,
DOI 10.1007/978-1-4614-5134-1_25,

1 Introduction

The multiple traveling salesman problem (MTSP) is a generalization of the well known traveling salesman problem (TSP), in which there are m salesmen who start from and terminate to the depot (cf. [6]). The multi-depots TSP is a also a generalization of the TSP. It is well-known that the TSP has received important attention by the OR and CS communities; on the other hand, the study of the generalized multi depots and other variants of the problem remains limited. In this paper, we address the three depots Hamiltonian path problem, namely 3DHPP, which can be viewed as a variant of the well-known Hamiltonian path problem with multiple depots (cf., [3]). For the 3DHPP, we show the existence of a $\frac{3}{2}$-approximation algorithm for a broad family of metric cases. The proposed algorithm is mainly based on extending the construction scheme already used by Rathinam et al. [7]. The aforementioned result is established when costs are symmetric and they satisfy the triangle inequality.

Let $D = \{d_1, d_2, d_3\}$ be the set of vertices representing the three distinct depots, $U = \{1, 2, 3, \ldots, n\}$ be the set of vertices denoting n destinations such that $n \geq 2$ and $V = D \bigcup U$. The edge (i, j) joining vertices i and j has a cost $C_{ij} \in Q^+$, where Q^+ denotes the set of all positive rational numbers. Assume that all costs are symmetric (i.e., $\forall\ (i, j) \in V,\ Cij = Cji$) and that they satisfy the following triangle inequality: $C_{ik} \leq C_{ij} + C_{jk},\ \forall\ (i, j, k) \in V$. A path for salesman l may be denoted by an ordered sequence of vertices $P_l = (d_l, v_1^l, v_2^l, \ldots, v_{k_l}^l)$, where $l = 1, 2, 3$ and k_l represents the number of vertices visited by the lth salesman and $\forall j \in \{1, \ldots, k_l\}$, $v_{lj} \in U$. The cost of the path P_l traveled by the lth salesman is defined as follows:

$$C(P_l) = \begin{cases} C(d_l, v_1^l) + \Sigma_{j=1}^{k_l - 1} C(v_j^l, v_{j+1}^l) & \text{if } k_l > 0 \\ 0 & \text{otherwise, when } l = 1, 2, 3. \end{cases}$$

The objective of the problem is to find the paths P_1, P_2, and P_3 such that:

1. Each destination of U is visited exactly once by any salesman,
2. Each salesman visits at least one destination, and
3. The sum of the costs corresponding to the salesman, i.e., $\Sigma_{l=1}^{3} C(P_l)$, is minimum.

For the rest of the paper, we shall denote the studied problem by 3DHPP, for *3-Depots Hamiltonian Path Problem*. The remainder of the paper is organized as follows: first, in Sect. 2 we discuss the main steps of the approach used for tailoring the approximation algorithm for the 3DHPP. In Sect. 3 an example is given in order to illustrate the main steps of the proposed algorithm. Finally, the main results is announced in Sect. 4 for which the sketch of the proof is given along the example.

2 An Approximation Algorithm for the 3DHPP

Herein, we will show that 3DHPP has an approximation algorithm, when the following steps are used. Indeed, the proposed algorithm can be viewed as a five-step algorithm in which a series of decomposition, construction, and reconstructions are applied. These steps are described as follows:

1. Find a *minimum cost constrained forest*, namely F, such that, there are three trees in the forest, where

   (a) Each tree is characterized by its depot.
   (b) The degree of each depot is equal to 1.

   Notice that the number of odd-degree, the set of destination vertices in each tree, is odd.

   Let $E(F)$ be the edges in the constrained forest and O_i, for $i = 1,2,3$, be the odd-degree (i.e., the destination vertices in the ith tree of F).
2. Let consider $(V_O, E(V_O))$ be a graph defined as follows:

   (a) V_O denotes the set of the odd-degree destination vertices of F,
   (b) $E(V_O)$ contains the edges lying any two vertices in V_O.

   Note that, because $|O_1|$, $|O_2|$ and $|O_3|$ are odd, then

   $$|V_O| = |O_1| + |O_2| + |O_3| \text{ is odd.}$$

   Now, find a *partial matching* M, i.e., a set of edges $E(M) \subseteq E(V_O)$ whose cardinality is equal to $\frac{|V_O|-3}{2}$, which also matches $|V_O| - 3$ vertices in V_O. Of course, such a matching realizes the *partial matching with a minimum cost.* Hence, let m_1, m_2 and m_3 denote the three destination vertices that are note matched.
3. Add the edges of the minimum cost, associated with the *partial matching* provided by step (2) to the *minimum cost constrained forest* obtained by step (1), and connect both depots vertices d_1 and d_2 by using an edge whose cost is equal to zero cost, and also the destinations m_2 and m_3.

   Consider the following new provided multigraph:

   $$G_n = (V, E(M) \bigcup E(F) \bigcup \{(d_1, d_2, d_3)\}).$$

   We can observe that the degree of each vertex of G_n is even except for the two vertices m_1 and d_3.
4. Find an Eulerian path, namely E in G_n, such that the path starts from the vertex m_1, terminates with the vertex d_3 and visits each of the edges in G_n exactly once.
5. Apply a shortcut phase of the edges in the Eulerian path to get a Hamiltonian path such that each vertex in V is visited exactly once. After such a shortcutting phase, let's denote the obtained Hamiltonian path as follows:

   $$E_{sc} = (m_1, \ldots, d_1, d_2, \ldots, m_2, m_3, \ldots, d_3).$$

Then, the sequence of destinations which is:

(a) To assign to the second salesman the subsequence of the provided Hamiltonian path which starts from d_2 and terminates at m_2.
(b) To assign to the first salesman the reverse of the remaining subsequence of the provided Hamiltonian path which starts from d_1 and terminates at m_1.
(c) To assign to the third salesman the reverse of the remaining subsequence of the Hamiltonian path which has d_3 as the starting vertex and m_3 its terminal vertex.

is the returned solution by the algorithm.

3 Example

In this section, we show how the different steps of the algorithm can be applied in order to construct the final solution for the approximation algorithm.

Indeed, according to the first step, Fig. 1 can show how the three trees can be computed and how each one of these trees can be relied to the depot vertex, i.e., d_1, d_2, and d_3, respectively.

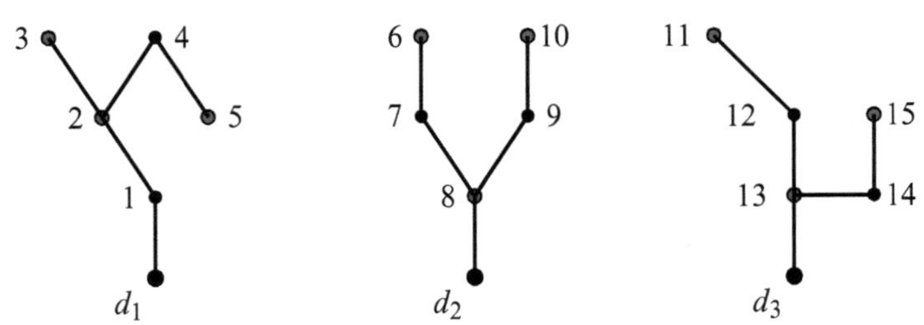

Fig. 1 Step 1:(b) The odd-degree destination vertices $|V_0| = |O_1 \bigcup O_2 \bigcup O_3| = 9$

Of course, we can observe that such a forest (noted F) is a minimum cost constrained one. We also can remark that the construction respects the conditions (a) and (b) of step 1.

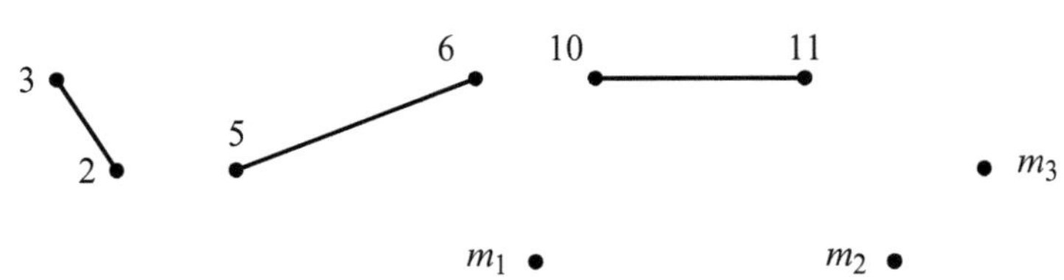

Fig. 2 A minimum cost partial matching of cardinality equal to $\frac{|V_0|-3}{2} = 3$

We recall that step 2 serves to find a partial matching of a special cardinality. Indeed, Fig. 2 illustrates such a partial matching satisfying the cardinality of 3 fixed at step 2.

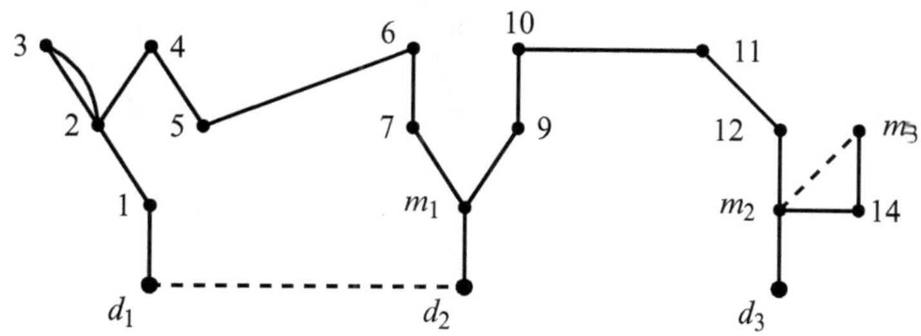

Fig. 3 Step 3 and 4: Add the edges of M to F. Also add a zero cost edges (d_1,d_2) and (m_2,m_3); Eulerian path=$\{m_1,7,6,5,4,2,3,2,1,d_1,d_2,m_1,9,10,11,12,m_2,m_3,14,m_2,d_3\}$

Following both steps 3 and 4, we can observe (cf. Fig. 3) how we use the adding-phase in order to get the Eulerian path, namely E.

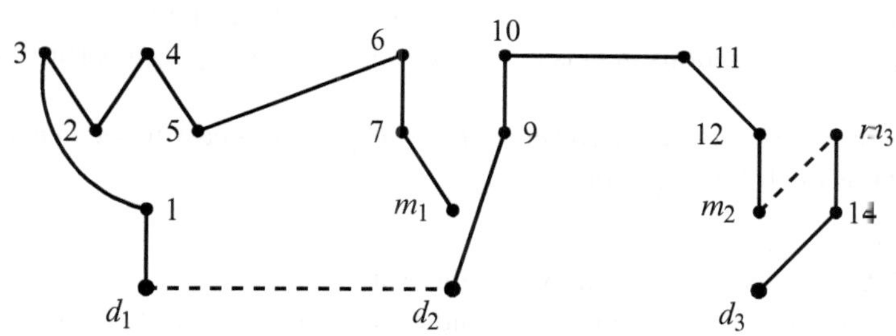

Fig. 4 Step 5: Shortcut the edges of the Eulerian path to find a Hamiltonian path $=\{m_1,7,6,5,4,2,3,1,d_1,d_2,9,10,11,12,m_2,m_3,14,d_3\}$

By applying the shortcutting phase of step 5, we can observe the result (cf. Fig. 4) the new Hamiltonien path, namely H, provided from the previously Eulerian path E.

Finally, according to the last three points of step 5, which corresponds to the removing phase, Fig. 5 illustrates the final solution provided by the algorithm.

Theorem 1. *The algorithm achieves a $\frac{3}{2}$ approximation for a three-depot Hamiltonian path problem (3DHPP)for a broad family of instances where the costs are symmetric and satisfy the triangle inequality, with complexity $O(n^3)$ steps.*

Proof. (a) Show that the algorithm is of complexity $O(n^3)$:

In the algorithm, we seek a forest, a matching, Eulerian path and Hamiltonian path.

The complexity of the algorithm is dominated by two steps: finding a forest of minimum cost and finding a partial matching of the minimum cost.

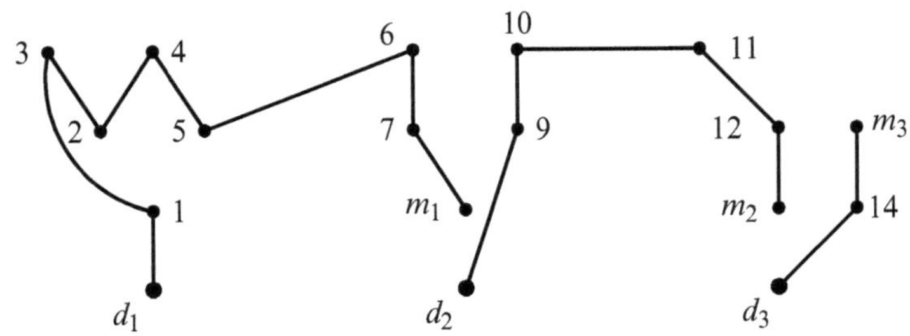

Fig. 5 Step 5: Remove the zero cost edges to find the paths for the 3 salesmen:$\{d_1,1,3,2,4,5,6,7,m_1\}$, $\{d_2,9,10,11,12,m_2\}$,$\{d_3,14,m_3\}$

Finding a forest can be seen as a problem of intersection of two matroids. Ended to see this:

Let V be the set of all vertices: $V = n+3$, and E be the set of edges joining any two vertices of V. Let F_1 a family of subsets such that each $F \in F_1$: the graph $G = (V, F_1)$ is acyclic and there is no path connecting depots in G. $M_1 = (E, F_1)$ is a graphic matroid.

Consider F_2 to be a family of subsets of E such that for every $F \in F_2$, the degree of each depot in $G = (V, E)$ is at most 1 and the number of edges joining every two destinations in G is at most $n-3$. $M_2 = (E, F_2)$ is a partition matroid [2,5].

If one considers each basic element $I \in F_1 \bigcap F_2$, it is easy to see that it satisfies contains the following properties:

- I is acyclic.
- I does not contain a path joining deposits.
- The degree of each deposit is at most 1 and the number of edges joining two destinations is at most $n-3$.

I contains n edges, so these properties mean that every depot is of degree 1, and the problem of minimum cost of forest can be seen as a problem of intersection of two matroids M_1 and M_2, thus it can be found in $O(n^3)$ using the algorithm Brezovec [1]

In step 2, we look for a partial matching of the minimum cost, this can be solved in $O(n^3)$ by applying some classic matching algorithm, for instance the algorithm of Edmonds [4]

The complexity of step (4) and (5) is $O(n)$, respectively, hence the complexity of the algorithm is $O(n^3)$.

(b) We now show that the approximation ratio of this algorithm is $\frac{3}{2}$ for a broad family of metric instances.

It is clear that the solution S produced by the algorithm is bounded by the sum of the cost(F) and cost(M) because the algorithm determines a forest of minimum cost and a minimum cost matching, then applies shortcuts of the edges in the Eulerian path:

$\text{cost}(S) \leq \text{cost}(F) + \text{cost}(M)$.
We know that $\text{cost}(F) \leq \text{cost}(opt)$, it is then enough to show that the cost(M) is bounded by cost(opt).

The idea of proof is to show that any optimal solution of 3DHPP consists of at least two partial pairwise disjoint matchings of cardinality $\frac{|V_0|-3}{2}$ for the vertices of V_0.
To see this, consider any optimal solution of the 3DHPP. In this solution, there are three pairwise disjoint paths P_1^*, P_2^* and P_3^* corresponding to salesman 1, 2, and 3, respectively.

We define $z_l = \{v : v \in V_0, v \in P_l^*\}$ for l = 1,2,3. Every vertex of z_l is a destination visited by the lth salesman in the optimal solution that also is an odd degree vertex of the minimum cost spanning forest found by the algorithm.

Note that $z_1 \bigcap z_2 \bigcap z_3 = \emptyset$, $z_1 \bigcup z_2 \bigcup z_3 = V_0$ and $|z_1| + |z_2| + |z_3| = |V_0|$ is odd.
Shortcut the edges in the optimal solution such that the paths P_1, P_2 and P_3 are only incident on the vertex present in $z_1, z_2,$and z_3. We now show that it is possible to decompose the set of edges present in paths into two disjoint set of matchings edges as follows:

Case 1. $Z(P(d_1))$ **odd**, $Z(P(d_2)) = Z(P(d_3)) = 0$

The chain that connects the vertices of $Z(P(d_1))$ in order of their occurrence to $P(d_1)$ is the cost or less for $P(d_1)$ So at cost (opt) it can be decomposed into two matchings that leave three vertices of $Z(P(d_1))$ non-paired as follows:
Let $Z(P(d_1)) = \{u_1, u_2, \ldots, u_{2p+1}\}$, $|P(d_1)| = 2p+1 \geq 5$; in this case we consider the following decomposition:
$E_1 = \{(u_2,u_3),(u_4,u_5),\ldots,(u_{2p-4},u_{2p-3}),(u_{2p-2},u_{2p-1})\}$. E_1 does not cover the vertices u_1, u_{2p} and u_{2p+1}.
$E_2 = \{(u_3,u_4),(u_5,u_6),\ldots,(u_{2p-1},u_{2p})\}$, E_2 does not cover the vertices u_1, u_2 and u_{2p+1}.
E_1 and E_2 are two disjoint matchings of cardinality $\frac{|V_0|-3}{2}$, each cost less than the cost (M).

Case 2. $Z(P(d_1))$ **odd**, $Z(P(d_2))$ **even not bold** $Z(P(d_3)) = 0$

Same decomposition in two matchings that leave one vertex of $Z(P(d_1))$ uncoupled and two of $Z(P(d_2))$.
Let $Z(P(d_1)) = \{u_1,u_2,\ldots,u_{2p+1}\}$, $Z(P(d_2)) = \{v_1,v_2,\ldots,v_{2k}\}$; in this case we consider two disjoint sets E_1, E_2 with:
$E_1 = \{(u_1,u_2)(u_3,u_4),\ldots,(u_{2p-1},u_{2p})\} \bigcup \{(v_2,v_3)(v_4,v_5),\ldots,(v_{2k-2},v_{2k-1})\}$ E_1 does not cover the vertices u_{2p+1}, v_1 and v_{2k}.
$E_2 = \{(u_2,u_3)(u_4,u_5),\ldots,(u_{2p},u_{2p+1})\} \bigcup \{(v_2,v_3)(v_4,v_5),\ldots,(v_{2k-2},v_{2k-1})\}$,
E_2 does not cover the vertices u_1, v_1 and v_{2k}.
E_1 and E_2 are two disjoint matchings of cardinality $\frac{|V_0|-3}{2}$, so each one costs less than the cost (M).

Case 3. $Z(P(d_1))$ **is odd**, $Z(P(d_2))$ **and** $Z(P(d_3))$ **are even**

Same decomposition in two matchings: one that leaves one vertex of $Z(P(d_1))$ uncoupled and no $Z(P(d_2))$ and the other leaves a $Z(P(d_1))$ and two $Z(P(d_2))$

Let $Z(P(d_1)) = \{u_1, u_2, \ldots, u_{2p+1}\}$, $Z(P(d_2)) = \{v_1, v_2, \ldots, v_{2k}\}$ and $Z(P(d_3)) = \{w_1, w_2, \ldots, w_{2r}\}$; in this case we consider two disjoint sets E_1, E_2 with:

$E_1 = \{(u_1,u_2)(u_3u_4), \ldots, (u_{2p-1}, u_{2p})\} \bigcup \{(v_1,v_2)(v_3,v_4), \ldots, (v_{2k-1}, v_{2k})\} \bigcup \{(w_2,w_3)(w_4,5), \ldots, (w_{2r-2}, w_{2r-1})\}$.

E_1 does not cover the vertices u_{2p+1}, w_1, and w_{2r}.

$E_2 = \{(u_2,u_3)(u_4u_5), \ldots, (u_{2p}, u_{2p+1})\} \bigcup \{(v_2,v_3)(v_4v_5), \ldots, (v_{2k-2}, v_{2k-1})\} \bigcup \{(w_1,w_2), (w_3,4), \ldots, (w_{2r-1}, w_{2r})\}$.

E_2 does not cover the vertices u_1, v_1, and v_{2k}.

E_1 and E_2 are two disjoint matchings of cardinality $\frac{|V_0|-3}{2}$, each cost less than the cost (M).

Case 4. $Z(P(d_1)), Z(P(d_2))$ **and** $Z(P(d_3))$ **are odd**

Same decomposition in two matchings, each one leaving a vertex not coupled to each $Z(P(d_i))$.

Let $Z(P(d_1)) = \{u_1, u_2, \ldots, u_{2p+1}\}$, $Z(P(d_2)) = \{v_1, v_2, \ldots, v_{2k+1}\}$ and $Z(P(d_3)) = \{w_1, w_2, \ldots, w_{2r+1}\}$, in this case we consider two disjoint sets E_1, E_2 with:

$E_1 = \{(u_1,u_2), (u_3,u_4), \ldots, (u_{2p-1}, u_{2p})\} \bigcup \{(v_1,v_2)(v_3,v_4), \ldots, (v_{2k-1}, v_{2k})\} \bigcup \{(w_1,w_2)(w_3,4), \ldots, (w_{w2r-1}, w_{2r})\}$.

E_1 does not cover the vertices u_{2p+1}, v_{2k+1} and w_{2r+1}.

$E_2 = \{(u_2,u_3), (u_3,u_4), \ldots, (u_{2p}, u_{2p+1})\} \bigcup \{(v_2,v_3)(v_4,v_5), \ldots, (v_{2k}, v_{2k+1})\} \bigcup \{(w_2,w_3)(w_4,5), \ldots, (w_{2r}, w_{2r+1})\}$.

E_2 does not cover the vertices u_1, v_1 and w_1.

E_1 and E_2 are two disjoint matchings of cardinality $\frac{|V_0|-3}{2}$, each cost less than the cost (M).

Hence, $C(S) \leq \frac{3}{2}C(opt)$.

This analysis applies to nearly all metric instance of 3DHPP that can be input of the algorithm of Sect. 2, except for the case where the multigraph G_n consists of two connected components, one with no destination vertices of odd degree and a second with three uncoupled odd degree destination vertices. In this case, we should add one more edge to couple two among these vertices. It is straightforward to see that, even in this case, the added edge can never be of cost such that the returned solution is ≥ 2 times the cost of the optimal.

4 Conclusion

To our knowledge, there is no approximation algorithm better than 2 for metric 3DHPP. Herein, we propose an algorithm which admits an approximation ratio of $\frac{3}{2}$ for a broad family of metric cases and guarantees a ratio $r < 2$ in the general metric case.

References

1. Bresovec, C., Cornuejols, G., Glover, F.: A matroid algorithm and its application to the efficient solution of two optimization problems on graphs. Math. Program. **42**, 471–487 (1998)
2. Cerdeira, J.O.: Matroids and a forest cover problem. Math. Program. **66**, 403–405 (1994)
3. Demange, M.: Algorithme d'approximation: un petit tour en compagnie d'un voyageur de commerce. Lagazette des Mathématiques. **102**, 51–90 (2004)
4. Edmonds, J.: Maximum matching and a polyhedron with 0,1-vertices. J. Res. Natl. Bur. Stand. **69B**, 125–130 (1965)
5. Lawler, E.L.: Combinatorial Optimisation: Networks and Matroids. Dover Publication, New York (2001)
6. Malik, W., Rathinam, S., Darbha, S.: An approximation algorithm for a symmetric generalized multiple depot, multiple traveling salesman problem. Oper. Res. Lett. **35**, 747–753 (2007)
7. Rathinam, S., Sengupta, R.: $\frac{3}{2}$-approximation algorithm for two variants of 2-depot Hamiltonian path problem. Oper. Res. Lett. **38**, 63–68 (2010).

MIX
Papier aus verantwortungsvollen Quellen
Paper from responsible sources
FSC® C105338

If you have any concerns about our products,
you can contact us on
ProductSafety@springernature.com

In case Publisher is established outside the EU,
the EU authorized representative is:
Springer Nature Customer Service Center GmbH
Europaplatz 3, 69115 Heidelberg, Germany

Printed by Libri Plureos GmbH
in Hamburg, Germany